ELEMENTARY COURSE

OF

CIVIL ENGINEERING

FOR THE USE OF

CADETS OF THE UNITED STATES MILITARY ACADEMY.

BY

J. B. WHEELER,

Professor of Civil and Military Engineering in the United States Military Academy at West Point, N. Y., and Brevet-Colonel U. S. Army.

FIFTH EDITION.

THIRD THOUSAND.

NEW YORK:
JOHN WILEY AND SONS,
53 EAST TENTH STREET,
1893

Press of J. J. Little & Co.,
Astor Place, New York.

PREFACE.

THIS text-book is prepared especially for the cadets of the United States Military Academy, to be used while pursuing their studies in the course of Civil Engineering laid down for them.

The object of the book is to state concisely the principles of the science of Civil Engineering, and to illustrate these principles by examples taken from the practice and writings of civil engineers of standing in their profession.

These principles and facts are widely known and are familiar to all well-informed engineers; they will, however, be new to the beginner.

The present edition differs slightly from the one that has been used for the past seven years. The modifications in the text are simply those that have been suggested by the use of the book in the class-room. The differences between the two editions are, however, not sufficiently great to prevent a simultaneous use of both old and new in the same class.

J. B. W.

WEST POINT, N. Y., July, 1884.

ANALYTICAL TABLE OF CONTENTS.

CONTENTS.

CHAPTER V.—PRESERVATIVES.

PART II.

Strength of Materials.

CHAPTER VI.—STRAINS.

PART IV.

CHAPTER IX.—MASONRY.

Mechanics of Masonry.

CHAPTER X.—MASONRY CONSTRUCTION.

PART V.

CHAPTER XI.—FOUNDATIONS.

Foundations on Land.

CHAPTER XII.—FOUNDATIONS IN WATER.

PART VI.

CHAPTER XIII.—BRIDGES.

CHAPTER XIV.—TRUSSED BRIDGES.

CHAPTER XV.—TUBULAR AND IRON PLATE BRIDGES.

CHAPTER XVI.—ARCHED BRIDGES.

PART VIII.

CHAPTER XXI.—ROADS.

CHAPTER XXII.—LOCATION AND CONSTRUCTION OF ROADS.

CHAPTER XXIII.—RAILROADS.

CHAPTER XXIV.—CANALS.

INTRODUCTORY CHAPTER.

I. **Engineering** is defined to be "**the science and art of utilizing the forces and materials of nature.**"

It is divided into two principal branches, **Civil** and **Military Engineering.**

The latter embraces the planning and construction of all defensive and offensive works used in military operations.

The former comprises the designing and building of all works intended for the comfort of man, or to improve the country either by beautifying it or increasing its prosperity.

In this branch the constructions are divided into two classes, according as the parts of which they are made are to be relatively at rest or in motion. In the former case they are known as **structures**, and in the latter as **machines.**

II. It is usual to limit the term **civil engineering** to the planning and construction of works of the first class, and to use the term **mechanical** or **dynamical engineering** when the works considered are machines.

It is also usual to subdivide civil engineering into classes, according to the prominence given to some one or more of its parts when applied in practice, as *topographical engineering*, *hydraulic engineering*, *railway engineering*, etc. By these divisions, greater progress toward perfection is assured. Notwithstanding this separation into branches and subdivisions, there are certain general principles common to them all.

III. The object of the following pages is to give in regular order these elementary principles, common to all branches of engineering, which the student should learn, so that he may understand the nature of the engineer's profession, and know how to apply these principles in practice.

IV. A **structure** is a combination of portions of solid materials so arranged as to withstand the action of any external forces to which it may be exposed, and still to preserve its form. These portions are called **pieces**, and the surfaces where they touch and are connected are called **joints**. The term **solid** here used is applied to a *body that offers an appreciable resistance to the action of the different forces to which it may be subjected.*

V. That part of the solid material of the earth upon which the structure rests is called the **foundation**, or **bed of the foundation**, of the structure.

VI. In planning and building a structure, the engineer should be governed by the following conditions:

The structure should possess the **necessary strength**; should **last the required time**; and its cost must **be reasonable.** In other words, the engineer in projecting and executing a work should duly consider the elements of **strength, durability**, and **economy**.

VII. The permanence of a structure requires that it should possess **stability, strength**, and **stiffness**. It will possess these when the following conditions are fulfilled:

When all the external forces, acting on the whole structure, are in equilibrium;

When those, acting on each piece, are in equilibrium;

When the forces, acting on each of the parts into which a piece may be conceived to be divided, are in equilibrium; and

When the alteration in form of any piece, caused by the external forces, does not pass certain prescribed limits.

A knowledge, therefore, of the forces acting on the structure, and of the properties of the materials to be used in its construction, is essential.

VIII. The designing and building of a structure form three distinct operations, as follows:

1. The conception of the project or plan;
2. Putting this on paper, so it can be understood, and
3. Its execution.

The first requires a perfect acquaintance with the locality where the structure is to be placed, the ends or objects to be attained by it, and the kind and quantity of materials that can be supplied at that point for its construction.

The **second** requires that the projector should know something of drawing, as it is only by drawings and models accompanied by descriptive memoirs, with estimates of cost, that the arrangement and disposition of the various parts, and the expense of a proposed work, can be understood by others.

The drawings are respectively called the **plan, elevation,** and **cross-section,** according to the parts they represent. A symmetrical structure requires but few drawings; one not symmetrical, or having different fronts, will require a greater number.

These, to be understood, must be accompanied by written specifications explaining fully all the parts.

The estimate of cost is based upon the cost of the materials, the price of labor, and the time required to finish the work.

The **third** may be divided into three parts:

1. The field-work, or laying out the work;
2. The putting together the materials into parts; and
3. The combining of these parts in the structure.

This requires a knowledge of surveying, levelling, and other operations incident to laying out the work;

A knowledge of the physical properties of the materials used;

The art of forming them into the shapes required; and

How they should be joined together to best satisfy the conditions that are to be imposed upon the structure.

ELEMENTARY COURSE

OF

CIVIL ENGINEERING.

PART I.

BUILDING MATERIALS.

1. The materials in general use by civil engineers for their constructions may be arranged in three classes:

1st. Those which constitute the more solid components of structures; as **Wood, Stone,** and the **Metals.**

2d. Those which unite the solid parts together; as **Glue, Cements, Mortars, Mastics,** etc.

3d. Those mixtures and chemical preparations which are employed to protect the structure from the action of the weather and other causes of destructibility; as **Paints, Solutions of Salts, Bituminous Substances,** etc.

CHAPTER I.

WOOD.

2. The abundance and cheapness of this material in the United States, the ease with which it could be procured and worked, and its strength, lightness, and durability, under favorable circumstances, have caused its very general use in every class of constructions.

Timber, from the Saxon word *timbrian,* to build, is the term applied to wood of a suitable size, and fit for building purposes. While in the tree it is called **standing timber;** after the tree is felled, the portions fit for building are cut into proper lengths and called **logs or rough timber;** when the latter have been squared or cut into shape, either to be used in this form or cut into smaller pieces, the general term **timber** is applied to them; if from the trunk of the tree, they are known as **square** or **round, hewn** or **sawed**, according to the form of cross-section and mode of cutting it; if from the branches or roots, and of crooked shape, they are called **compass timber.** The latter is used in ship-building.

The logs, being sawed into smaller pieces, form **lumber,** and the latter is divided into classes known as joists, scantlings, strips, boards, planks, etc., and, when sawed to suit a given bill; as dimension stuff.

3. The trees used for timber are **exogenous**—that is, they grow or increase in size by formation of new wood in layers on its outer surface.

If the trunk of a tree is cut across the fibres, the cut will show a series of consecutive rings or layers.

These layers are of annual growth in the temperate zones, and, by counting them, the approximate age of the tree may be determined.

The trunk of a full-grown tree presents three distinct parts: the **bark,** which forms the exterior coating; the **sap-wood,** which is next to the bark; the **heart,** or inner part, which is easily distinguishable from the sap-wood by its greater density, hardness and strength, and oftentimes by its darker color.

The heart embraces essentially all that part of the trunk which is of use as a building material. The sap-wood possesses but little strength, and is subject to rapid decay, owing to the great quantity of fermentable matter contained in it. The bark is not only without strength, but, if suffered to remain on the tree after it is felled, it hastens the decay of the sap-wood and heart.

VARIETIES OF TIMBER-TREES IN THE UNITED STATES.

4. The forests of our own country produce a great variety of the best timber for every purpose. For use in construction, trees are divided into two general classes, **soft wood** and **hard wood trees.**

The first includes all coniferous trees, like the pines, and also some few varieties of the leaf-wood trees; and the other includes most of the timber trees that are non-coniferous, like the oaks, etc.

The soft wood trees generally contain turpentine, and are distinguished by straightness of fibre and by the regularity of form of the tree. The timber made from them is more easily sawed or split along the grain, and much more easily broken across the grain, than that of the second class.

The hard-wood, or non-coniferous timber, contains no turpentine, and, as a class, is tough and strong.

Examples of Soft-wood Trees.

5. **Yellow Pine** (*Pinus mitis*).—This tree is, perhaps, in this country the most widely distributed of all the pines, being found in all the States from New England to the Gulf of Mexico. In the Southern States it is called the **Spruce Pine**, and the **Short-leaved Pine.**

The heart-wood is fine grained and moderately resinous. Its sap-wood decays rapidly when exposed to the weather. The tree grows mostly in light clay soils and furnishes a strong and durable timber extensively used in house and ship building.

Long-leaf Pine (*Pinus australis*).—This tree is found from southeastern Virginia to the Gulf, and is the principal tree where the soil is sandy and dry. Inferior growths of it are frequently called *Yellow Pine*. It has but little sap-wood. The heart-wood is fine grained, compact, and has the resinous matter very uniformly distributed.

The timber made from it is strong and durable, being considered superior to that of the other pines. Its quality depends, however, on the kind of soil in which the tree grows, being less resinous in rich soils.

Red Pine (*Pinus resinosa*).—This tree is found in Canada and the northwestern parts of the United States, and is often wrongly called "**Norway Pine.**" It furnishes good, strong and durable timber.

White Pine (*Pinus strobus*).—This tree is found in Canada and New England, and along the Alleghanies as far south as Georgia, and frequently called **Northern Pine.** Its timber is light, soft, free from knots, slightly resinous, easily worked, and durable when not exposed to the weather. It is used in a great variety of ways for building purposes and for joiners' work.

6. **Fir.**—The genus **Fir** (*Abies*), commonly known as **Spruce,** furnishes large quantities of timber and lumber which are extensively used throughout the Northern States. The lumber made from it has the defects of twisting and splitting on exposure to the weather and of decaying rapidly in damp situations. The **common fir** (*Abies alba* and *Abies nigra*), the **spruce fir** found in Northern California, and the

Oregon fir [*Pinus* (*Abies*) *Douglasii*] which grows to an enormous size, all furnish timber much used in building.

7. **Hemlock** (*Abies Canadensis*) is a well-known species, used throughout the Northern States as a substitute for pine when the latter is difficult or expensive to procure. It is very perishable in damp situations or when subjected to alternate wetness and dryness. It has been used in considerable quantities in positions where it is entirely submerged in fresh water. Hemlock timber has the defects of being shaky, full of knots, and more difficult to work than pine.

8. **Cedar.**—The **White Cedar**, called **Juniper**, and the **Cypress** are celebrated for furnishing a very light timber of great durability when exposed to the weather; on this account it is much used for shingles and other exterior coverings. The shingles made of it will last, so it is said, for 40 years. These two trees are found in great abundance in the swamps of the Southern States.

9. The foregoing kinds of timber, especially the pines, are regarded as valuable building materials, on account of their strength, their durability, the straightness of the fibre, the ease with which they are worked, and their applicability to almost all the purposes of constructions in wood.

Examples of Hard-wood Trees.

10. **White Oak** (*Quercus alba*).—The bark of this tree is light, nearly white; the leaf is long, narrow, and deeply indented; the wood is compact, tough, and pliable, and of a straw color with a pinkish tinge.

It is largely used in ship-building, the trunk furnishing the necessary timber for the heavy frame-work, and the roots and large branches affording an excellent quality of compass-timber. Boards made from it are liable to warp and crack. This tree grows throughout the United States and Canada, but most abundantly in the Middle States. Proximity to salt air during the growth of the tree appears to improve the quality of the timber. The character of the soil has a decided effect on it. In a moist soil, the tree grows to a larger size, but the timber loses in firmness and durability.

Live Oak (*Quercus virens*).—The wood of this tree is of a yellowish tinge; it is heavy, compact, and of a fine grain; it is stronger and more durable than that of any other species, and on this account is considered invaluable for the purposes of ship-building, for which it has been exclusively reserved.

The live oak is not found farther north than the neighborhood of Norfolk, Virginia, nor farther inland than from fifteen to twenty miles from the sea-coast.

Post Oak (*Quercus obtusiloba*).—This tree seldom attains a greater diameter than about fifteen inches, and on this account is mostly used for posts, from which use it takes its name. The wood has a yellowish hue and close grain; is said to exceed white oak in strength and durability, and is therefore an excellent building material for the lighter kinds of frame-work. This tree is found most abundantly in the forests of Maryland and Virginia, and is there frequently called *Box White Oak* and *Iron Oak*. It also grows in the forests of the Southern and Western States, but is rarely seen farther north than the southern part of New York.

Chestnut White Oak (*Quercus prinus palustris*).—This tree is abundant from North Carolina to Florida. The timber made from it is strong and durable, but inferior to that of the preceding species.

Water Oak (*Quercus aquatica*).—This tree gives a tough but not a durable timber. It grows in the Southern country from Virginia to as far south as Georgia and Florida.

Red Oak (*Quercus rubra*).—This tree is found in all parts of the United States. The wood is reddish, of a coarse texture, and quite porous. The timber made from it is generally strong, but not durable.

11. **Black Walnut** (*Juglans nigra*).—The timber made from this tree is hard and fine-grained. It has become too valuable to be used in building purposes, except for ornamentation.

Hickory (*Carya tomentosa*).—The wood of this tree is tough and flexible. Its great heaviness and liability to be worm-eaten have prevented its general use in buildings.

12. There are a number of other trees, belonging to both hard and soft woods, that produce timber inferior to those named. They may possibly in the future be used to some extent to furnish timber for building purposes. The Red Cedar, Chestnut, Ash, Elm, Poplar, American Lime or Basswood, Beech, Sycamore, Tamarack, etc., have all been used to a limited extent in constructions when the other kinds were not to be obtained.

PREPARATION OF TIMBER.

13. **Felling.**—Trees should not be felled for timber until they have attained their mature growth, nor after they ex-

hibit symptoms of decline; otherwise the timber will not possess its maximum strength and durability. Most forest trees arrive at maturity in between fifty and one hundred years, and commence to decline after one hundred and fifty or two hundred years. When a tree commences to decline, the extremities of its older branches, and particularly its top, exhibit signs of decay. The age of a tree can, in most cases, be approximately ascertained either by its external appearances or by cutting into the centre of its trunk and counting the rings or layers of the sap and heart.

Trees should not be felled while the sap is in circulation; for this substance is of such peculiarly fermentable nature, that if allowed to remain in the fallen timber, it is very productive of destruction of the wood. The best authorities on the subject agree that the tree **should be felled in the winter season.**

The practice in the United States accords with the above, not so much on account of the sap not being in circulation, as for the reason that the winter season is the best time for procuring the necessary labor, and the most favorable for removing the logs, from where they are cut, to the points where they are to be made into rafts.

As soon as the tree is felled, it should be stripped of its bark and raised from the ground. A short time only should elapse before the sap-wood is taken off and the timber reduced nearly to its required dimensions.

14. **Measuring Timber.**—Timber is measured by the cubic foot, or by *board measure;* the unit of the latter is a board one foot square and one inch thick.

Appearances of Good Timber.

15. Among trees of the same species, that one which has grown the slowest, as shown by the narrowness of its annual rings, will in general be the strongest and most durable.

The grain should be hard and compact, and if a cut be made across it, the fresh surface of the cut should be firm and shining.

And, in general, other conditions being the same, the strength and durability of timber will increase with its weight, and darkness of color.

Timber of good quality should be straight-grained, and free from knots. It should be free from all blemishes and defects.

Defects in Timber.

16. **Defects** arise from some peculiarity in the growth of the tree, or from the effects of the weather.

Strong winds oftentimes injure the growing tree by twisting or bending it so as to partially separate one annual layer from another, forming what is known as **rolled timber** or **shakes.**

Severe frosts sometimes cause cracks radiating from the centre to the surface.

These defects, as well as those arising from worms or age, may be detected by examining a cross-section of the log.

SEASONING OF TIMBER.

17. Timber is said to be **seasoned** when by some process, either natural or artificial, the moisture in it has been expelled so far as to prevent decay from internal causes.

The term **seasoning** means not only the drying of the timber, but also the removal or change of the albuminous substances in it. These substances are fermentable, and when present unchanged in the timber are ever ready to promote decay.

The seasoning of timber is of the greatest importance, not only to its own durability, but to the solidity of the structure for which it may be used; for, if the latter, when erected, contained some pieces of unseasoned or **green** timber, their after-shrinking might, in many cases, cause material injury, if not complete destruction, to the structure.

Natural Seasoning consists in exposing the timber freely to the air, but in a dry place, sheltered from the sun and high winds.

This method is preferable to any other, as timber seasoned in this way is both stronger and more durable than when prepared by any artificial process. It will require, on an average, about two years to season timber thoroughly by this method. For this reason, artificial methods are used to save time.

Water Seasoning.—The simplest artificial method consists in immersing the timber in water as soon as cut, taking care to keep it entirely submerged for a fortnight, and then to remove it to a suitable place and dry it. The water will remove the greater portion of the sap, even if the timber is full when immersed. This method doubtless weakens the timber to some extent, and therefore is not recommended where strength in the timber is the most important quality.

Boiling and **Steaming** have both been used for seasoning, but are open to the same objection as the last method; viz.; the impairing of the elasticity and strength of the timber.

Hot-air Process.—This consists in exposing the timber in a chamber, or oven, to a current of hot air, whose temperature varies according to the kind and size of the timber to be seasoned. This is considered the best of the artificial methods. The time required for sufficient seasoning depends upon the thickness of the timber, ordinary lumber requiring from one to ten weeks.

DURABILITY AND DECAY OF TIMBER.

18. Timber lasts best when kept, or used, in a dry and well-ventilated place. Its durability depends upon its protection from decay and from the attacks of worms and insects.

The **wet** and **dry rot** are the most serious causes of the decay of timber.

Wet Rot is a slow combustion, a decomposition of moist organic matter exposed to the air, without sensible elevation of temperature. The decay from wet rot is communicated by contact, and requires the presence of moisture.

To guard against this kind of rot, the timber must not be subjected to a condition of alternate wetness and dryness, or even to a slight degree of moisture if accompanied by heat and confined air.

Dry Rot is a decay arising from the decomposition of the fermentable substances in the timber; it is accompanied by the growth of a fungus, whose germs spread in all directions, finally converting the wood into a fine powder. The fungus is not the cause of decay; it is only a morbid growth due to the decaying fibres of the wood.

Dry rot derives its name from the effect produced and not from the cause, and although it is usually generated in moisture, it is frequently found to be independent of extraneous humidity. Externally, it makes its first appearance as a mildew, or a white or yellowish vegetation of like appearance. An examination under a microscope of a section of a piece of wood attacked by dry rot shows minute white threads spreading and ramifying throughout the substance.

Dry rot only attacks wood which is dead, whereas wet rot may seize the tree while it is still alive and standing. Timber, not properly seasoned, used where there is a want of free circulation of air, decays by dry rot even if there be only a small amount of moisture present. It will also decay by dry rot, if covered while unseasoned by a coat of paint, or similar substance.

Durability under certain Conditions, and Means of Increasing it.

19. Timber may be subjected to the following conditions:

It may be kept constantly dry, or at least practically so
It may be kept constantly wet in fresh water.
It may be constantly damp.
It may be alternately wet and dry.
It may be constantly wet in sea-water.

20. **Timber kept constantly dry** in well-ventilated positions, will last for centuries. The roof of Westminster Hall is more than 450 years old. In Stirling Castle are carvings in oak, well preserved, over 300 years old; and the trusses of the roof of the Basilica of St. Paul, Rome, were sound and good after 1000 years of service. The timber dome of St. Mark, at Venice, was in good condition 850 years after it was built.

It would seem hardly worth while to attempt to increase the durability of timber when under these conditions, except where it may be necessary to guard against the attacks of insects, which are very destructive in some localities. Damp lime hastens the decay of timber; the latter should therefore, in buildings, be protected against contact with the mortar.

21. **Timber kept constantly wet in fresh water,** under such conditions as will exclude the air, is also very durable.

Oak, elm, beach, and chestnut piles and planks were found beneath the foundation of Savoy Place, London, in a perfect state of preservation, after having been there 650 years.

The piles of the old London Bridge were sound 800 years after they were driven. In the bridge built by Trajan, the piles, after being driven more than 1600 years, were found to have a hard exterior, similar to a petrifaction, for about four inches, the rest of the wood being in its ordinary condition.

We may conclude that timber submerged in fresh water will need no artificial aid to increase its durability, although in time it may be somewhat softened and weakened.

22. **Timber in damp situations.**—Timber in damp situations is in a place very unfavorable for durability, and is liable, as previously stated, to decay rapidly. In such situa-

tions only the most lasting material is to be employed, and every precaution should be taken to increase its durability.

23. **Timber alternately wet and dry.**—The surface of all timber exposed to alternations of wetness and dryness gradually wastes away, becoming dark-colored or black. This is wet rot, or simply "*rot.*"

Density and resinousness exclude moisture to a great extent; hence timber possessing these qualities should be used in such situations. Heart-wood, from its superior density, is more durable than sap-wood; oak, than poplar or willow. Resinous wood, as pine, is more durable than the non-resinous, as ash or beech, in such situations.

24. **Timber constantly wet in sea-water.**—The remarks made about timber placed in fresh water apply equally to this case, as far as relate to decay from rot. Timber immersed in salt water is, however, liable to the attacks of two of the destructive inhabitants of our waters, the **Limnoria terebrans** and **Teredo navalis**; the former rapidly destroys the heaviest logs by gradually eating in between the annual rings; and the latter, the well-known *ship-worm*, converts timber into a perfectly honeycombed state by its numerous perforations. They both attack timber from the level of the mud, or bottom of the water, and work to a height slightly above mean low water. The timber, for this distance, must be protected by sheathing it with copper, or by thickly studding the surface with broad-headed iron nails, or other similar device. Resinous woods resist their attacks longer, most probably on account of the resin in the wood. The resin after a time is washed or dissolved out, and the timber is then speedily attacked.

An examination of piles in the wharf at Fort Point, San Francisco harbor, where these agents are very destructive, showed that piles which were driven without removing the bark, resisted to a certain extent, their destructive attacks.

Timber saturated with dead oil by the process known as creosoting is said to offer an effective resistance.

PRESERVATION OF TIMBER.

25. The necessity of putting timber into damp places has caused numerous experiments to be made as to the best method of increasing its durability under such circumstances.

There are three means which may be used to increase the durability of timber placed in damp situations, viz:

1st. To season it thoroughly.
2d. To keep a constant circulation of air about it.
3d. To cover it with a preservative.

The cellulose matter of the woody fibre is very durable when not acted upon by fermentation, and the object of seasoning is to remove or change the fermentable substances, as well as to expel the moisture in the timber, thus protecting the cellulose portion from decay. Even if the timber be well seasoned, thorough ventilation is indispensable in damp situations. The rapid decay of sills and lower floors is not surprising where there are neither wall-gratings nor ventilating flues to carry off the moisture and the foul gases rising from the earth under them. The lower floors would last nearly as long as the upper ones if the earth were removed to the bottom of the foundation and the space filled in with dry material, as sand, plaster, rubbish, etc., or the bottom covered with a concrete floor to exclude the moisture, and arrangements made to allow a free circulation of air under the sills.

An external coating of paint, pitch, or hot oil increases the durability of well-seasoned timber, but such a coating upon the surface of **green** timber produces just the opposite effect. The coating of paint closes the pores of the outer surface, and prevents the escape of the moisture from within, thus retaining in the wood the elements of decay.

It is not always practicable to employ the foregoing means in damp places to preserve the timber, and other methods have to be used. These methods are based upon the principle of expelling the albuminous substances and replacing them by others of a durable nature, or on that of changing the albuminous substances into insoluble compounds by saturating the timber with salts of an earthy or metallic base which will combine with the albuminous matter and make it inert.

Some of the methods which have been proposed, or used, are as follows:

Kyanizing.—Kyan's method is to saturate the timber with a solution of mercuric chloride, one pound of chloride to four gallons of water.

The complete injection of the liquid is obtained either by long immersion in the liquid in open vats, or by great pressure upon both solution and wood in large wrought-iron tanks.

The expensiveness of the process, and its unhealthiness to those employed in it, forbid its extensive use.

Burnettizing.—Burnett's process is to use a solution of chloride of zinc, one pound of the chloride to ten gallons of water; the solution being forced into the wood under a pressure of 150 pounds to the square inch.

Earle's Process consisted in boiling the timber in a solution of one part of sulphate of copper to three parts of the sulphate of iron; one gallon of water being used with every pound of the salts. A hole was bored through the whole length of the piece; the timber was then immersed from two to four hours, and allowed to cool in the mixture.

Ringold and Earle invented the following process: A hole from ½ to 2 inches in diameter was made the whole length of the piece, and the timber boiled from two to four hours in lime-water. After the piece was dried, the hole was filled with lime and coal-tar. Neither of the last two methods was very successful.

Common Salt is known in many cases to be a good preservative. According to Mr. Bates's opinion this method often answers a good purpose if the pieces so treated are not too large.

Boucherie's Process employs a solution of sulphate of copper or pyrolignite of iron. One end of the green stick is enclosed in a close-fitting collar, to which is attached a water-tight bag communicating through a flexible tube with an elevated reservoir containing the solution. Hydrostatic pressure soon expels the sap. When the solution issues in a pure state from the opposite end of the log, the process is complete.

It was found that the fluid will pass a distance of twelve feet along the grain under less pressure than is necessary to force it across the grain three-fourths of an inch. The operation is performed upon green timber with great facility.

In 1846, 80,000 railroad ties of the most perishable woods, impregnated, by Boucherie's process, with sulphate of copper, were laid down on French railways. After nine years' exposure they were found as perfect as when laid. This experiment was so satisfactory that most of the railways of that country at once adopted the process. It has been suggested to wash out the sap with water, which would not coagulate the albumen, and then to use the solution.

Bethel's Process.—The timber is placed in an air-tight cylinder of boiler-iron, and the air partially exhausted. **Dead oil** is then admitted at a temperature of 120° Fahr., and a pressure of about 150 pounds to the square inch is then applied, and maintained from five to eight hours, according to the size of the timbers under treatment. The oil is then drawn off, and the timber is removed.

The **Seeley Process** consists in subjecting the wood, while immersed in dead oil, to a temperature between 212° and 300° Fahr. for a sufficient length of time to expel any moisture present; the water being expelled, the hot oil is quickly replaced by cold, thus condensing the steam in the pores of the timber, forming a vacuum into which oil is forced by atmospheric pressure and capillary attraction. In this process from six to twelve pounds of oil is expended for each cubic foot of wood.

The theory of this process is that the first part of the operation seasons the wood, destroys or coagulates the albumen, and expels the moisture; and that the second part fills the wood-cells with a material that is an antiseptic and resists destructive agents of every kind.

Robbins's Process consists in treating timber with coal-tar in the form of vapor.

The wood is placed in an air-tight iron chamber, with which is connected a still or retort, over a furnace. The furnace is then fired and the wood kept exposed to the heated vapors of the coal tar from six to twelve hours; the operation is then considered complete.

The most improved of all these methods is **Seeley's**; this is a modification and an improvement of Bethel's process, and is generally known as "**creosoting.**"

It is thought that the ancient Egyptians knew of some process of preserving wood. Old cases, supposed to have been 2,000 years old, apparently of sycamore impregnated with bitumen, have been found to be still perfectly sound and strong.

CHAPTER II.

STONE.

26. **The qualities required** in stone for building purposes are so various that no very precise directions can be given to exactly meet any particular case. What would be required for a sea-wall would not be suited to a dwelling-house. In most cases the choice is limited by the cost. The most essential properties of stone as a building material are **strength, hardness, durability,** and **ease of working.** These properties are determined by experience or actual experiment.

27. The term **Stone**, or **Rock**, is applied to any aggregation of several mineral substances; as a building material, stones may be either **natural** or **artificial.**

Natural Stones may be subdivided into three classes; the **silicious**, the **argillaceous**, and the **calcareous**, according as silica, clay, or lime is the principal constituent.

Artificial Stones are imitations of natural stone, made by consolidating fragmentary solid material by various means; they may be subdivided into classes as follows:

1st. Those in which two or more kinds of solid materials are mixed together and consolidated by baking or burning; as **brick, tiles,** etc.

2d. Those in which the solid materials are mixed with some fluid or semi-fluid substance, which latter, hardening afterwards by chemical combinations, binds the former firmly together; as **ordinary concrete, patent stone,** etc.

3d. Those in which the solid materials are mixed with some hot fluid substance which hardens upon cooling; as **asphaltic concrete,** etc.

I. NATURAL STONES.

GENERAL OBSERVATIONS ON THE PROPERTIES OF STONE AS A BUILDING MATERIAL.

28. **Strength, hardness, durability,** and **ease of working** have already been mentioned as essential properties to be considered in selecting stone for building purposes.

It is not easy to judge of the qualities from external appearances. In most cases stone, which has one of the three properties first named, will have also the other two. In general, when the texture is uniform and compact, the grain fine, the color dark, and the specific gravity great, the stone is of good quality. If there are cracks, cavities, presence of iron, etc., even though it belong to a good class of stone, it will be deficient in some of these essential qualities, and should be rejected. A coarse stone is ordinarily brittle, and is difficult to work; it is also more liable to disintegrate than that of a finer grain.

29. **Strength.**—Among stones of the same kind, the strongest is almost always that which has the greatest heaviness.

As stone is ordinarily to be subjected only to a crushing force, it will only be in particular cases that the resistance to this strain need be considered, the strength of stone in this respect being greater than is generally required of it. If its durability is satisfactorily proved, its strength, as a rule, may be assumed to be sufficient.

30. **Hardness.**—This property is easily ascertained by actual experiment and by a comparison made with other stones which have been tested. It is an essential quality in stone exposed to wear by attrition. Stone selected for paving, flagging and for stairs, should be hard and of a grain too coarse to admit of becoming very smooth under the action to which it is submitted.

By the absorption of water, stones become softer and more friable.

31. **Durability.**—By this term is meant the power to resist the wear and tear of atmospheric agencies, the capacity to sustain high temperature, and the ability to resist the destructive action of fresh and salt water.

The appearances which indicate probable durability are often deceptive.

As a general rule, among stones of the *same kind*, those which are fine-grained, absorb least water, and are of greatest specific gravity, are also most durable under ordinary exposures. The weight of a stone, however, may arise from a large proportion of metallic oxide—a circumstance often unfavorable to durability.

The various chemical combinations of iron, potash, and alumina, when found in considerable quantities in the silicious rocks, greatly affect their durability. The decomposition of the feldspar by which a considerable portion of the silica is removed when the potash dissolves, leaves an excess of aluminous matter behind. The clay often absorbs water, becomes soft, and causes the stone to crumble to pieces.

32. **Frost**, or rather the alternate action of freezing and thawing, is the most destructive agent of nature with which the engineer has to contend. Its effects vary with the texture of stones; those of a fissile nature usually split, while the more porous kinds disintegrate, or exfoliate at the surface. When stone from a new quarry is to be tried, the best indication of its resistance to frost may be obtained from an examination of any rocks of the same kind, within its vicinity, which are known to have been exposed for a long period. Submitting the stone fresh from the quarry to the direct action of freezing would seem to be the best test of it, if it were not that there are some kinds of stone that are much affected by frost when they are first quarried due to the moisture present in the stone, which moisture is lost by exposure to the air, and is never reabsorbed to the same amount.

A test for ascertaining the probable effects of frost on stone was invented by M. Brard, a French chemist, and may be used for determining the probable comparative durabilities of specimens. It imitates the disintegrating action of frost by means of the crystallization of sodium sulphate. The process may be stated briefly as follows: Let a cubical block, about two inches on the edge, be carefully sawed from the stone to be tested. A cold saturated solution of the sodium sulphate is prepared, placed over a fire, and brought to the boiling-point. The stone, having been weighed, is suspended from a string, and immersed in the boiling liquid for thirty minutes. It is then carefully withdrawn, the liquid is decanted free from sediment into a flat vessel, and the stone is suspended over it in a cool cellar. An efflorescence of the salt soon makes its appearance on the stone, when it must be again dipped in the liquid. This should be frequently done during the day, and the process be continued for about a week. The earthy sediment found at the end of this period in the vessel is carefully weighed, and its quantity will give an indication of the like effect of frost. This process is given in detail in Vol. XXXVIII. *Annales de Chémie et de Physique.*

This test, having corresponded closely with their experience, has received the approval of many French architects and engineers. Experiments, however, made by English engineers on some of the more porons stones, by exposing them to the alternate action of freezing and thawing, gave results very different from those obtained by Brard's method.

33. **The Wear of Stone** from ordinary exposure is very variable, depending not only upon the texture and constituent elements of the stone, but also upon the locality, and the position, it may occupy in a structure, with respect to the prevailing driving rains. This influence of locality on the durability of stone is very marked. Stone is observed to wear more rapidly in cities than in the country, and exhibits signs of decay soonest in those parts of a building exposed to the prevailing winds and rains.

The disintegration of the stratified stones placed in a wall is materially affected by the position of the strata or laminæ with respect to the exposed surface, proceeding faster when the faces of the strata are exposed, as is the case when the stones are not placed with their laminæ lying horizontally.

Stones are often exposed to the action of high temperatures, as in the case of great conflagrations. They are also used to protect portions of a building from great heat, and sometimes to line furnaces. Those that resist a high degree of heat are termed **fire-stones**. A good fire-stone should be infusible, and not liable to crack or exfoliate from heat. Stones that contain lime or magnesia are usually unsuitable. Also, silicates containing an oxide of iron.

Their durability under such circumstances should be considered when selecting them for building.

The only sure test, however, of the durability of any kind of stone is its wear, as shown by experience.

34. **Expansion of Stone from Heat.**—Experiments have been made in this country and Great Britain to ascertain the expansion of stone for every degree of Fahrenheit, and the results have been tabulated. Within the ordinary ranges of temperature the stone is too slightly affected by expansion or contraction to cause any perceptible change. Professor Bartlett's experiments, however, showed that in a long line of coping the expansion was sufficiently great to crush mortar between the blocks.

35. **Preservation of Stone.**—To add to the durability of stone, especially of that naturally perishable or showing signs of decay, various processes have been tried or proposed. All have the same end in view; viz., to fill the exposed pores of the stone with some substance which shall exclude the air and moisture. Paints and oils are used for this purpose. Great results have been expected from the use of soluble glass (silicate of potash), and also from silicate of lime. The former, being applied in a state of solution in water, gradually hardens, partly through the evaporation of its water, and partly through the removal of the potash by the carbonic acid in the air. The latter is used by filling the pores with a solution of silicate of potash, and then introducing a solution of calcium chloride or lime nitrate; the chemical action produces silicate of lime, filling the pores of the natural stone. Time and experience will show if the hopes expected from the use of these silicates will be realized.

36. **Ease of Working the Stone.**—This property is to a certain extent the inverse of the others. The ease with which stone can be cut or hammered into shape implies either softness or else a low degree of cohesiveness between its particles.

It often happens that its hardness may prevent a stone, in every other way suitable, from being wrought to a true surface and from receiving a smooth edge at the angles. Moreover, the difficulty of working will increase very materially the cost of the finished stone.

It requires experience and good judgment to strike a medium between these conflicting qualities.

37. **Quarrying.**—If the engineer should be obliged to get out his own stone by opening a new quarry, he should pay particular attention to the best and cheapest method of getting it out and hauling it to the point where it is to be used. In all cases he will, if possible, open the quarry on the side of a hill, and arrange the roads in and leading to it with gentle slopes, so as to assist the draught of the animals employed. The stone near the surface, not being as good as that beneath, is generally discarded. The mass or bed of stone being exposed, a close inspection will discover the natural joints or fissures along which the blocks will easily part from each other. When natural fissures do not exist, or smaller blocks are required, a line of holes is drilled at short regular intervals, or grooves are cut in the upper surface of a bed. Then blunt steel wedges or pins, slightly larger than the holes, are inserted, and are struck sharply and simultaneously with hammers until the block splits off from the layer.

If large masses of stone be required, resort is had to **blasting**. This operation consists in boring the requisite number of holes, loading them with an explosive compound, and firing them. The success of blasting will depend upon a judicious selection of the position and depth of the holes and upon the use of the proper charges.

Instead of trusting, as is too often done, to an empirical rule, or to no rule at all, it is well, by actual experiments on the particular rock to be quarried, to ascertain the effect of different charges, so as to determine the amount required in any case, to produce the best result.

VARIETIES OF BUILDING STONES IN GENERAL USE.

SILICIOUS STONES.

38. **Silicious Stones** are those in which silica is the principal constituent. With a few exceptions, their structure is crystalline-granular, the grains being hard and durable. They emit sparks when struck with a steel, and do not generally effervesce with acids.

Some of the principal silicious stones used in building are **Syenite, Granite, Gneiss, Mica Slate, Hornblende Slate, Steatite,** and the **Sandstones**. For their composition, particular description, etc. see any of the manuals of mineralogy.

Syenite, Granite, and **Gneiss.**—These stones differ but little in the qualities essential to a good building material, and from the great resemblance of their external characters and physical properties are generally known to builders by the common term granite.

Granite (*Syenite*, *Granite*, and *Gneiss*).—This stone ranks high as building material, in consequence of its superior strength, hardness, and durability, and furnishes a material particularly suitable for structures which require great strength. It does not resist well very high temperatures, and its great hardness requires practised stone-cutters to be employed in working it into proper shapes. It is principally used in works of magnitude and importance, as light-houses, sea-walls, revetment-walls of fortifications, large public buildings, etc. Only in districts where it abounds is it used for ordinary dwelling-houses. It was much used by the ancients, especially by the Egyptians, some of whose structures, as far as the stone is concerned, are still remaining in good condition, after 3,000 years' exposure. Granite occurs in extensive beds, and may be obtained from the quarries in blocks of almost any size required. Gneiss, in particular, having the mica more in layers, presents more of a stratified appearance, and admits of being broken out into thin slabs or blocks. A granite selected for building purposes should have a fine grain, even texture, and its constituents uniformly disseminated through the mass. It should be free from pyrites or any iron ore, which will rust and deface, if not destroy the stone on exposure to the weather. The feldspathic varieties are the best, and the syenitic are the most durable. An examination of the rock in and around the quarry may give some idea of its durability.

Mica Slate has in its composition the same materials as gneiss, and breaks with a glistening or shining surface. The compact varieties are much used for flagging, for door and hearth stones, and for lining furnaces, as they can be broken out in thin, even slabs. It is often used in ordinary masonry work, in districts where it abounds.

Hornblende Slate resembles mica slate, but is tougher, and is an excellent material for flagging.

Steatite, or Soapstone, is a soft stone easily cut by a knife, and greasy to the touch. From the ease with which it is worked, and from its refractory nature, it is used for fire-stones in furnaces and stoves, and for jambs in fire-places. Being soft, it is not suitable for ordinary building purposes.

Sandstone is a stratified rock, consisting of grains of silicious sand, arising from the disintegration of silicious stones, cemented together by some material, generally a compound of silica, alumina, and lime. It has a harsh feel, and every dull shade of color from white, through yellow, red, and brown, to nearly a black. Its strength, hardness, and durability vary between very wide limits; some varieties being little inferior to good granite as a building-stone, others being very soft, friable, and disintegrating rapidly when exposed to the weather. The least durable sand-stones are those which contain the most argillaceous matter; those of a feldspathic character also are found to withstand poorly the action of the weather. The best sandstone lies in thick strata, from which it can be cut in blocks that show very faint traces of stratification; that which is easily split into thin layers, is weaker. It should be firm in texture, not liable to peel off when exposed, and should be free from pyrites or iron-sand, which rust and disfigure the blocks. It is generally porous and capable of absorbing much water, but it is comparatively little injured by moisture, unless when built with its layers set on edge. In this case the expansion of water between the layers in freezing makes them split or "scale" off. It should be placed with the strata in a horizontal position, so that any water which may penetrate between the layers may have room to expand or escape. Most of the varieties of sandstone yield readily under the chisel and saw, and split evenly; from these properties it has received from workmen the name of **free-stone.** It is used very extensively as a building-stone, for flagging, for road material; and some of its varieties furnish an excellent fire-stone.

Other varieties of silicious stones besides those named, as **porphyry, trap** or **greenstone, basalt, quartz-rock (cobble-stone), buhr-stone,** etc., are used for building and engineering purposes, and are eminently fit, either as cut-stone or rubble, as far as strength and durability are concerned.

ARGILLACEOUS STONES.

39. **Argillaceous** or **Clayey Stones are** those in which clay exists in sufficient quantity to give the stone its characteristic properties. As a rule, the natural argillaceous stones, excepting roofing slate, are deficient in the properties of hardness and durability, and are unfit for use in engineering constructions.

Roofing Slate is a stratified rock of great hardness and density, commonly of a dark dull blue or purplish color. To be a good material for roofing, it should split easily into even slates, and admit of being pierced for nails without being fractured. It should be free from everything that can on exposure undergo decomposition. The signs of good quality in slate are compactness, smoothness, uniformity of texture, clear dark color; it should give a ringing sound when struck, and should absorb but little water. Being nearly impervious to water, it is principally used for covering of roofs, linings of water-tanks, and for other similar purposes.

CALCAREOUS STONES.

40. **Calcareous Stones** are those in which **lime** (*calcium monoxide*) is the principal constituent. It enters either as a sulphate or carbonate.

Calcium Sulphate, known as **gypsum** in its natural state, when burnt and reduced to a powder, is known as **plaster-of-Paris.** A paste made of this powder and a little water, soon becomes hard and compact. Gypsum is not used as a building-stone, being too soft. The plaster, owing to its snowy whiteness and fine texture, is used for taking casts, making models, and for giving a *hard finish* to walls. Care must be taken to use it only in dry and protected situations, as it absorbs moisture freely, then swells, cracks, and exfoliates rapidly.

Calcium Carbonates, or **Limestones**, furnish a large amount of ordinary building-stone, ornamental stone, and form the source of the principal ingredient of cements and mortars.

They are distinguished by being easily scratched with a knife, and by effervescing with an acid. In texture they are either **compact** or **granular**; in the former case the fracture is smooth, often conchoidal; in the latter it has a crystalline-granular surface, the fine varieties resembling loaf-sugar.

The limestones are generally impure carbonates, and we are indebted to their impurities for some of the most beautiful as well as the most invaluable materials used for constructions. Those stones which are colored by metallic oxides, or by the presence of other minerals, furnish the numerous colored and variegated marbles; while those which contain a certain proportion of impurities as silica, alumina, etc., yield, on calcination, those cements which, from possessing the property of hardening under water, have received the names of hydraulic lime, hydraulic cement, etc.

Limestones that can be made to have a smooth surface and take a polish are known as **marbles**; the coarser kinds are called **common limestones**, and form a large class of much value for building purposes.

41. **Marbles.**—Owing to the high polish of which they are susceptible, and their consequent value, the marbles are mostly reserved for ornamental purposes.

They present great variety, both in color and appearance, and the different kinds have generally received some appropriate name descriptive of their use or appearance.

Statuary Marble is of the purest white, finest grain, and is free from all foreign minerals. It receives a delicate polish, without glare, and is, therefore, admirably adapted to the purposes of the sculptor, for whose uses it is mostly reserved.

Conglomerate Marble.—This consists of two varieties; the one termed **pudding** stone, composed of rounded pebbles embedded in compact limestone; the other termed **breccia**, consisting of angular fragments united in a similar manner. The colors of these marbles are generally variegated, making the material very handsome and ornamental.

Bird's-eye Marble.—The name of this stone is descriptive of its appearance after sawing or splitting, the *eyes* arising from the cross-sections of a peculiar fossil (*fucoides demissus*) contained in the mass.

Lumachella Marble.—This is a limestone having shells embedded in it, and takes its name from this circumstance.

Verd Antique.—This is a rare and costly variety, of a beautiful green color, the latter being caused by veins and blotches of *serpentine* diffused through the limestone.

There are many other varieties that receive their name either from their appearance or the localities from which they are obtained.

Many of these are imitated by dealers, who, by processes known to themselves, stain the common marbles so successfully that it requires a close examination to distinguish the false from the real.

Common Limestone.

42. This class furnishes a great variety of building-stones, which present great diversity in their physical properties. Some of them seem as durable as the best silicious stones, and are but little inferior to them in strength and hardness; others decompose rapidly on exposure to the weather; and some kinds are so soft that, when first quarried, they can be scratched with the nail and broken between the fingers. The durability of limestones is materially affected by the foreign minerals they may contain; the presence of clay injures the stone for building purposes, particularly when, as sometimes happens, it runs through the bed in very minute veins; blocks of stone having this imperfection soon separate along

these veins on exposure to moisture. Ferrous oxide, sulphate and carbonate of iron, when present, are also very destructive in their effects, frequently causing by their chemical changes rapid disintegration.

Among the varieties of impure carbonates of lime are the **magnesian limestones**, called **dolomites**. They are regarded in Europe as a superior building material; those being considered the best which are most crystalline, and are composed of nearly equal proportions of the carbonates of lime and magnesia. The magnesian limestone obtained from quarries in New York and Massachusetts is not of such good quality; the stone obtained being, in some cases, extremely friable.

II.—ARTIFICIAL STONES.

1st.—BRICK.

43. A **brick** is an artificial stone, made by moulding *tempered* clay into a form of the requisite shape and size, and hardening it, either by baking in the sun or by burning in a kiln or other contrivance. When hardened by the first process, they are known as *sun-dried*, and by the latter as *burnt-brick*, or simply **brick**.

44. **Sun-dried Brick.**—Sun-dried bricks have been in use from the remotest antiquity, having been found in the ruins of ancient Babylon. They were used by the Greeks and Romans, and especially by the Egyptians. At present they are seldom employed.

They were ordinarily made in the spring or autumn, as they dried more uniformly during those seasons; those made in the summer, drying too rapidly on the exterior, were apt to crack from subsequent contraction in the interior.

It was not customary to use them until two years after they had been made.

Walls, known as *adobes*, made of earth hardened in a similar way, are found in parts of our country and in Mexico. They furnish a simple and economical mode of construction where the weights to be supported are moderate, and where fuel is very scarce and expensive. This mode, however suitable for a southern, is not fit for our climate.

45. **Burnt Brick.**—Bricks may be either *common* or *pressed*, *hand* or *machine made*.

The qualities of a brick are dependent upon the kind of

earth used, the *tempering* of this earth, the *moulding* of the raw brick, and the *drying* and *burning* processes.

46. **Common Brick.**—The size and form of common bricks vary but little. They are generally rectangular parallelopipedons, about 8½ inches long, 4 inches broad, and 2⅝ inches thick, the exact size varying with the contraction of the clay.

Kinds of Earth.—The argillaceous earths suitable for brick-making may be divided into three principal classes, viz.:

Pure Clays, those composed chiefly of aluminum silicate, or one part of alumina and two of silica, combined with a small proportion of other substances, as lime, soda, magnesia, ferrous oxide, etc.;

Loams, which are mechanical mixtures of clay and sand; and

Marls, which are mechanical mixtures of clay and carbonate of lime.

Pure clay, being made plastic with water, may be moulded into any shape, but will shrink and crack in drying, however carefully and slowly the operation be conducted. By mixing a given quantity of sand with it, these defects may be greatly remedied, while the plastic quality of the clay will not be materially affected.

The loams oftentimes have too much sand, and are then so loose as to require an addition of clay or other plastic material to increase their tenacity.

Earth is frequently found containing the proper proportions of clay and sand suitable for making bricks; but, if it be not naturally fit for the purpose, it should be made so by adding that element which is lacking. The proportion of sand or clay to be added should be determined by direct experiments.

Silicate of lime, if in any considerable quantity in the earth, makes it too fusible. Carbonate of lime, if present in any considerable quantity in the earth, would render it unfit, since the carbonate is converted, during the burning, into lime, which absorbs moisture upon being exposed, would cause disintegration in the brick.

Preparation of the Earth.—The earth, being of the proper kind, is first dug out before the cold weather, and carried to a place prepared to receive it. It is there piled into heaps and exposed to the weather during the winter, so as to be mellowed by the frosts, which break up and crumble the lumps.

In the spring the earth is turned over with shovels, and the stones, pebbles, and gravel are removed; if either clay or sand be wanting, the proper amount is added.

Tempering.—The object of tempering is to bring the earth into a homogeneous paste for the use of the moulder. This is effected by mixing it with about half its volume of water, and stirring it and kneading it either by turning it over repeatedly with shovels and treading it over by horses or men until the required plasticity is obtained, or by using the pug-mill or a similar machine.

The plastic mass is then moulded into the proper forms by hand or machinery.

By Hand.—In the process by hand the mould used is a kind of box, without top or bottom, and the tempered clay is dashed into it with sufficient force to completely fill it, the superfluous clay being removed by *striking* it with a straight-edge. The newly-made brick is then turned out on a drying-floor, or on a board and carried to the place where it is to dry.

47. **By Machines.**—Bricks are now generally moulded by machines. These machines combine the pug-mill with an apparatus for moulding. This apparatus receives the clay as discharged from the pug-mill, presses it in moulds, and pushes the brick out in front ready to be removed from the frames and carried to the drying-floor.

48. **Drying.**—Great attention is necessary in this part of the process of manufacture. The raw bricks are dried in the open air or in a drying-house, where they are spread out on the ground or floor, and are frequently turned over until they are sufficiently hard to be handled without injury. They are then piled into stacks under cover for further drying.

In drying bricks, the main points to be observed are to protect them from the direct action of the sun, from draughts of air, from rain and frost, and to have each brick dry uniformly from the exterior inwards. The time allowed for drying depends upon the climate, the season of the year, and the weather.

49. **Burning.**—The next stage of manufacture is the burning. The bricks are arranged in the kiln so as to allow the passage of the heat around them; this is effected by piling the bricks so that a space is left around each. This arrangement of the bricks, called *setting* the kiln, is to allow the heat to be diffused equally throughout, to afford a good draught, and to keep up a steady heat with the least amount of fuel.

A very moderate fire is next applied under the arches of the kiln to expel any remaining moisture from the raw brick; this is continued until the smoke from the kiln is no longer black. The fire is then increased until the bricks of the arches attain a white heat; it is then allowed to abate in some degree, in order to prevent complete vitrifaction; and it is thus alternately raised and lowered until the burning is complete, as ascertained by examining the bricks at the top of the kiln. The bricks should be slowly cooled; otherwise they will not withstand the effects of the weather. The cooling is done by closing the mouths of the arches and the top and sides of the kiln, in the most effectual manner, with moist clay and burnt brick, and by allowing the kiln to remain in this state until the heat has subsided. The length of time of burning varies, but is often fifteen days or thereabouts.

50. **General Qualities and Uses.**—Bricks, when properly burnt, acquire a degree of hardness and durability that renders them suitable for nearly all the purposes to which stone is applicable; for, when carefully made, they are in strength, hardness, and durability but little inferior to the ordinary kinds of building-stone. They remain unchanged under the extremes of temperature, resist the action of water, set firmly and promptly with mortar, and, being both cheaper and lighter than stone, are preferable to it for many kinds of structures, as for the walls of houses, small arches, etc.

The Romans employed bricks in the greater part of their constructions. The scarcity of stone in Holland and the Netherlands led to their extensive use, not only in private but in their public buildings, and these countries abound in fine specimens of brick-work.

51. **Characteristics of good Bricks.**—Good bricks should be regular in shape, with plane surfaces and sharp edges; the opposite faces should be parallel, and adjacent faces perpendicular to each other.

They should be free from cracks and flaws; be hard; possess a regular form, and uniform size; and, where exposed to great heat, infusibility.

They should give a clear, ringing sound when struck; and when broken across, they should show a fine, compact, uniform texture, free from air-bubbles and cracks.

They should not absorb more than $\frac{1}{15}$ of their weight of water.

52. From the nature of the process of burning, it will be evident that in the same kiln must be found bricks of very different qualities. There will be at least three varieties: 1, bricks which are burned too much; 2, those, just enough; and, 3, those, not enough. The bricks forming the arches and adjacent to the latter, being nearer the fire, will be burnt to great hardness, or perhaps vitrified; those in the interior will be well burnt; and those on top and near the exterior will be under-burned. The first are called **arch** brick; the second, **body, hard,** or, if the clay had contained ferrous-oxide, **cherry red**; and the third, **soft, pale,** or **sammel** brick.

The arch bricks are very hard but brittle, and have but slight adhesion with mortar; the soft or sammel, if exposed to the weather, have not requisite strength or durability, and can, therefore, be used only for inside work.

53. **Pressed Brick.**—Pressed brick are made by putting the raw bricks, when nearly dry, into moulds of proper shape, and submitting them to a heavy pressure by machinery. They are heavier than the common brick. All machine-made bricks partake somewhat of the nature of pressed brick.

54. **Fire-bricks.**—Fire-bricks are made of refractory clay which contains no lime or alkaline matter, and remains unchanged by a degree of heat that would vitrify and destroy common brick. They are *baked* rather than burnt, and their quality depends upon the fineness to which the clay has been ground and the degree of heat used in making them.

They are used for facing fireplaces, lining furnaces, and wherever a high degree of temperature is to be sustained.

Bricks light enough to float in water were known to the ancients. During the latter part of the last century M. Fabbroni, of Italy, succeeded in making floating bricks of a material known as *agaric mineral*, a kind of calcareous tufa, called fossil meal. Their weight was only one-sixth that of common brick; they were not affected by the highest temperature, and were bad conductors of heat.

55. Brick-making was introduced into England by the Romans, and arrived at great perfection during the reign of Henry VIII.

The art of brick-making is now a distinct branch of the useful arts, and the number of bricks annually made in this country is very great, amounting to thousands of millions.

The art of brick-making does not belong to that of the engineer. But as the engineer may, under peculiar circumstances, be obliged to manufacture brick, the foregoing outline has been given.

Tiles.

56. **Tiles** are a variety of brick, and from their various uses are divided into three classes, viz.: **roofing, paving,** and **draining** tiles.

Their manufacture is very similar to that of brick, the principal differences arising from their thinness. This requires the clay to be stronger and purer, and greater care to be taken in their manufacture.

Their names explain their use.

2d.—CONCRETES.

57. **Concrete** is the term applied to any mixture of mortar with coarse solid materials, as gravel, pebbles, shells, or fragments of brick, tile, or stone.

The term **concrete** was formerly applied to the mixture made with common lime mortar; **béton**, to the mixture when the mortar used was hydraulic, *i. e.*, will harden under water.

The proportions of mortar and coarse materials are determined by the following principle: *that the volume of cementing substance should always be slightly in excess of the volume of voids of the coarse materials to be united.* This excess is added as a precaution against imperfect manipulation.

Concrete is mixed by hand or by machinery.

One method, by hand, used at Fort Warren, Boston Harbor, was as follows: The concrete was prepared by first spreading out the gravel on a platform of rough boards, in a layer from eight to twelve inches thick, the smaller pebbles at the bottom and the larger on the top, and then spreading the mortar over it as uniformly as possible. The materials were then mixed by four men, two with shovels and two with hoes, the former facing each other, always working from the outside of the heap to the centre, then stepping back, and recommencing in the same way, and continuing the operation until the whole mass was turned. The men with hoes worked each in conjunction with a shoveller, and were required to *rub well into the mortar* each shovelful as it was turned and spread. The heap was turned over a second time, this having been usually sufficient to make the mixture complete, to cover the entire surface of each pebble with mortar, and to leave the mass of concrete ready for use.

Various machines have been devised to effect the thorough mixing of the materials. A pug-mill, a cylinder in an inclined position revolving around its axis, a cubical box revolving eccentrically, and various other machines, have been used.

58. **Uses of Concrete.**—Concrete has been generally used in confined situations, as foundations, or as a backing for massive walls. For many years it has been extensively employed in the construction of the public works throughout the United States, and is now extended in its application, not only to foundations, but even to the building of exterior and partition walls in private buildings. It has of recent years had quite an extensive application in harbor improvements in Europe. There are evidences of its extensive use in ancient times

in Rome; many public buildings, palaces, theatres, aqueducts, etc., being built of this material. It has been asserted that the pyramids of Egypt are built of artificial stone composed of small stone and mortar.

It is especially suitable as a building material when dryness, water-tightness, and security against vermin are of consequence, as in cellars of dwelling-houses, magazines on the ground, or underneath, for storage of provisions, etc.

59. **Remarks.**—In order to obtain uniformly a good concrete by the use of hydraulic lime or cement, or both, it is essential—

1. That the amount of water be just sufficient to form the cementing material into a viscous paste, and that it be systematically applied;
2. That each grain of sand or gravel be entirely covered with a thin coating of this paste; and
3. That the grains be brought into close and intimate contact with each other.

These conditions require more than the ordinary methods and machinery used in making mortars, especially if a superior article be desired.

Patent Stones.

60. Various attempts from time to time have been made to make an imitation which, possessing all the merits, and being free from the defects, of the most useful building-stones, would supplement, if not supersede, them. These imitations are generally artificial sandstones.

Béton Aggloméré.

61. **Béton aggloméré**, or **Coignet-Béton**, is an artificial sandstone, made by M. François Coignet, of Paris, France, in which the grains of sand are cemented together by a lime paste possessing hydraulic properties.

It is made by placing the hydraulic cement with about one-third its volume of water into a mill, and mixing until a plastic and sticky paste is formed. This paste and perfectly dry sand, in suitable proportions, are then put into a powerful mill and mixed together until a pasty powder is formed. The pasty powder is placed in layers of from one and a half to two inches thick, in strong moulds, and rammed by repeated blows of an iron-shod rammer until each layer of material is reduced to about one-third of its original thickness. The upper surface is struck with a straight-edge, and smoothed off with a trowel. The mould is turned over on a bed of sand, and detached from the block. If the block be small, it may be handled after one day; larger pieces should have a longer time to harden.

In common practice, the cement and the sand in a dry state are mixed with shovels, spread out on the floor, and then sprinkled with the proper amount of water. The dampened mixture is shovelled into the mill and thoroughly mixed.

The proportions of sand and lime will vary according to the probable uses of the stone; 6 volumes of sand to 1 of hydraulic lime in powder; or, 5 of sand, 1 of hydraulic lime, and 1 of Portland cement, are sometimes used.

The distinctive features of this béton are the very small proportion of water used, the thorough mixing of the materials, and the consolidation effected by ramming the layers.

If too much water be used, the mixture cannot be suitably rammed; if too little, it will be deficient in strength.

Béton aggloméré is noted for its strength, hardness, and durability, and has had quite an extensive application in France; aqueducts, bridges, sewers, cellars of barracks, etc., have been built with it.

Ransome's Patent Stone.

62. Among other artificial stones that are offered to the builder are several bearing the name of Ransome, an English engineer. The patent silicious stone, Ransome's apœnite, and Ransome's patent stone, are all artificial sandstones, in which the cement is a silicate of lime. They differ mostly in the process of making.

A patent stone has been made in San Francisco and in Chicago, and employed to some extent in those cities.

Principles of Manufacture.—Dry sand and a solution of silicate of soda, about a gallon of the silicate to a bushel of sand, are thoroughly mixed in a suitable mill, and then moulded into any of the forms required. These blocks or forms are then saturated by a concentrated solution of calcium chloride, which is forced through the moulded mass by exhaustion of the air, by gravity, or by other suitable means. The chemical reactions result in the formation of an insoluble silicate of lime, which firmly unites all the grains of the mass into one solid, and a solution of sodium chloride (common salt). The latter is removed by washing with water.

Remark.—The artificial stone thus formed is uniform and homogeneous in its texture, and said to be free from liability to distortion or shrinkage. It is also claimed that it is not affected by variations of climate or temperature.

3D.—ASPHALTIC CONCRETE.

63. **Asphaltic Concrete** is a concrete in which the solid materials are united by **mastic**, a mixture of powdered limestone, or similar material, with artificial or natural combinations of bituminous or resinous substances.

The manufacture of mastics will be described under the head of UNITING MATERIALS; the manufactured product may be bought in blocks ready for use.

Asphaltic concrete is made as follows:

The mastic is broken into small pieces, not more than half a pound each, and placed in a caldron, or iron pot, over a fire. It is constantly stirred to prevent its burning, and as soon as melted there is gradually added two parts of sand to each one of the mastic, and the whole mass is constantly stirred until the mixture will drop freely from the implement used in stirring.

The ground having been made perfectly firm and smooth, covered with ordinary concrete, or otherwise prepared, the mixture is applied by pouring it on the surface to be coated, taking care to spread it uniformly and evenly throughout. A square or rectangular strip is first laid, and then a second, and so on, until the entire surface is completely covered, the surface of each square being smoothed with the float. Before the concrete hardens a small quantity of fine sand is sifted over it and is well rubbed in with a trowel or hand-float.

The thickness of the coating will depend upon its situation, being less for the capping of an arch than for the flooring of a room, and less for the latter than for a hall or pavement that is to be in constant use.

Care is taken to form a perfect union between edges of adjoining squares, and, where two or more thicknesses are used, to make them break joints.

A mixture of coal tar is frequently used as a substitute for mastic.

Uses.—The principal uses of asphaltic concrete are for paving streets, side-walks, floors of cellars, etc.

4TH.—GLASS.

64. **Glass** is a mixture of various insoluble silicates. Its manufacture depends upon the property belonging to the alkaline silicates, when in a state of fusion, of dissolving a large quantity of silica. The mixture hardens on cooling, and is destitute of crystalline structure.

Uses.—Glass is extensively used in building, as a roof-covering for conservatories, ornamental buildings, railroad depots, and other structures for which the greatest possible light or the best-looking material is required. Other uses, as for windows, sky-lights, doors, etc., are familiar to every one.

65. **Glazing** is the art of fixing glass in the frames of windows. The panes are secured with putty, a composition of whiting and linseed-oil with sometimes an addition of white lead. Large panes should be additionally secured by means of small nails or brads.

CHAPTER III.

METALS.

66. The metals used in engineering constructions are **Iron, Steel, Copper, Zinc, Tin, Lead,** and some of their alloys.

IRON AND STEEL.

67. **Iron** has the most extensive application of all the metals used for building purposes. It is obtained from the ore by smelting the latter in a blast-furnace. When the fuel used is coal, the blast is generally of hot-air; in this process, known as the *hot-blast*, the air, before being forced into the furnace, is heated high enough to melt lead.

When the metal has fused, it is separated from the other substances in the ore, and is allowed to combine with a small amount of carbon, from 2 to 5 per cent., forming a compound known as *cast-iron*.

A sufficiency of cast-iron having accumulated in the furnace, the latter is tapped, and the molten metal running out is received in sand in long straight gutters, which have numerous side branches. This arrangement is called the *sow* and *pigs;* hence the name of pig-iron.

The iron in the pig is in a shape to be sent to market, and in suitable condition to be remelted and cast into any required form, or to be converted into **wrought or malleable** iron.

Impurities.—The strength and other good qualities of the iron depend mainly on the *absence* of *impurities*, and especially of those substances known to cause brittleness and weakness, as sulphur, phosphorus, silicon, calcium, and magnesium.

CAST-IRON.

68. **Cast-Iron** is a valuable building material, on account of its great strength, hardness, and durability, and the ease with which it can be cast or moulded into the best forms for the purposes to which it is to be applied

Varieties of Cast-Iron.—Cast-iron is divided into six varieties, according to their relative hardness. This hardness seems to depend upon the proportion and state of carbon in the metal, and apparently not so much on the total amount of carbon present in the specimen, as on the proportionate amounts in the respective states of mechanical mixture and of chemical combination. Manufacturers distinguish the different varieties by the consecutive whole numbers from 1 to 6.

No. 1 is known as **gray cast-iron**, and No. 6 as **white cast-iron.** They are the two principal varieties.

Gray Cast-Iron, of good quality, is slightly malleable when cold, and will yield readily to the action of the file if the hard outside coating is removed. It has a brilliant fracture of a gray, sometimes bluish gray, color. It is softer and tougher, and melts at a lower temperature, than white iron.

White Cast-Iron is very brittle, resists the file and chisel, and is susceptible of high polish. Its fracture presents a silvery appearance, generally fine-grained and compact.

The intermediate varieties, as they approach in appearance to that of No. 1 or No. 6, partake more or less of the properties characteristic of the extreme varieties.

Numbers 2 and 3, as they are designated, are usually considered the best for building purposes, as combining strength and pliability.

Appearances of Good Cast-iron.

69. A medium-sized grain with a close compact texture indicates a good quality of iron. The color and lustre presented by the surface of a recent fracture are good indications of its quality. A uniform dark-gray color with a high metallic lustre is an indication of the best and strongest iron. With the same color, but less lustre, the iron will be found to be softer and weaker. No lustre with a dark and mottled color indicates the softest and weakest of the gray varieties.

Cast-iron, of a light-gray color and high metallic lustre, is usually very hard and tenacious. As the color approaches to white, and as the metallic changes to a vitreous lustre, hardness and brittleness of the iron become more marked; when the extreme, a dull or grayish white color with a very high vitreous lustre, is attained, the iron is of the hardest and most brittle of the white variety.

70. **Test of its Quality.**—The quality of cast-iron may be tested by striking a smart stroke with a hammer on the edge of a casting. If the blow produces a slight indentation, without any appearance of fracture, the iron is shown to be slightly malleable, and therefore of a good quality; if, on the contrary, the edge is broken, there is an indication of brittleness in the material, and consequent want of strength.

71. **Strength.**—The strength of cast-iron varies with its density, and the density depends upon the temperature of the metal when drawn from the furnace, the rate of cooling, the head of metal under which the casting is made, and the bulk of the casting.

From the many causes by which the strength of iron may be influenced, it is very difficult to judge of the quality of a casting by its external characters; however, a uniform appearance of the exterior devoid of marked inequalities of surface, generally indicates uniform strength; and large castings are generally proportionally weaker than small ones.

WROUGHT OR MALLEABLE IRON.

72. **Wrought**, or **Malleable Iron**, in its perfect condition, is simply *pure iron*.

It generally falls short of such condition to a greater or less extent, on account of the presence of the impurities referred to in a previous paragraph. It contains ordinarily more than one-quarter of one per cent. of carbon.

It may be made by direct reduction of the ore, but it is usually made from cast-iron by the process called *puddling.*

Wrought-iron is tough, malleable, ductile and infusible in ordinary furnaces. At a white heat it becomes soft enough to take any shape under the hammer, and admits of being *welded.* In order to weld two pieces together, each surface should be free from oxide. If there be any oxide present, it is easily removed by sprinkling a little sand or dust or borax over the surfaces to be joined; either of these forms with the rust a fusible compound, which is readily squeezed out by the hammering or rolling.

Appearances of good Wrought-iron.

73. The fracture of good wrought-iron should have a clear gray color, metallic lustre, and a fibrous appearance. A crystalline structure indicates, as a rule, defective wrought-iron. *Blisters, flaws,* and *cinder-holes* are defects due to bad manufacture.

Strength.—The strength of wrought-iron is very variable, as it depends not only on the natural qualities of the metal, but also upon the care bestowed in forging, and upon the greater or less compression of its fibres when it is rolled or hammered into bars.

Forms.—The principal forms in which wrought-iron is sent to market are **Bar-iron, Round-iron, Hoop** and **Sheet-iron,** and **Wire.**

Bar-iron comes in long pieces with a rectangular cross-section, generally square, and is designated as 1 inch, 1½ inch, 2 inch, according to its dimensions. It is then cut and worked into any shape required.

Bars receive various other forms of cross-section, depending upon the uses that are to be made of them. The most common forms are the T, H, I, and L, cross-sections, called T-iron, H-iron, etc., from their general resemblance to these letters, and one whose section is of this shape, ⊔, called channel iron. The section like an inverted U is frequently seen.

Round iron comes in a similar form, except the cross-section is circular, and it is known, in the same way, as 1 inch, 2 inch, etc.

Hoop and Sheet-iron are modifications of bar-iron, the thickness being very small in comparison with the width.

Corrugated iron is sheet-iron of a modified form, by which

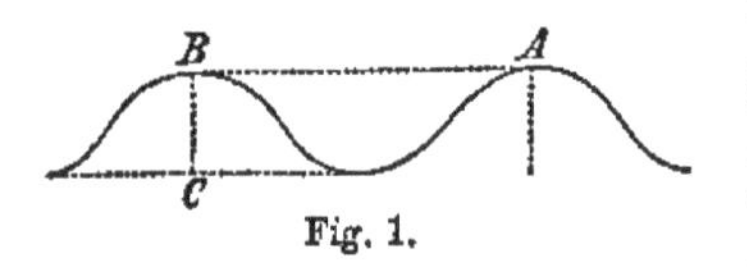

Fig. 1.

its strength and stiffness are greatly increased. The distance between the corrugations, A B, (Fig. 1.) varies, being 3, 4, or 5 inches; the depth, B C, being about one-fourth A B.

Iron Wire.—The various sizes of wire might be considered as small sizes of round-iron, distinguished by numbers depending on the dimensions of cross-section, except that wire is *drawn* through circular holes in a metal plate, while round-iron is *rolled*, to obtain the requisite cross-sections.

The numbers run from 0 to 36; No. 0 wire has a diameter equal to one-third of an inch, and No. 36 one equal to .004 of an inch; the other numbers being contained between these, and the whole series being known as the Birmingham Wire Gauge.

A series in which the numbers run from 0 to 40, the extremes being nearly the same as that just given, is sometimes used. It is known as the American Gauge.

STEEL.

74. Steel, the hardest and strongest of the metals, is a chemical combination of iron and carbon, standing between wrought and cast-iron.

No sharp dividing line can be drawn between wrought-iron and steel, based on the proportions of carbon present in the product. The differences in their physical properties are largely due to the process of manufacture. Many of the properties peculiar to wrought-iron have been found to disappear upon melting the iron, showing that they were the result of the manipulation to which the iron was subjected.

The term **steely-iron, or semi-steel,** has been applied when the compound contains less than 0.5 per cent. of carbon; **steel,** when containing more than this, and less than 2 per cent.; but when 2 per cent. or more is present, the compound is termed cast-iron, as before stated.

75. Steel is made from iron by various processes, which are of two general classes; the one in which carbon is added to malleable iron; the other in which a part of the carbon is abstracted from cast-iron. Like iron, steel is seldom pure, but contains other substances which, as a rule, affect it injuriously. There are, however, some foreign substances which, introduced into the mass during manufacture, have a bene-

ficial effect upon the steel by increasing its hardness and tenacity and making it easier to forge and weld.

76. Steel, used for building purposes, is made generally by one of three processes:

1. By fusion of blister steel in crucibles; as cast-steel;
2. By blowing air through melted cast-iron; as Bessemer steel; or—
3. By fusion of cast-iron on the open hearth of a reverberatory furnace, and adding the proper quantities of malleable iron or scrap steel; as Siemens-Martin steel.

77. The different kinds of steel are known by names given them either from their mode of manufacture, their appearance, from some characteristic constituent, or from some inventor's process; such are *German-steel, blister-steel, shear-steel, cast-steel, tilted-steel, puddled-steel, granulated-steel, Bessemer-steel,* etc.

German-steel is produced direct from certain ores of iron, by burning out a portion of the carbon in the cast-iron obtained by smelting the ore. It is largely manufactured in Germany, and is used for files and other tools. It is also known as **natural** steel.

Blister-steel is made by a process known as "*cementation*," which produces a direct combination of malleable iron and carbon. The bars, after being converted into steel, are found covered with *blisters*, from which the steel takes its name. It is brittle, and its fracture presents a crystalline appearance. It sometimes receives the name of *bar-steel.*

Shear-steel is made by putting bars of blister-steel together, heating and welding them under the forge-hammer, or between rolls; the product is called "Shear-steel," "Double," "Single," or "Half," from the number of times the bars have been welded together. It is used for tools.

Cast-steel, known also as *crucible*-steel, is made by breaking blistered steel into small pieces, and melting it in close crucibles, from which it is poured into iron moulds. The resulting ingot is then rolled or hammered into bars.

Its fracture is of a silvery color, and shows a fine, homogeneous, even, and close grain. It is very brittle, acquires extreme hardness, and is difficult to weld without a flux.

This is the finest kind of steel, and the best adapted for most purposes in the arts; but, from its expensiveness, it is not much used in building.

Tilted-steel is made from blistered steel by moderately heating the latter and subjecting it to the action of a tilt or trip-hammer; by this means the tenacity and density of the steel are increased.

Puddled-steel is made by puddling pig-iron, and stopping the process at the instant when the proper proportion of carbon remains.

Granulated-steel is made by allowing the melted pig-iron to fall into water, so that it forms into grains or small lumps; the latter are afterwards treated so as to acquire the proper proportion of carbon, and are then melted together.

Bessemer-steel, which takes its name from the inventor of the process, is made by direct conversion of cast-iron into steel. This conversion is effected either by decarbonizing the melted cast-iron until only enough of carbon is left to make the required kind of steel, or, by removing all the carbon, and then adding to the malleable iron remaining in the furnace the necessary proportion of carbon; the resulting product is then immediately run into large ingots.

Siemens-Martin steel is another variety of steel obtained directly from the cast-iron, and takes its name from the inventors of the process. In this process, the carbon is not removed by a blast of atmospheric air, as in the Bessemer process, but by the oxygen of the iron ore or iron scales, etc., the oxygen being freed as a gas during combustion.

In each of the last two processes, the temperature is so great as to melt wrought-iron with ease.

There are other kinds of steel, possessing certain characteristics peculiar to themselves or claimed for them, but whose process of manufacture is not publicly known.

78. **Hardening and Tempering.**—Steel is more granular than iron, and is much more easily melted, but the great difference between them is the capability of the steel to become extremely hard and elastic when *tempered*. The quality of the steel depends in a great measure on the operation of hardening and tempering.

It is hardened by being heated to a cherry-red color, and then being suddenly cooled by being plunged into some cold liquid. In this way it is rendered very brittle, and so hard as to resist the hardest file. To give elasticity, it is tempered; this is done by heating the hardened steel to a certain degree, and cooling it quickly; the different degrees of heat will depend upon the use to which the steel is to be put.

These qualities of hardness and elasticity adapt it for various uses, for which neither cast nor wrought-iron would be suitable.

DURABILITY OF IRON AND STEEL.

79. Constructions in these metals are, like those in wood, subject to the same general conditions. They may be ex-

posed to the air in a dry place, or in a damp place, be kept alternately wet and dry, or be entirely immersed in fresh or salt water.

Their exposure to the air or moisture, especially if an acid be present, is followed by rusting which proceeds with rapidity after it once begins. The corrosion is more rapid under exposure to alternate wetness and dryness than in either of the other cases.

Cast-iron is usually coated with a film of graphite and ferrous silicate, produced by the action of the sand of the mould on the melted iron; this film is very durable, and, if not injured, the casting will last a long time without rusting.

Iron kept in a constant state of vibration rusts less rapidly than in a state of rest.

Iron completely imbedded in brick-work or masonry is preserved from rust, and in cathedrals and other ancient buildings it has been found in good condition after six hundred years. In these cases the iron was probably protected by the lime in the mortar, the latter being a good preservative.

The rapid deterioration of iron-work when exposed to the air and to moisture makes its protection, so as to increase its durability, a matter of great importance.

PROTECTION OF IRON-WORK.

80. The ordinary method, used to protect iron from rust, is to cover its surface with some material that withstands the action of the air and moisture, even if it be for a limited time.

The following are some of the methods:

By **painting.**—The surface of the iron is covered with a coat of paint. Red and white lead paints, ochreous or iron oxide paints, silicate paints, and bituminous paints, all are used. For this purpose, the value of the paint depends greatly upon the quality of the oil with which it is mixed. The painting must be renewed from time to time.

By **japanning.**—The iron being placed in a heated chamber, or furnace, the paint is there applied, and is to some extent absorbed by the iron, forming over it a hard, smooth, varnish-like coating.

By the use of **coal-tar.**—The iron is painted with coal-tar alone or mixed with turpentine or other substances; another method consists in first heating the iron to about 600° Fahr., and then boiling it in the coal-tar.

By the use of **linseed oil.**—The iron is heated, and the surface while hot is smeared over with cold linseed-oil.

By **galvanizing.**—This term, "galvanized iron," is applied to articles of iron coated with zinc. The iron, being thoroughly cleaned and free from scale, is dipped into a bath of melted zinc, and becomes perfectly coated with it. This coating protects the iron from direct action of the air and moisture, and as long as it lasts intact the iron is perfectly free from rust.

COPPER.

81. This metal possesses great durability under ordinary exposure to the weather, and from its malleability and tenacity is easily manufactured into thin sheets and fine wire.

When used for building purposes, its principal application is in roof-coverings, gutters, and leaders, etc. Its great expense, compared with the other metals, forms the chief objection to its use.

ZINC.

82. This metal is used much more than copper in building, as it is much cheaper and is exceedingly durable. Though zinc is subject to oxidation, the oxide does not scale off like that of iron, but forms an impervious coating, protecting the metal under it from the action of the atmosphere, thus rendering the use of paint unnecessary.

In the form of sheets, it can be easily bent into any required shape.

The expansion and contraction caused by variations of temperature are greater for zinc than iron, and when zinc is used for roof-coverings, particular attention must be paid to seeing that plenty of *play* is allowed in the laps.

Zinc, before it is made into sheets or other forms, is called *spelter*.

TIN.

83. This metal is only used, in building, as a coating for sheet-iron or sheet-copper, protecting their surfaces from oxidation.

LEAD.

84. This metal was at one time much used for roof-covering, lining of tanks, etc. It is now almost entirely superseded by the other metals.

It possesses durability, but is wanting in tenacity; this requires the use of thick sheets, which increase both the expense and the weight of the construction.

ALLOYS.

85. An alloy is a compound of two or more metals, mixed while in a melted state. **Bronze, gun-metal, bell-metal, brass, pewter,** and the various **solders** are some of the alloys that have a limited application to building purposes.

CHAPTER IV.

UNITING MATERIALS.

86. Structures composed of wood and iron have their different portions united principally by means of straps and pins made of solid materials; in some cases, especially in the smaller structures, a cementing material is used, as glue, etc.

The use of straps, pins, and like methods of fastenings will be described under the head of FRAMING.

Structures composed of stone have their different portions united principally by cementing materials, as **limes, cements, mortars,** etc.

GLUE.

87. **Glue** is a hard, brittle, brownish product obtained by boiling to a jelly the skins, hoofs, and other gelatinous parts of animals, and then straining and drying it.

When gently heated with water, it becomes viscid and tenacious, and is used as a uniting material. Although possessing considerable tenacity, it is so readily impaired by moisture that it is seldom used in engineering constructions, except for **joiner's** work.

LIMES AND CEMENTS.

LIMES.

88. If a limestone be calcined, the carbonic acid will be driven off in the process, and the substance obtained is generally known as **lime.**

This product will vary in its qualities, depending on the amount and quality of the impurities of the limestone. As a building material, the products are divided into three principal classes:

1. **Common or fat lime.**
2. **Hydraulic lime.**
3. **Hydraulic cement.**

Common lime is sometimes called **air-lime,** because a paste made from it with water will harden only in the air.

Hydraulic lime and cement are also called **water limes and cements,** because a paste made from either of them with water has the valuable property of hardening under water.

The principal use of the limes and cements in the engineer's art is as an ingredient in the mortars and concretes.

Varieties of Limestone.

89. The majority of limestones used for calcination are not pure carbonates, but contain various other substances, the principal of which are silica, alumina, magnesia, etc.

If these impurities be present in sufficiently large quantities, the limestone will yield on calcination a product possessing **hydraulic properties.**

Limestones may therefore be divided into two classes, **ordinary** and **hydraulic,** according as the product obtained by calcination does or does not possess hydraulic properties.

90. **Ordinary Limestone.**—A limestone which does not contain more than ten per cent. of these impurities, produces common lime when calcined. White chalk, and statuary marble, are specimens of pure limestone.

91. **Hydraulic Limestones.**—Limestones containing more than ten per cent. of these impurities are called hydraulic limestones, because they produce, when properly calcined, a lime having hydraulic properties.

The hydraulic limestones are subdivided into **silicious, argillaceous, magnesian** and **argillo-magnesian,** according to the nature of the predominating impurity present in the stone.

Physical Characters and Tests of Hydraulic Limestones.

92. The simple external characters of a limestone, as color, texture, fracture, and taste, are insufficient to enable a person to decide whether it belongs to the hydraulic class.

Limestones are generally of some shade of drab or of gray, or of a dark grayish blue; have a compact texture, even or conchoidal fracture, a clayey or earthy smell and taste. Although the hydraulic limestones are usually colored, still the stone may happen to be white, from the combination of lime with a pure clay.

The difficulty of pronouncing upon the class to which a limestone belongs renders necessary a resort to chemical analysis and experiment.

To make a complete chemical analysis of a limestone requires more skill in chemical manipulations than engineers usually possess; but a person who has the ordinary elementary knowledge of chemistry can ascertain the quantity of clay or of magnesia contained in a limestone, and (knowing this) can pronounce, with tolerable certainty, as to the probabilities of its possessing hydraulic properties after calcination.

Having from the proportions ascertained that the stone will probably furnish a lime with hydraulic properties, a sample of it should be submitted to experiment. The only apparatus required for this purpose is a crucible that will hold about a pint, and a mortar and pestle. The bottom as well as the top or cover of the crucible should be perforated to give an upward current of air and allow the carbonic acid to escape. The stone to be tested is broken into pieces as nearly the same size as possible, not exceeding three-fourths of an inch cube, and placed in the crucible. When more than one specimen is to be tried, and a comparison between them made, there should be several crucibles. Access being had to an anthracite coal-fire in an open grate, or to any other steady fire, the crucibles are embedded in and covered with glowing coals, so that the top and bottom portions of their contents will attain simultaneously a bright-red heat, each crucible containing as nearly as possible the same quantity of stone. If there be only one crucible, two or three of the fragments are removed in forty-five minutes after the stone has reached a red heat; in forty-five minutes afterwards two or three more are taken out, and this repeated for four and a half and perhaps six hours, which time will be sufficient to expel all the carbonic acid. If there be several crucibles, they themselves may be removed in the same order. By this means we will have some samples of the stone that are burnt too much, some not enough, and some of a class between them.

The specimen, if a cement, will not slake when sprinkled with water. By reducing it to a powder in the mortar, mixing it to a stiff paste with water, immersing it in fresh or salt water, and noting the time of setting and the degree of hardness it attains, an approximate value of the cement may be obtained.

Calcination of Limestones.

93. As the object in burning limestone is to drive off the water and carbonic acid from the limestone, many devices have been used to effect it. A pile of logs burning in the open air, on which the limestone or oyster-shells are thrown, has been frequently used to obtain common lime. It is, however, generally manufactured by burning the limestone in a kiln suitably constructed for the purpose.

94. Kilns are divided into two classes: 1st, the **intermittent** kilns, or those in which the fuel is all at the bottom, and the limestone built up over it; and, 2d, the **perpetual** or **draw** kiln, in which the fuel and the limestone are placed in the kiln in alternate layers. The fuel used is either wood or coal. In the first class one charge of lime is burned at a time, and, when one burning is complete, the kiln is completely cleared out previous to a second; while in the latter class fresh layers of fuel and limestone are added at the top as the lime is drawn out at the bottom.

The shapes given to the interiors of kilns are very different. The object sought is to obtain the greatest possible uniform heat with the smallest expenditure of fuel, and for this purpose thick walls are necessary to prevent loss of heat by radiation.

95. **Intermittent Kilns.**—The simplest form of kiln is that represented in Fig 2, in which wood is used for fuel. It has a circular horizontal cross-section, and is made of hammered limestone without mortar.

The cut represents a vertical section through the axis and arched entrance communicating with the interior of a kiln for burning lime with wood; *c, c, c,* large pieces of limestone forming the arch upon which the mass of limestone rests; A, arched entrance communicating with the interior.

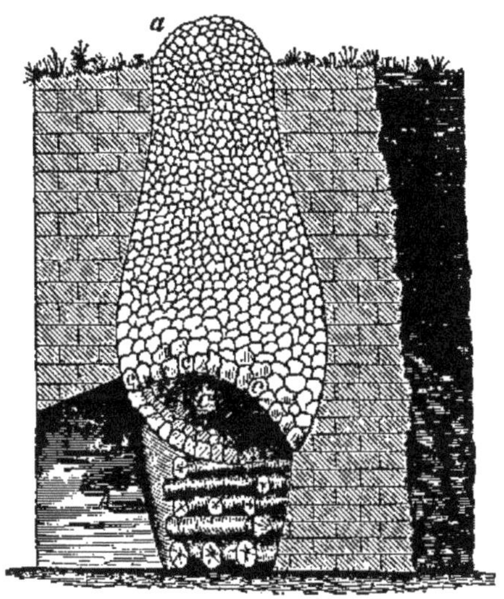

FIG. 2.

It is usually placed on the side of a hill, so that the top may be accessible for charging the kiln.

The largest pieces of the limestone to be burned are formed into an arch, *c*, *c*, *c*, and above this the kiln is filled by throwing the stone in loosely from the top, the largest stones first and smaller ones afterwards, heaping them up, as shown in the figure. The fuel is supplied through the arched entrance, A.

The circular seems the most suitable form for the horizontal sections of a kiln, both for strength and for economy of heat. Were the section the same throughout, or the form of the interior of the kiln cylindrical, the strata of stone, above a certain point, would be very imperfectly burned when the lower strata were calcined just enough, owing to the rapidity with which the inflamed gases arising from the combustion are cooled by coming into contact with the stone. To procure, therefore, a temperature which shall be nearly uniform throughout the heated mass, the horizontal sections of the kiln should gradually decrease from the point where the flame rises, which is near the top of the dome of broken stone, to the top of the kiln. This contraction of the horizontal section from the bottom upward should not be made too rapidly, as the draught would be thereby injured and the capacity of the kiln too much diminished; and in no case should the area of the top opening be less than about one-fourth the area of the section taken near the top of the dome. The proportions between the height and mean horizontal section will depend on the texture of the stone, the size of the fragments into which it is broken for burning, and the greater or less ease with which it vitrifies.

A better kiln than the one shown in Fig. 2 will be obtained by giving an ovoidal shape to the interior, lining it with fire-brick, substituting for the arch of limestones a brick arch with openings to admit a free circulation of air, so as to secure the necessary draught, and arranging it with a fire-grate.

The management of the burning is a matter of experience. For the first eight or ten hours the fire should be carefully regulated, in order to bring the stone gradually to a red heat. By applying a high heat at first, or by any sudden increase of it before the mass has reached a nearly uniform temperature, the stone is apt to shiver, and to choke the kiln by stopping the voids between the courses of stone which form the dome. After the stone is brought to a red heat, the supply of fuel should be uniform until the end of the calcination. Complete calcination is generally indicated by the diminution which gradually takes place in the mass, and which, at this stage, is about one-sixth of the primitive volume; by the broken appearance of the stone which forms the dome, and by the interstices being choked up with fragments of the burnt stone; and by the ease with which an iron bar may be forced down through the burnt stone in the kiln. When these indications of complete calcination are observed, the kiln should be closed for ten or twelve hours to confine the heat and finish the burning of the upper strata.

The defects of the intermittent kilns are the great waste of fuel, and that the stone nearest the fire is liable to be injured by over-burning before the top portions are burnt enough.

96. **Perpetual Kilns.**—Perpetual kilns are intended to remedy these defects, especially the waste of heat. A simple form of a kiln of this class is shown in Figs. 3 and 4. The interior is an inverted frustum of a cone from five to five and a half feet in diameter at bottom, and nine or ten at top, and thirteen or fourteen high. It is arranged with three arched entrances, *a*, *a*, *a*, for drawing the lime, and they are arranged with doors for regulating the draught.

Fig. 3 represents a horizontal section made near the base, and Fig. 4, a vertical section on A B, through the axis of the kiln.

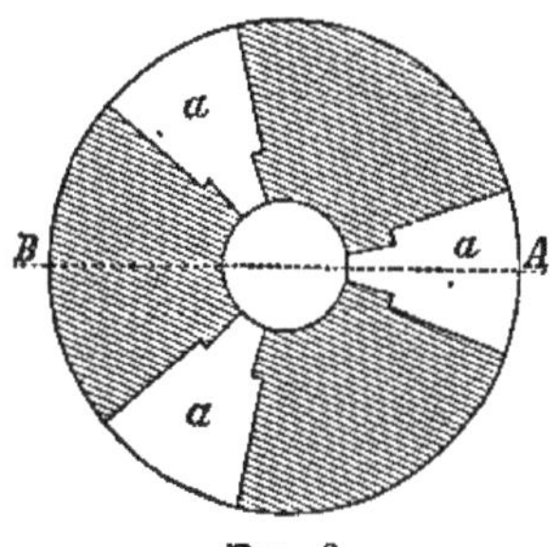

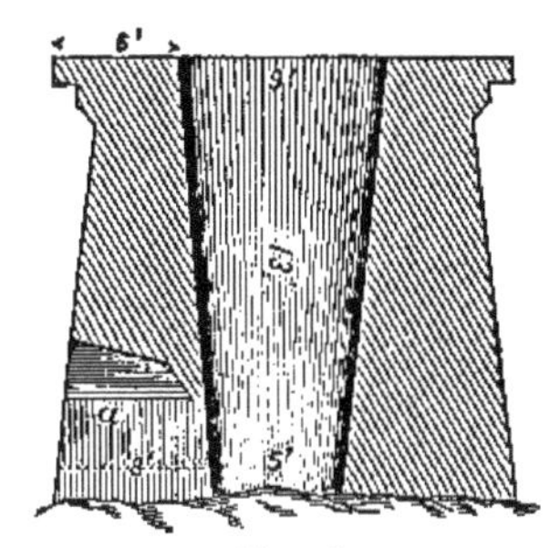

FIG. 3. FIG. 4.

These kilns are arranged for burning by first placing a layer of light wood at the bottom, then a layer of coal, and then a layer of limestone. Layers of coal and limestone follow alternately until the kiln is filled. The lower layer is ignited, and as the burnt mass settles down, and the lime near the bottom is sufficiently burnt, the drawing commences.

Wood is not as convenient a fuel as coal for this kiln, the principal objections being the difficulty of obtaining the pieces always the same size and of distributing it uniformly in the layers.

The **perpetual** kiln is more economical than the **intermittent** in the use of fuel, but requires more skill and caution in its management.

The perpetual kiln invented by Mr. C. D. Page, of Rochester, N. Y., is extensively used in the western part of New York and in Maine. It is known as a perpetual **flame** or **furnace** kiln, is arranged for either wood or coal, anthracite or bituminous, and avoids the defects arising from mixing the fuel and stone together.

The foregoing are types of the kilns used for burning limestones, whether the product is to be common lime or hydraulic cement. The perpetual kiln is generally used for burning limestone for cement.

Figures 5 and 6 represent vertical sections through the axis of the kiln and draw-pit of the ordinary perpetual kilns used in the United States for burning lime-stone for cement.

Fig. 5 represents the section of the kiln used in Maryland

and Virginia; and Fig. 6 of those preferred in New York and Ohio.

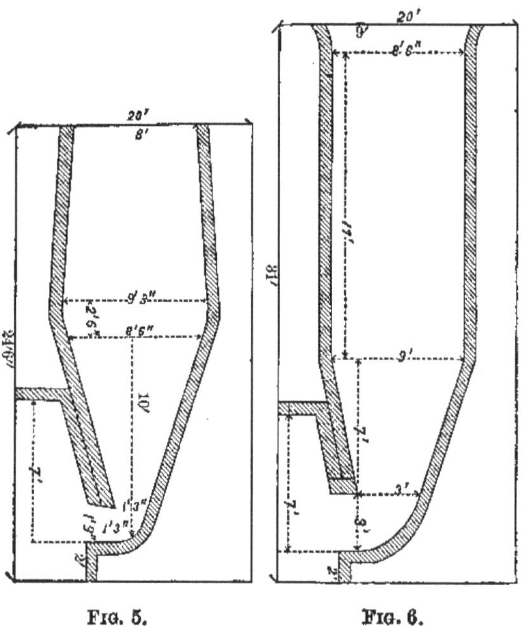

FIG. 5. FIG. 6.

97. The great object of a kiln is to give a cement of good and homogeneous quality with economy of fuel. This uniformity of product is not obtained from either the intermittent or perpetual kilns ordinarily used; some of the stone being over-burnt, while other portions, usually the largest fragments, are under-burnt, in some cases partly raw inside. Both over and under-burnt pieces are difficult to reduce to powder, and materially affect the quality of the cement. It is very evident that dissimilar stones should not be burned together in the same kiln.

Various kilns have been devised to remedy all defects, and still be economical of fuel. The perpetual flame or furnace kiln of Page, before named, and the annular or ring kiln, of which the Hoffman is a type, are noted examples.

Products of Calcination of Limestones.

98. The products obtained by calcination have been divid-

ed into common lime, hydraulic lime, and hydraulic cement.

COMMON LIME.

99. **Lime, common lime, air-lime, quick-lime, caustic lime** (synonymous terms) is a calcium monoxide, produced whenever any variety of pure or nearly pure limestone is calcined with a heat of sufficient intensity and duration to expel the carbonic acid [*carbon dioxide*]. It is amorphous, infusible, somewhat spongy, highly caustic, has a specific gravity of 2.3, and possesses great avidity for water. On being mixed with an equivalent of water, the water is rapidly absorbed with evolution of great heat; the lime swells, bursts into pieces, and finally crumbles into a fine white powder, of which the volume is from two and a half to three and a half times that of its original bulk. In this condition the lime is said to be **slaked** and ready for use in making mortar.

The limestones which furnish the lime of commerce are seldom pure, the impurities amounting sometimes to nearly ten per cent. The purer the limestone, the larger is the increase of volume or the *growth* of the lime in slaking, and the more unctuous to the sight and touch is the paste made therefrom. For this reason the limes made from the purer stones are often called **fat** or **rich** limes, as distinguished from those known as **poor** or **meagre** limes, and which are made from stones containing considerable impurity.

The poor limes are seldom reduced to an impalpable, homogeneous powder by slaking, and are characterized by less *growth*. They yield a thin paste, and are principally used as fertilizers. If it be necessary to use them for building purposes, they should be reduced to a fine powder by grinding; however, they should never be used if it be possible to avoid so doing.

HYDRAULIC LIMES.

100. These occupy an intermediate place between the common limes and the hydraulic cements. They are obtained by calcining limestones in which the impurities, silica, alumina, magnesia, etc., range from *ten* to *twenty* per cent. When ten to twenty per cent. of impurity is chiefly clay, and is homogeneously mixed with the carbonate of lime, the stones are known as argillaceous hydraulic limestones; and when this proportion of impurity is chiefly of silica, they are called silicious hydraulic limestones.

Hydraulic lime, upon being mixed with water, slakes more slowly than the meagre limes, suffers a slight elevation of temperature accompanied by little or no vapor, and an increase of volume rarely exceeding one-third of its original bulk. A paste made from this lime after it has been slaked, hardens under water.

It is not manufactured in the United States, nor is it known if there be in the United States any deposits of the argillaceous hydraulic limestones capable of furnishing good hydraulic lime.

Hydraulic lime, made from the argillaceous limestone, is manufactured in several localities in France, notably at Seilley, about seventy miles from Paris.

The best type of hydraulic lime from the silicious limestone is that known as the **hydraulic lime of Teil**, from the quarries of Teil on the Rhone, Department of Ardèche, France.

HYDRAULIC CEMENT.

101. If the limestone contain more than 20 per cent. and less than 40 of the impurities before named, the product obtained by calcination is an **hydraulic cement.**

Hydraulic cement will not slake, and a paste made from it with water will harden or *set* under water. The rapidity of setting and the degree of hardness will vary with the homogeneous character of the stone, the proportions into which the clay and lime enter, and the intensity and duration of the burning.

The effect of heat on lime-stones varies with the constituent elements of the stone. The pure limestones, and those in which the only impurity is not more than 22 per cent. of clay, will stand a high degree of temperature, losing their carbonic acid and water without fusing, while the others become more or less vitrified when the temperature much exceeds a red heat.

102. There are two general classes of hydraulic cements, the **slow** and the **quick** setting.

If the limestone contain at least 20, and not more than 22 per cent. of clay, and is burned at high heat, the product is a heavy, slow-setting cement.

If there be from 27 to 30 per cent. of clay, aud even as high as 35 in some cases, and the burning be moderate, the result is a light, quick-setting cement.

The stone that might, with proper burning, have yielded a slow-setting cement, will, if burned at a moderate heat, produce a light, quick-setting cement. The Roman cement, that of Vassy, and the hydraulic cements ordinarily made in the United States, are examples of the quick-setting class.

The proportion existing between the impurities and the lime exercises a controlling influence on the properties of the hydraulic cements, and, when the proportion of lime is less than 40 per cent., the stone will, upon calcination, produce neither lime, hydraulic lime, nor hydraulic cement.

POZZUOLANAS.

103. If clay be present in excess in the limestone, the product obtained by calcination is known as **calcareous pozzuolana**, and when there is 10 per cent. of lime or less, simply **pozzuolana.**

Pozzuolana, which gives the name to this class, is a kind of tufa, of volcanic origin, containing about 9 per cent. of lime, 45 of silica, 15 of alumina, and the rest of other impurities, and is found near Rome, in Italy.

It was originally discovered at the foot of Mount Vesuvius, near the village of Pozzuoli, whence its name.

It sometimes exists in a coherent form, but more frequently in powder of coarse, sharp, and angular grains, generally brown in color, running to reddish. If lime be added to supply the deficiency, hydraulic properties can be imparted to the mortar made from it. This fact has been known for centuries, and Vitruvius and Pliny both speak of its high qualities and its use by the Romans in the marine constructions of their time.

104. **Artificial Pozzuolanas.**—They may be prepared by grinding well-burnt bricks to powder, or by burning brick-clay and grinding it.

Trass or Terras.

105. This substance resembles pozzuolana, is used in the same manner, and possesses the same properties. It is used in Holland, being principally obtained from Bonn and Andernach, on the Rhine, below Coblentz. If any deposits exist in the United States, they are not known.

MANUFACTURE OF LIMES AND CEMENTS.

106. Common lime is obtained, as already stated, by the calcination of limestones, in which there is less than ten per cent.

of impurities; the limestone is burnt in kilns, and in the manner already described.

Manufacture of Hydraulic Limes.

107. Hydraulic lime is not manufactured in the United States.

In France it is made by burning the stone in a suitable kiln at a heat sufficient to drive off the carbonic acid. While still warm from the kiln, the stone is sprinkled with from 15 to 20 per cent. of its own weight of water, care being taken not to use enough to convert any portion of it into paste. The slaking soon begins, and the stone falls to pieces. The mass in then thrown together in large heaps, and left undisturbed for six or eight days. It is then screened with sieves of 25 to 30 fine wires to the lineal inch.

The portion which passes the screen is hydraulic lime.

Manufacture of Hydraulic Cements.

108. The hydraulic cements produced at a low heat are light in weight and quick-setting, and the mortars and concretes made from them never attain the strength and hardness of those made from the heavy and slow-setting cements produced by burning with heat of great intensity and duration.

Hydraulic Cements from Argillaceous Limestones.

109. **Heavy, Slow-setting Cements.**—The best example of this class is the Portland cement, which is made from argillaceous limestones, containing from 20 to 22 per cent. of clay, or from an artificial mixture of carbonate of lime and clay in similar proportions. Nineteen-twentieths of all the Portland cement of the present day is artificial. It is manufactured extensively throughout Europe, either by the **wet process**, as in England, or the **dry process**, as in Germany.

The Wet Process.

110. **The wet process**, as practised by the works near London, is as follows: The carbonate of lime is furnished by the

chalks, and the clay is from the shores of the Medway and Thames and adjoining marshes; both the chalk and clay are practically pure.

First. The clay and chalk in the proper proportions, about one to three by weight, are mixed together in a circular wash-mill, so arranged as to thoroughly pulverize the chalk and convert the whole into a semi-fluid paste.

Second. When the thorough mixture is effected, the liquid, resembling whitewash in appearance, is drawn off into reservoirs, where it is left to settle. The heavier material, or *raw cement*, settles to the bottom, and then the surplus water which is clear is removed. Samples are taken from the reservoirs from time to time and tested. If any error be discovered in the proportions, it is corrected.

Third. When by evaporation the mixture has attained the consistency of hard butter or stiff clay, it is removed from the reservoirs to rooms artificially heated, and is spread out for further drying.

Fourth. After it has dried sufficiently, it is burned in suitable kilns at a white heat, just below the point of vitrifaction.

Fifth. The product is then ground between ordinary mill-stones to a powder of the necessary fineness. It is then ready for use.

The Dry Process.

111. **The dry process**, as practised in Germany, is as follows: The carbonate of lime and clay are first kiln-dried at the temperature of 212° Fahr., then mixed together in the proper proportions, between 20 and 23 per cent. of clay to between 80 and 77 per cent. of the carbonate of lime, and reduced to a fine powder. This powder is then made into a stiff paste, and then into blocks about the size of bricks. These bricks are dried and then burnt at a high heat in a kiln, and then ground to powder as in the preceding case.

112. It is an easy matter to pulverize the materials, either wet or dry, mix them, and then grind the burnt stone to a powder. The difficult part is the proper application and management of the heat in burning. The mysterious conversion which takes place in the kiln under a heat of sufficient intensity to make glass, is to some extent beyond our control, and to a great extent beyond our knowledge.

In whatever manner apparently homogeneous limestones may be exposed to burning at a high temperature, it is impossible to avoid the vitrifaction of some layers containing an excess of silica, and to prevent others not having enough clay from producing cements having lime in excess. For this reason an artificial mixture of clay and carbonate of lime is generally relied upon for Portland cement.

The superior quality of Portland cement appears to depend greatly upon the presence of the **double silicate of lime and alumina,** which is formed only at a high heat.

If an argillaceous limestone does not contain at least 20 per cent. of clay, the carbonate of lime is in excess, and the high heat necessary to produce a heavy, slow-setting cement fails to produce the semi-fusion which is the characteristic of such a cement.

113. **Light, Quick-Setting Cements.**—If the limestone contain more than 23 per cent. of clay, as great as 30 per cent. and exceptionally as high as 35 per cent., and the calcination be kept below the point of vitrifaction, it will yield a light, quick-setting cement. The result appears to be **silicate and aluminate of lime** with uncombined clay, but more especially silica, which, being inert, adulterates and injures the cement.

A cement of this kind sets quickly under water, but is far inferior to the Portland cement in hardness and final strength. Those of Vassy, Grenoble, etc., in France, and the English and French Roman cements made from nodules of septaria, belong to this class.

This kind of cement may be made artificially, and was quite extensively used before the superior qualities of the Portland cement were known.

If the limestone contain more than 23 per cent. of clay homogeneously distributed through the mass, and is burnt with a heat of great intensity and duration, similar to that required to produce Portland cement, it generally fuses into a species of slag or glass, and is worthless as a cement.

Hydraulic Cements from Argillo-Magnesian Limestones.

114. The natural hydraulic cements of the United States are made from the limestones whose principal ingredients are carbonate of lime, carbonate of magnesia, and clay.

The usual process of manufacture is to break the stone into pieces not exceeding twelve or fifteen pounds in weight, and burn them in an ordinary kiln, either intermittent or perpetual, the latter being generally used when coal is the fuel. After being burnt, the fragments are crushed by suitable machinery, and then reduced to a powder by grinding. The powder is then packed in barrels and sent to market.

These limestones cannot be burned with the intensity and duration of heat necessary to make Portland cement, without fusing into a slag destitute of hydraulic properties. Like those argillaceous limestones which have more than 23 per cent. of clay, they will, if properly burned, produce a light, quick-setting cement, which is a **silicate and aluminate of lime and magnesia.**

The cements from the valley of Rondout Creek, Ulster County, N. Y., known as Rosendale cement; from near Shepherdstown, Va.; Cumberland, Md.; Louisville, Ky.; Sandusky, Ohio; Utica, Ill.; and other localities in the U. S., are made from this stone, and belong to this class of cements.

The Rosendale cement, which is the most valuable of them, will, under favorable circumstances, attain about one-third of the ultimate strength and hardness of the Portland cement.

Hydraulic Cements from Magnesian Limestones.

115. Pure carbonate of magnesia, known as **magnesite,** when burned at a cherry-red heat, reduced to powder, and made in a paste, possesses hydraulic properties. If the powder be mixed in a paste with magnesium chloride—or, a very good substitute for it, *bittern*, the residue of sea-water after the salt has been separated by crystallization—a cement is made superior in strength and hardness to any other known, not excepting even the Portland. This calcined magnesite has been patented under the name of Union cement.

The dolomites, or magnesian limestones, when burned properly and reduced to a powder, will give a mortar with hydraulic properties; and in general any magnesian limestone containing as high as 60 per cent. of carbonate of magnesia, if properly burned, will yield an hydraulic cement, whether clay be present or not.

Scott's Hydraulic Cement.

116. This is a cement invented by Major Scott, of the Royal Engineers, British Army, and is referred to, not for any marked advantages it possesses, but for the peculiarity of its mode of manufacture.

The limestone is calcined in the usual manner, producing common lime. It is then, in layers of one and a half to two feet thick, laid over the arches of a perforated oven, and brought to a dull glow. The fire is then raked out, and iron pots containing coarse, unpurified sulphur (about fifteen pounds to each cubic yard of lime) are pushed in on the grate-bars, and the sulphur ignited. The oven is closed, so as to prevent the escape of the sulphurous vapor. After the sulphur has been consumed, the mass is allowed to cool, and is then ground to a powder like other cements.

Why lime treated in this manner should acquire hydraulic properties is not fully known.

TESTS FOR LIMES AND CEMENTS.

117. The manufacture of limes and cements having become a special branch of industry in the United States and Europe, the engineer can easily obtain the kinds required for his purposes, and will rarely, if ever, be placed in a position requiring him to make them. He will be more particularly concerned in knowing how to test the samples furnished him, so as to be able to make a judicious selection.

Test for Rosendale Cement.—Rosendale cement should be ground fine enough so that 90 per cent. of it can pass a No. 30 wire sieve of thirty-six wires to the lineal inch both ways; should weigh not less than sixty-eight pounds to the struck bushel, loosely measured; and when made into a stiff paste without sand, and formed into bars, should, when seven days old, sustain, without rupture, a tensile strain of sixty pounds to the square inch of cross-section, the sample having been six days in water.

Test for Portland Cement.—Portland cement should possess the same degree of fineness as just given; should weigh one hundred and six pounds to the struck bushel, loosely measured; and under the same conditions should sustain a tensile strain of one hundred and seventy-eight pounds to the square inch of cross-section.

Test for other varieties.—The relative value of other varieties of cements can be determined by subjecting them to similar tests and comparing the results.

Wire Test.—The wire test was formerly used to determine the hydraulic activity of samples. It is as follows: The paste is made into cakes of one and a quarter inches in diameter and five-eighths of an inch thick, and is immersed in water of an established temperature (65° F.); the times are then noted which are required before the cakes will support, without depression, the point of a wire one-twelfth of an inch in diameter loaded to weigh one-quarter of a pound, and of another wire one-twenty-fourth of an inch in diameter weighing one pound. This test is still used to some extent, especially by the French.

The wire test, when applied to cement pastes without sand, does not give a correct indication of the values of their hydraulic properties.

STORAGE OF LIMES AND CEMENTS.

118. Hydraulic limes and cements deteriorate by exposure to the air. If liable to be kept on hand for several months, they should be stored in a tight building free from draughts of air, and the casks should be raised several inches above the floor, if stone or earthen.

Cements, that have been injured by age or exposure, may have their original energy restored by recalcination. Samples have been restored by being submitted to a red heat of one hour's duration.

Common lime, for the same reasons, should be preserved in tight vessels. It is usually sent to market in barrels, and is reduced to powder by slaking. The fineness of the powder, its growth, the phenomena of slaking, and the degree of unctuousness of the paste made with water, are the tests for good lime.

MORTAR.

119. **Calcareous Mortar**, ready for use, is a mixture, in a plastic condition, of lime, sand, and water. It is used to bind together the solid materials in masonry constructions, and to form coatings for the exterior surfaces of the walls and interior of buildings.

It may be divided into two principal classes—**common mortar** when made of common lime, and **hydraulic mortar** when hydraulic lime or cement is used.

When mortar is thin-tempered or in a fluid state, it is known as **grout.**

Hardened Mortar is simply an artificial stone, and should fulfil the essential conditions already given for stone—viz., should possess *strength*, *hardness*, and *durability*. These qualities vary with the quality of the lime or cement employed, the kind and quantity of sand, the method and degree of manipulation, and the position, with respect to moisture or dryness, in which the mortar is subsequently placed.

Common mortar will harden only partially in damp places excluded from free circulation of air, and not at all under water. These places are, on the contrary, favorable to the induration of hydraulic mortars.

Slaked Lime.

120. Before the lime is mixed with sand to form mortar, it must first be **slaked**.

The methods of slaking lime are classed under three heads: 1, **drowning**; 2, **immersion**; and 3, **spontaneous or air** slaking.

The **first** is to throw on the lumps of lime, just as they come from the kiln, enough water to reduce them to paste. The workmen are apt to throw on more water than is required; hence the name.

The **second** is to break the lumps of lime into pieces not exceeding an inch through, then to place them in a basket or other contrivance, and to immerse them in water for a few seconds, withdrawing them before the commencement of ebullition. A modification of this method is to form heaps of the proper size of these broken lumps, and then to sprinkle a certain quantity of water upon the lime, the amount of water being from one-fourth to one-third the volume of the lime, the rose of a watering-pot being used in sprinkling.

The **third** is to allow the lime to slake spontaneously by absorbing moisture from the surrounding atmosphere.

The first method is the one most generally used in the United States.

The lumps of lime are collected together in a layer from six to eight inches deep, in a water-tight box, or a basin of sand coated over with lime-paste to make it hold water, and then the amount of water sufficient to reduce the lime to a paste is poured over them. This amount of water is approximately determined by a trial of a small quantity of lime beforehand. It is important that all the water necessary should be added at the beginning. After an interval of five or ten minutes the water becomes heated to the boiling-point, and all the phenomena of slaking follow.

The workmen are apt to use too much water in the beginning, or, not using enough, to add more when the slaking is in progress. In the first case the resulting paste will be too thin, and in the latter the checking of the slaking will make the product lumpy.

As soon as the water is poured on the lime, it is recommended to cover the mass with canvas or boards, or with a layer of sand of uniform thickness after the slaking is well under way. Another recommendation is, that the lime be not stirred while slaking.

Writers disagree as to the relative values of these three methods of slaking lime. Supposing that in the first process *all the water required to produce a stiff paste, and no more than this, is poured on at the beginning*, these modes may be arranged in their order of superiority, as follows:

For fat limes: 1, drowning, or the ordinary method; 2, spontaneous slaking; and, 3, immersion. For hydraulic limes: 1, ordinary method; 2, immersion; and, 3, spontaneous slaking.

In the matter of cost, the first mode has a decided advantage over the others. The second is not only expensive from the labor required, but difficult from the uncertainty of the period of immersion at the hands of the workmen. The third involves the expense of storage-rooms or sheds and time, a period from twenty days to even a year being necessary to complete the slaking.

Preservation of the Lime after being Slaked.

121. The paste obtained by the first mode may be preserved any length of time if kept from contact with the air. It is usual to put it in tight casks, or in reservoirs; to put it in trenches and cover it with sand will be sufficient for its preservation.

The powder, from the second and third modes, may be preserved for some time, by placing it in casks or bins with covers, or in dry sheds in heaps, covered over with cloth or dry sand.

General Treussart thought that lime should be used immediately after it was slaked. In this country such is the ordinary practice. The general opinion of engineers is however adverse to this practice, and in some parts of Europe it is the custom to slake the lime the season before it is used.

Sand.

122. **Sand** is the granular product arising from the disintegration of rocks. It may therefore, like the rocks from which it is derived, be divided into three principal varieties—the silicious, the calcareous, and the argillaceous.

Sand is sometimes named from the locality where it is obtained, as **pit-sand**, which is procured from excavations in inland deposits of disintegrated rock; **sea-sand** and **river-sand**, which are taken from the shores of the sea or rivers.

Builders again classify sand according to the size of the grain. The term **coarse sand** is applied when the grain varies between $\frac{1}{8}$ and $\frac{1}{16}$ of an inch in diameter; the term **fine sand**, when the grain is between $\frac{1}{16}$ and $\frac{1}{24}$ of an inch in diameter; and the term **mixed sand** is used for any mixture of the two preceding kinds.

The usual mode of determining the size of sand is to screen it by passing it through sieves of various degrees of fineness. The sieves are numbered according to the number of openings in a square inch of the wire gauze of which they are made.

The silicious sands, arising from the quartzose rocks, are the most abundant, and are usually preferred by builders. The calcareous sands, from hard calcareous rocks, are more rare, but form a good ingredient for mortar. Some of the argillaceous sands are valuable, as when mixed with common lime they impart to it hydraulic properties.

The property, which some argillaceous sands possess, of forming with common or slightly hydraulic lime a compound which will harden under water, has long been known in France, where these sands are termed *arènes*. The sands of this nature are usually found in hillocks along river valleys. These hillocks sometimes rest on calcareous rocks or argillaceous tufas, and are frequently formed of alternate beds of sand and pebbles. The sand is of various colors, such as yellow, red, and green, and seems to have been formed from the disintegration of clay in a more or less indurated state. They form, with common lime, an excellent mortar for masonry, exposed either to the open air or humid localities, as the foundations of edifices.

Pit-sand has a rougher and more angular grain than river or sea sand, and on this account is generally preferred by builders for mortar to be used in brick or stone work.

River and sea sand are by some preferred for plastering,

because they are whiter and have a finer and more uniform grain than pit-sand.

The sand used in common mortar should be clean, sharp, and neither too coarse nor too fine.

Its cleanliness may be known by its not soiling the fingers when rubbed between them; and its sharpness can be told by filling the hand and closing it firmly, listening to the sounds made by the particles when rubbed against each other.

Dirty sand, as well as sea sand, should before using be washed, to free it from impurities.

Sand enters mortar as a mechanical mixture, and is used to save expense by lessening the quantity of lime, to increase the resistance of the mortar to crushing, and to lessen the amount of shrinking during the drying of the mortar.

It injures the tenacity of mortar, and if too much be used the mortar will crumble when dry.

PROPORTIONS OF INGREDIENTS.

123. The quantity or proportion of sand to the lime varies with the quality of the lime and the uses to be made of the mortar.

Vicat gives for common mortar the proportion of 2.4 parts of sand to one of pure slaked lime in paste, by measure.

The practice of the United States Corps of Engineers in making hydraulic mortars has been to add from 2.5 to 3.5 in bulk of compact sand to one of lime and cement, or cement alone, in thick paste.

THE METHOD AND DEGREE OF MANIPULATION.

124. The ingredients of mortar are incorporated either by manual labor or by machinery; the latter method gives results superior to the former. The machines used for mixing mortar are the ordinary pug-mill (Fig. 7), like those employed by brickmakers for tempering clay, the grinding-mill (Fig. 8), or mill of any other pattern suitable for the work. The grinding-mill is a better machine for this purpose than the pug-mill, because it not only reduces the lumps found in the most carefully-burnt stone after the slaking is apparently complete, but it brings the lime to the state of a uniform stiff paste, in which condition it should be before the sand is incorporated with it.

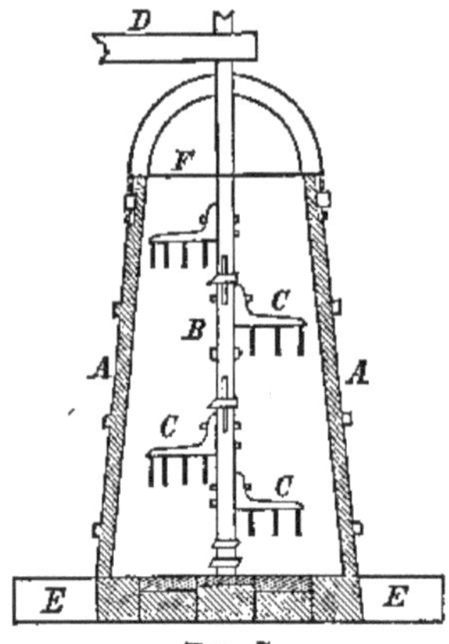

FIG. 7.

Fig. 7 represents a vertical section through the axis of a **pug-mill** for mixing or tempering mortar. This mill consists of a hooped vessel of the form of a conical frustum, which receives the ingredients, and of a vertical shaft, to which arms with teeth resembling an ordinary rake, are attached for the purpose of mixing the ingredients.

A, A, section of sides of the vessel.

B, vertical shaft, to which the arms C are affixed.

D, horizontal bar for giving a circular motion to the shaft B.

E, sills of timber supporting the mill.

F, wrought-iron support, through which the upper part of the shaft passes.

Fig. 8 represents a part of a **mortar mill** for crushing lime and tempering mortar.

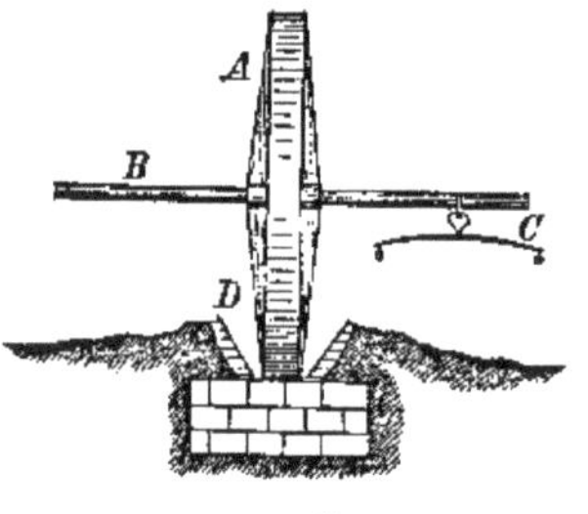

FIG. 8.

A, a heavy wheel of timber or cast iron.

B, a horizontal bar passing through the wheel, fixed to a vertical shaft, and arranged at the other end, C, with the proper gearing for a horse.

D, a circular trough which receives the ingredients to be mixed. The trough is of trapezoidal cross-section, from 20 to 30 feet in diameter, about 18 inches wide at top, 12 inches deep, and is built of hard brick, stone, or timber laid on a firm foundation.

A good example of a grinding-mill is given on page 98 of Lieut. W. H. Wright's "Treatise on Mortars," in describing the mill used at Fort Warren, Boston Harbor.

The **steam mortar-mill**, in which the wheels or stones revolved on edge, and which was used at Fort Taylor, Key West, Florida, the **mortar mill** of Greyveldinger, used in Paris, in which a revolving screw performs the mixing, as also the Fort Warren mortar-mill above alluded to, are described in Gillmore's "Treatise on Limes, Cements, and Mortars."

125. **Process of making Mortar with the Mill.**—The lime-paste is first put in the circular trough, and to this is added by measurement about one-half of the sand required for the batch. The mill is set in motion, and the ingredients thoroughly incorporated. The remainder of the sand is then added, and as much water as may be necessary to bring the mass to the proper consistency.

If common mortar is to be rendered hydraulic by adding hydraulic cement, the latter should be added to the lime-paste just before the mill is set in motion; a very quick-setting cement should not be added until the last portions of sand are thrown in.

126. **Process by Hand.**—The measure of sand required for the batch is placed on the floor and formed into a basin, in which the unslaked lime is placed, the lumps being broken to the proper size. The necessary quantity of water is poured on by a hose, watering-pots, or ordinary buckets, and the lime stirred as long as vapor is evolved. The ingredients are well mixed together with the shovel and hoe, a little water being added occasionally if the mass be too stiff. It is customary then to heap the mortar compactly together, and allow it to remain until ready for use.

The rule in mixing mortar, either by machinery or hand, is to see that *the lime and sand be thoroughly incorporated.*

SETTING OF MORTARS.

127. A mortar has **set** when it has become so hard that its form cannot be altered without fracture. The **set** is determined by the wire test. If the mortar supports the point of the wire without depression or penetration, it is assumed that the mortar has set.

Theory of Setting of Mortars.

128. Common mortar slowly hardens in the air, from the surface towards the interior, by drying and by the absorption of carbonic acid. The process is slow, but in time, under favorable circumstances, a hard material is produced. The carbonic acid, absorbed by the mortar, combines with the lime, forming a carbonate with an excess of base, and the hardening is due to this reaction and to pressure.

Hydraulic mortars, and paste made with hydraulic cement, harden by a species of crystallization that takes place when the silicates of lime, alumina and magnesia, which are anhydrous after calcination, become hydrates upon being mixed with water.

The compounds which are formed by burning the limestone fit to produce Portland cement at a high heat require but *three* equivalents of water for their hydration, while those formed at a low heat take *six*. This is probably the cause of the superior strength and hardness attained by the Portland cement.

In the cements obtained from the argillo-magnesian limestones the presence of the silicate of magnesia is given as the reason why these cements are more durable for constructions in the sea, as the silicate of magnesia resists the action of sea-water better than the silicates of lime and alumina, unless other ingredients introduce adverse conditions.

ADHERENCE OF MORTAR.

129. The force with which mortars, in general, adhere to other materials depends on the nature of the material, its texture, and the state of the surface to which the mortar is applied.

In applying mortars, the materials to be joined should be thoroughly moistened — a point too often neglected — and the surfaces made clean. Precautions should be taken to prevent too rapid drying, and the mortar should be as stiff as it can be used, still being in a plastic condition.

Mortar adheres more strongly to brick, and more feebly to wood, than to any other material. Among stones of the same class it generally adheres better to the porous and coarse-grained than to the compact and fine-grained. Among surfaces it adheres more strongly to the rough than to the smooth.

The adhesion of common mortar to brick and stone, for the first few years, is greater than the cohesion of its own particles. The contrary is the case with hydraulic cement.

From experiments made by Rondelet on the adhesion of common mortar to stone, it appears that it required a force varying from 15 to 30 pounds to the square inch, applied perpendicular to the plane of the joint, to separate the mortar and stone after six months' union; whereas only 5 pounds to the square inch were required to separate the same surfaces when applied parallel to the plane of the joint.

HARDNESS, STRENGTH, AND DURABILITY OF MORTARS.

130. The same general rules for determining these qualities in stone are applicable in mortars, and, as with stone, experience is the best test.

The principal causes of deterioration and decomposition of mortars are:

1. Changes of temperature, producing expansions and contractions.

2. Alternations of freezing and thawing, producing exfoliations and disintegrations of the parts exposed to their influence.

Common mortars, which have had time to harden, resist the action of severe frosts very well, if they are made rather *poor*, or with an excess of sand. The proportions should be 2½ volumes, or over, of sand to one of the lime in paste.

Hydraulic mortars set equally well in damp situations and in the open air; and those which have hardened in the air will retain their hardness if afterwards immersed in water. They also resist well the action of frost, if they have had time to set before exposure to it; but, like common mortars, they require to be made with an excess of sand to withstand well atmospheric changes.

To ascertain the strength and compare the qualities of different mortars, experiments have been made upon the resistance offered by them to cross-strains.

The usual method has been to place small rectangular prisms of mortar, upon points of support at their extremities, and subject them to a cross-strain by applying a pressure at a point midway between the bearings.

131. Experiments made upon prisms a year old, which had been exposed to the ordinary changes of weather, gave the following as the average strength of mortars per square inch to resist rupture by a force of tension; the calculations being made from experiments on the resistance offered to a transverse strain:

Mortars of very strong hydraulic lime....	170	pounds.
" ordinary " "	140	"
" medium " "	100	"
" common lime...............	40	"
" " (bad quality)....	20	"

General Totten, late Chief of Engineers, U. S. Army, from his experiments deduced the following general results:

1. That mortar, of hydraulic cement and sand, is the stronger and harder as the quantity of sand is less.

2. That common mortar is the stronger and harder as the quantity of sand is less.

3. That any addition of common lime to a mortar of hydraulic cement and sand, weakens the mortar, but that a little lime may be added without any considerable diminution of the strength of the mortar, and with a saving of expense.

4. The strength of common mortars is considerably improved by the addition of an artificial pozzuolana, but more so by the addition of an hydraulic cement.

5. Fine sand generally gives a stronger mortar than coarse sand.

6. Lime slaked by sprinkling gave better results than lime slaked by drowning. A few experiments made on air slaked lime were unfavorable to that mode of slaking.

7. Both hydraulic and common mortar yielded better results when made with a small quantity of water than when made thin.

8. Mortar made in the mortar-mill was found to be superior to that mixed in the usual way with a hoe.

9. Fresh water gave better results than salt water.

USES OF MORTAR FOR STUCCO, PLASTERING, ETC.

132. The term **plastering is** ordinarily limited to the covering of interior walls and ceilings by coats of mortar, while the mortar covering exterior walls is called **stucco.** This latter term was originally applied to a species of plastering made to resemble marble, being quite hard and capable of receiving a polish. Outside plastering is used often to prevent the rain from penetrating the joints of the masonry, and in general when it is desired to have a smooth surface instead of a rough one.

Both inside and outside plastering, when properly done, require three coats to be used, the first known as the **scratch** coat, the second as the **brown,** and the third as **hard finish,** or stucco. The first coat is common-lime mortar, with a given quantity of bullock's hair mixed with it. It contains ordinarily a larger proportion of sand than common mortar does, so as to reduce the shrinkage to a minimum. When completed and partially dry, and still soft, it is with a pointed stick *scratched* in parallel scorings running diagonally across the surface at right angles to each other. When the first coat is dry enough, the brown coat is applied. This differs from the first in containing less hair in the mixture. This is followed by the third coat, which is hard finish for the inside, or stucco for the outside. The former is a paste of fine lime and plaster of Paris; the latter is a paste of fine lime made stiff with white sand.

If the outer plastering is to be exposed to the weather, it should be made of hydraulic mortar.

MASTICS.

133. **Mastic** is the term generally applied to a mixture of powdered limestone, or similar material, with artificial or natural combinations of bituminous or resinous substances.

It is used as a cement for other materials, or as a coating to render them water-proof.

The term **asphalt** is sometimes employed to designate the bituminous limestone, more generally the mastic after it has been moulded into blocks for transportation, frequently to the product obtained by mixing sand with the mastic, and by some to the raw bitumen or mineral tar. Calling the first **asphalt**, the second would be **asphaltic mastic**, the third **asphaltic concrete**, and the fourth **asphaltum**.

Bituminous Mastic.

134. **Bituminous mastic** is prepared by heating the mineral pitch or asphaltum in a large caldron or iron pot, and stirring in the proper proportion of the powdered limestone. This operation, although very simple in its kind, requires great attention and skill on the part of the workmen in managing the fire, as the mastic may be injured by too low or too high a degree of heat. The best plan appears to be to apply a brisk fire until the boiling liquid commences to give out a thin, whitish vapor. The fire is then moderated and kept at a uniform state, and the powdered stone is gradually added, and mixed in with the tar by stirring the two well together. If the temperature should be raised too high, the heated mass gives out a yellowish or brownish vapor. In this state it should be stirred rapidly, and be removed at once from the fire.

When the mixing is completed, the liquid mass is run into moulds, where it hardens into blocks of convenient shape and size.

The stone above used is a carbonate of lime naturally impregnated with bitumen, called sometimes Seyssel asphalt, from the place where it was quarried. The proportion of bitumen in the Seyssel stone is oftentimes as much as 17 per cent., and the amalgamation is more perfect than that of any artificial compound of the kind yet invented. To prepare it for the operation just described, the stone may be reduced to powder, either by roasting it in vessels over a fire, or by grinding it down in the ordinary mortar-mill. To be roasted, the stone is first reduced to fragments the size of an egg. These fragments are put into an iron vessel, heat is applied, and the stone is reduced to powder by stirring it and breaking it up with an iron instrument. This process is not only less economical than grinding, but the material loses a portion of the bitumen from evaporation, besides being liable to injury from too great a degree of heat. If to be ground, the stone is first broken as for roasting. Care should be taken, during the process, to stir the mass frequently, otherwise it may cake.

To use the mastic, the blocks are remelted, and the mixture, in this state or mixed with sand, is laid on the surface to be coated by pouring it on, generally in squares, care being taken to form a perfect union between edges, and to rub the surface smooth with an ordinary wooden float, especially if another layer is to be laid over the first.

135. **Proportions.**—The proportions for bituminous mastic are about 1 part of asphaltum to 7 or 8 by measure of the powdered limestone, according as the stone contains more or less bitumen.

Any petroleum or naphtha present in the stone must be removed; this is generally done by distillation. Clay in the limestone injures the mastic, and is oftentimes the cause of the cracks seen in asphaltic concrete after it has been laid.

Artificial Mastics.

136. **Artificial Mastics** have been formed by mixing coal-tar, vegetable tar, pitch, etc., with powdered limestone, powdered brick, litharge, etc.; but these mixtures are inferior to the bituminous mastic.

The impurities and volatile ingredients of coal-tar, mineral tar, and similar substances, render them less durable than mineral pitch, and the combinations made with them are inferior to those made with the latter, as might be expected. But, for certain purposes, the artificial mastics are extremely useful, as they are quite cheap and possess in a measure the advantages of bituminous mastic.

USES OF MASTICS.

137. The combinations of asphaltum were well known to the ancients, and a cement made of it is said to have been employed in the construction of the walls of Babylon.

The principal uses of mastic at the present day are for paving streets, sidewalks, floors, cellars, etc., and for forming water-tight coatings for cisterns, cappings of arches, terraces, and other similar roofings.

It has quite an extensive use in Europe at the present time. The principal sources of the asphalt are the Jurassic range in the Val de Travers, Pyrimont, Seyssel on the Rhone, and the neighboring localities, and Bechelbronn (or Lobsan), in Alsace.

Asphaltum alone has been frequently used for coatings, but in time it becomes dry and peels off. But made into mastic, evaporation is prevented and its durability increased.

The use of the mastic, for making asphaltic concrete, has already been described.

CHAPTER V.

PRESERVATIVES.

PAINTS.

138. **Paints** are mixtures of fixed and volatile oils, chiefly those of linseed and turpentine, with certain of the metallic salts and oxides, and with other substances; the latter are used either as pigments or stainers, or to give what is termed a **body** to the paint, and also to improve its drying properties.

Paints are mainly used, as protective agents, to secure wood and metals from the destructive action of air and water. As they possess only a limited degree of durability, they must be renewed from time to time. They are more durable in air than in water.

The principal materials used in painting are: Red and

white lead, red and **yellow ochre, prussian blue, verdigris, lamp-black, litharge, linseed-oil,** and **spirits of turpentine.**

By suitably combining the above, almost any color may be obtained. For example, a *lead* color is obtained by mixing a little lamp-black with the white lead, etc.

Linseed-oil, being boiled with the addition of a small quantity of litharge and sugar-of-lead, forms what is known as **drying oil.**

Spirits of turpentine is not generally used in the paints intended for external and finishing coats, as it does not stand exposure as well as oil.

139. In painting wood, the first thing to be done is to clean and smooth the surface to be covered. If the wood be resinous the knots must be *killed* before the paint is applied; this is done by applying a coat of red lead mixed with *sizing.* The surface being dry, the first coat, generally white lead mixed with linseed oil, is put on; this is called *priming.* This coat being dry, all holes, indentations, heads of nails, etc., should be filled and covered over with putty. The second coat of paint is then applied. If it be old work that is to be repainted, the entire surface should be scrubbed with soap and water, well scraped, and then rubbed down with sand-paper or pumice, in order to get rid of the old paint and to obtain an even, smooth surface.

JAPANNING.

140. **Japanning** is the name given to the process which forms over the surface of the material to be covered, a hard, smooth, varnish-like coating. [Art. 80.]

OILING.

141. **Oiling** is frequently used as a preservative. It may be done either while the surface to be protected is hot or cold. Linseed-oil is the material generally used.

VARNISHES.

142. **Varnishes** are made by dissolving resinous substances in alcohol, or in linseed-oil and spirits of turpentine, just as paints are made by similarly dissolving or mixing pigments.

Varnishes are used for the same purpose as paints, but especially when it is desired to give a clear, shining appearance to the surface on which they are laid.

COAL-TAR.

143. **Coal-tar** is much used as a preservative. It may be applied as a coating for the material, or it may be applied by the process known as "creosoting." [Art. 25.]

ASPHALTUM.

144. **Asphaltum** is used for the same purposes. Its uses are described in Art. 137.

METAL COVERINGS.

145. **Plating.**—Protection is frequently afforded by covering the material with a thin coating of a metal which is not affected, or to a very slight degree, by the destructive agencies to be guarded against.

Zinc applied to iron, by the process of "galvanizing," protects iron from direct action of the air and moisture as long as the coating is perfect. [Art. 82.]

Tin is used for the same purpose.

Nickel has been tried for brass.

OTHER PRESERVATIVES—BY CHEMICAL COMBINATIONS.

146. **Salts of Silica** have been tried for protection of building stones. [Art. 35.]

Various salts have been used to saturate timber, thus changing the albuminous substances in the timber into insoluble compounds by chemical action, and thus increasing its durability. [Art. 25.]

PART II.

STRENGTH OF MATERIALS.

CHAPTER VI.

STRAINS.

147. The materials in a structure are subjected to the action of various forces, according to the kind of construction of which they form a part, and the position they occupy in it.

In planning a structure, two general problems are to be considered.

I. The nature and magnitude of the forces which are to act on it; and,

II. The proper distribution and size of its various parts, so that they shall successfully resist the action of these forces.

In the former, if the intensities, directions, and points of application be known, the effect that the forces will exert may be determined.

In the latter, it is necessary to have a knowledge of the **strength of the materials** to be used in the structure.

148. Strength depends upon the internal organization of a body, and a material is said to have the requisite strength—to be strong enough—when, by reason of certain inherent physical properties, it possesses the ability to resist the action of an external force within limits.

All materials have not equal strength, nor does the same material resist equally the same force, when a change is made in its direction or point of application.

The degree of strength that a material possesses is determined by experience or experiment.

As it is not always practicable nor expedient to submit to the test of an actual experiment the piece to be used in a structure, its assumed degree of strength is obtained either by subjecting a piece of the same material, having the same dimensions, to conditions similar to those to which the former is to be submitted; or knowing the relations between the strengths of pieces of the *same* material of different dimensions, by deducing it. These relations are obtained by means of mathematics, and are confirmed by experience.

149. **Strains.**—Every **solid** body is supposed to be formed of molecules, infinitely small and infinitely close to each other, grouped together by certain laws. Each molecule is supposed to be so related to those surrounding it that its position cannot be changed except by the application of an extraneous force.

If a solid, which is not allowed to move from its place, be acted upon by an extraneous force the equilibrium of the internal forces acting between the molecules will be disturbed and variations caused in the distances between the molecules, and in the intensities of the forces that bind them together.

By these variations, an equilibrium between the external and internal forces is effected, and an alteration of the form of the solid is caused. This alteration of form is called a **strain**, and the force by which the molecules resist it is called a **stress**.

Since, for any section of a solid, the force developed in the body at the section is equal to the external force acting at that section to produce a strain, the term **stress** is frequently applied to the straining force acting at that section.

External forces, therefore, acting upon solids not free to move cause strains and develop stresses in the bodies so placed.

CLASSIFICATION OF STRAINS.

150. If a solid body like that of a beam (Fig. 9) be firmly fastened at one end so that it will not move, and then be acted upon by an extraneous force, this beam will be subjected to a strain. Stress will be developed in the beam to resist the strain and to establish an equilibrium between the external and internal forces.

When a beam, or any solid body, is strained, the elementary particles or fibres of which it is composed will have their figures and dimensions changed by the action of the straining force. If these particles be cubical in form before the application of the external force, they will, after the force has been applied, become parallelopipeds, either right or oblique. In considering the distortions of the elementary particle, the particle being assumed to be infinitely small, the curvature of the faces produced by the distortion may be regarded as inappreciable.

To examine the distortions to which a particle of the beam is exposed, let it be assumed that the beam is of a homogeneous material and its axis is parallel to one of the linear dimensions of the cubical elementary particle. Suppose the beam to be intersected by two planes perpendicular to its axis and infinitely near each other. Let A B and C D be the sec-

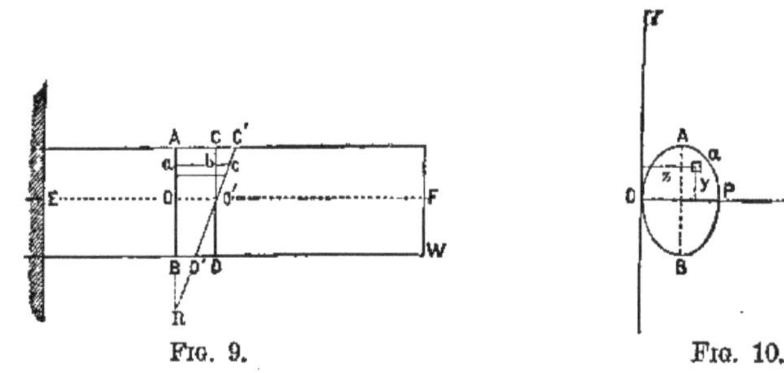

FIG. 9. FIG. 10.

tions cut from the beam by these planes. Suppose the section A B to be fixed and the section C D to take all the positions it can have with respect to the fixed section.

1. Let the action of the straining force be such that the section C D, while remaining parallel to A B, shall move away from it. This can be done only by lengthening the fibres connecting the two sections; and since the sections remain parallel, by lengthening all of them an equal amount.

This distortion of lengthening the fibres is called a **strain of tension,** the resistance offered by the fibre is called a **tensile stress,** and the external force producing the strain, a force of **extension.**

2. If the force acting on the section C D had been such as to make it approach A B, but still be parallel to it, the fibres would have been shortened. The strain would have been one of **compression**; the stress, **compressive**; and the straining force, a **crushing** one.

3. Suppose the section C D, under the action of the force, had taken a position as C′ D′, by turning around some line in its plane, as O′. This position could not have been assumed unless the fibres were deflected, and the fibres above the axis of rotation lengthened, and those below it shortened. The distortion of the fibre in this case is called a **cross strain**; the stress, a **transverse** one; and the straining force, a **bending** force, or force of **flexure.**

4. Suppose the section C D, under the action of the force, to be moved past A B, but the planes of the sections kept parallel. This position would require the fibres to be distorted as shown in Fig. 11, in which the fibre $a\,b$ takes a new position as $a\,b'$. Since the planes remain parallel, all the fibres connecting the sections are distorted equally. This distortion is called a **shearing** strain; the stress, a **shearing** one, or simply a **shear**; and the straining force, a **shearing** force.

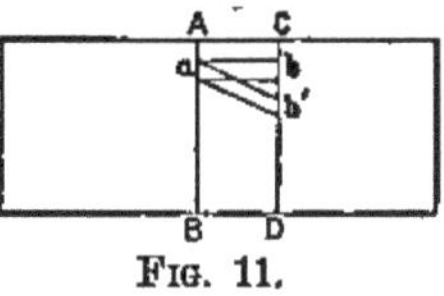

Fig. 11.

5. The section C D may by the action of the force be made to revolve around some line perpendicular to its plane. The fibres connecting the sections would become distorted, taking the form of oblique parallelopipeds with helical axes. This distortion is called a strain of **torsion**; the stress, **torsion**; and the straining force, one of **twisting.**

The section C D can be made to take other positions than the ones given, but on examination of any one of such positions, it will be found to be one of those just described, or one which can be separated into two or more of them. It follows, therefore, that every strain of an elementary fibre caused by an extraneous force will be one of those named, or a combination of two or more of them.

In considering the strains of the elementary particles of a body at a given section of the body, the sum of all the stresses developed in the fibres at this section is the stress developed at the section considered.

151. **Examples.**—Weights, either permanently or temporarily applied to a solid, form the extraneous forces that ordinarily strain a structure. The stresses developed are as follows:

1. **Compressive**; as the stress developed in a pillar when a load is placed on its top. The load tends to shorten the fibres, causing a strain of compression on the pillar.

2. **Tensile**; as the stress developed in a rod, chain, etc., fastened at one end and sustaining a weight at the other. The load tends to lengthen the fibres, causing a strain of tension on the rod, etc.

3. **Transverse**; as the stress developed by a load placed on a beam supported at its extremities. The action of the load is to bend the beam and cause a cross strain.

4. **Shearing**; as that where the effect of the load is to pull apart, in the direction of their lengths, two plates or bars of iron that are held together by rivets. The action of the force is such as to cause a shearing strain on the rivets.

5. **Torsion**; as that developed by a weight lifted by a windlass. The action of the force causes a strain of torsion on the axle. This strain is common in machinery but not in structures, as care is taken to distribute the loads over the latter so as to avoid developing a torsional stress in the material.

Each strain is accompanied by its corresponding stress, which is an increasing function of the strain. When the relation between a strain and its stress is known, the latter can be expressed in terms of the strain, equations formed, and the strength of the solid determined.

152. **Elasticity** is that property of a body by which the particles, when disturbed by an extraneous force, tend to return to their original positions upon the extraneous force ceasing to act.

When the displacements of the particles are very small, the particles upon the removal of the disturbing force resume their positions by the action of the elastic force, and the strain is said to be within the **limit of elasticity**.

The potential energy of elasticity of a particle while distorted is the work which it is capable of performing in returning to its original position.

Experiment shows that within the limit of elasticity the strains vary continuously, and are proportional to the forces causing them. The corresponding stresses being functions of the strains, may be represented by the strains multiplied by a constant quantity. This constant is called the **coefficient of elasticity**.

The coefficient of elasticity varies both with the kind of material of which the solid is composed and with the kind of strain, being different when the strains are alike and the material different, and different when the material is the same but the strains are unlike.

The general method used to obtain the relations existing between the strains of bodies and the corresponding stresses is to suppose the solid to be composed of elementary particles, each particle being a volume of regular geometrical form. The elementary particle is then referred to a system of rectangular co-ordinate planes with its linear dimensions parallel to the co-ordinate axes, and supposed to be subjected to any stress whatever within the limit of elasticity.

The stress is supposed to be separated into its *six* elementary ones—*three* acting in the direction of the linear dimensions of the particle to lengthen or to shorten the particle, and *three* along the faces to alter the angles between the faces of the particle. The first three are known as **normal**, and the latter three, as **tangential** stresses.

The strain caused in the particle by the action of the stress is also divided into its elementary ones. The strains affecting the length of the linear dimensions of the particle are known as the **direct**, and those affecting the angles between the faces as **transverse** strains.

The form of the particle having been assumed to be a cube, or a right parallelopiped, the equation of its surface referred to the co-ordinate axes is known. The displacements, or elementary strains caused by the elementary stress can be expressed in terms of the co-ordinates and the differentials entering the equation of the surface. The strains having been determined in extent and kind, the corresponding stresses can be expressed in terms of these strains and constants. Equations may then be formed, which being integrated will give the total strains and stresses in the solid.

Approximate methods are generally employed to find the relations between the strains and the corresponding stresses, and are considered sufficiently accurate for all practical purposes. They will be employed in the following pages.

The approximate method is to conceive the solid to be divided by a plane into two parts; then to find all the extraneous forces acting on one of these parts, on either side of the plane, to strain the body; then place the straining forces thus found equal to the entire stress developed in the body at the section made by the plane; assume the stress at this section to be distributed according to some law, deduced by experiment or theory, which is assumed to be true, or practically so, as regards the exact state of distribution; form equations expressing these conditions and applying to the particular cases under consideration. A discussion of the equations thus formed will give the stress on the unit of area of the section, the amount of strain, and the strength of the material necessary to resist the stress acting at the section.

CONSTANTS.

153. In discussing the equations deduced for determining the strength of building materials, certain constants are involved which depend for their value on the physical properties of the material under consideration. These constants

have been or are to be determined for each material by actual experiment.

There are four principal ones:

I. The **weight**, or specific gravity of the body;
II. The **limit of elasticity**;
III. The **coefficient of elasticity**;
IV. The **modulus of rupture.**

154. **The weight** enters as an element in all constructions; and to such an extent in some, as in masonry for example, that the moving or temporary loads to be borne may be disregarded, or considered as insignificant, in comparison with the weight of the structure itself.

155. **Limit of Elasticity.**—From a great number of experiments, made on a great variety of materials, it has been found that practically,

1st. All bodies are elastic.

2d. Within very small limits they may be considered as *perfectly elastic.*

3d. Within the elastic limit the amount of displacement is directly proportional to the force that produces it.

4th. Within a considerable distance beyond the elastic limit the amount of displacement is not exactly but nearly proportional to the force producing it.

The limit of elasticity of a body in any direction is determined by experiment, and its determination is a matter of great nicety; hence experimenters have paid more attention to determining the ultimate strength of materials; that is, to finding the limits beyond which any additional load will break the material.

If the material be strained beyond the elastic limit, the particles will not resume their former positions, and a permanent change of figure is the result. This permanent change is called a **set.** A set, when it is made, does not necessarily weaken the material, but it is better in most cases not to have it.

156. **Coefficient of Elasticity.**—The coefficients of elasticity vary with the material, and with the kind of stress developed.

Let it be required to determine the coefficient of elasticity for a homogeneous material strained only by a force of extension. Assume the material to be in the form of a straight bar of uniform cross-section, fastened at one end, and pulled by forces whose resultant acts along the axis of the bar. The intensity of the pull will be uniform on each cross-section.

Let W = the total pull, L = the length of the bar before it is strained, A = the area of its cross-section, and l = the elongation of the bar caused by the force W.

Then, $\frac{W}{A}$ = the force acting on a unit of cross-section, and $\frac{l}{L}$ = the amount of elongation for a unit of length of the bar.

Since the pull is uniformly distributed over the cross-section, it is assumed that the stress developed in the cross-section is so distributed, and that $\frac{W}{A}$ = the intensity of the stress on any unit of cross-section of the bar. But, the stress is equal to the strain multiplied by a constant, hence we have

$$\frac{W}{A} = \frac{l}{L} \times E,$$

in which E is the coefficient of elasticity. Whence,

$$E = \frac{W}{A} \div \frac{l}{L}, \quad \text{or} \quad E = \frac{W}{A} \times \frac{L}{l}. \quad . \quad (1)$$

By means of formula (1) the **coefficient of elasticity** for a homogeneous material strained by a force of extension can be obtained by experiment.

The following are some of the values of E, that have been obtained by experiment for various building materials, viz.:

Material.	Value of E.
Cast Iron	18,400,000 lbs.
Wrought Iron	24,000,000 "
Lead (cast)	720,000 "
Steel	29,000,000 "
Tin (cast)	4,608,000 "
Zinc (cast)	13,680,000 "
Ash	1,644,800 "
Fir	1,191,200 "
Pine, pitch	1,225,600 "
" yellow	1,600,000 "
Oak	1,451,200 "
Marble	2,520,000 "
Limestone (common)	1,533,000 "

157. Modulus of Rupture.—If the straining forces be continually increased in intensity, they will produce in time a *rupture*, or such a disfigurement of the solid as to make the material unfit for building purposes.

At the moment of rupture, or an instant before, the intensity of the stress developed in the material is greater than at any other period of the strain. This greatest intensity of the stress is known as the "**ultimate resistance**" of the material, and its value is obtained by experiment.

When the material is subjected to a strain of tension alone, the tensile stress is supposed to be distributed uniformly over its cross-section, and the stress on the unit of area is equal to the total stress divided by the area of cross-section. When this straining force is increased sufficiently to produce rupture of the material, the stress on the unit of area, at the moment rupture begins, is taken as the measure of its ultimate resistance. This stress on the unit, or the *force necessary to pull asunder a piece whose cross-section is unity* is called the **modulus of tenacity** for that material.

If the stress is a compressive one, or a shear, the corresponding stress on the unit of section, at the moment of rupture, is called the **modulus of crushing**, or **modulus of shearing**, as the case may be.

The values of these moduli are obtained by experiment, and are represented in the formulas by the letters T, C, and S.

When the strain is a cross one, the transverse stress developed at a given cross-section is not supposed to be distributed as just described, but is assumed to vary uniformly over the cross-section, being greatest on the units farthest from the axis of rotation. The stress on the unit of cross-section at the surface of the material when the fibres begin to tear apart, or to crush, is taken as the measure of ultimate resistance to cross strain, and is called the **modulus of rupture**, which is represented in the formulas by the letter R.

It would seem, since the rupture of a piece by a cross strain takes place by the fibres being torn apart or crushed, that the respective values of R, C, and T for the same material would be the same, or at least nearly equal, and that one symbol might be used to represent the respective values of the three. Experiment shows, however, that they are not equal, but vary considerably.

If the stress is one of torsion, the stress on the unit farthest from the axis is taken as the measure, is called the **modulus of torsion**, and is represented in the formulas by the letters T_t.

TENSION.

158. Extraneous forces acting on a piece fastened at one end, and in the direction of its axis, produce a strain of extension in the piece, if the direction of the resultant is from the fixed end, and of compression if the direction is toward it.

Let it be required **to determine the elongation produced in a straight bar,** of uniform cross-section, placed in a vertical position and fixed at one end, *by a system of forces whose resultant acts along the axis of the bar.*

Represent by (Fig. 12)
L, the original length of the bar,
W, the force applied to lengthen it,
l, the elongation due to W,
A, the area of the cross-section,
E, the coefficient of elasticity.
Then from eq. (1), we have

$$l = \frac{WL}{EA} \quad . \quad . \quad . \quad (2)$$

and,

$$W = EA\frac{l}{L} \quad . \quad . \quad (3)$$

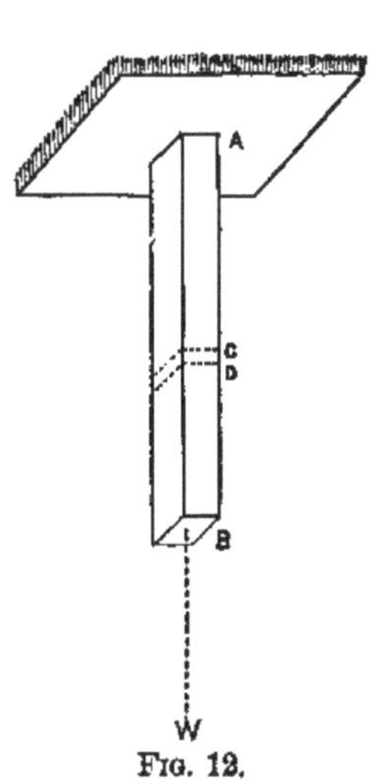

Fig. 12.

Eq. (2) shows that the elongation is directly proportional to the length of the bar and to the force itself, and inversely to the area of the cross-section and coefficient of elasticity; which is fully confirmed by experiment.

If in eq. (3) we make $A = 1$ and $l = L$, we shall have

$$W = E. \quad . \quad . \quad . \quad . \quad . \quad (4)$$

That is, the *coefficient* of *elasticity*, E, is the *force which*, applied to a bar, the *cross-section of which is a superficial unit*, would produce *an elongation equal to the original length of the bar*, supposing its *elasticity perfect up to this limit.*

This is a theoretical force; but as the law upon which it depends is true within the limits of elasticity, knowing W, A, and L, and determining l by measurement, the value that E would have if the elasticity remained perfect is easily found.

Divide W by A, and we have

$$\frac{W}{A} = \text{the stress on a unit of cross-section.}$$

If W′ be the force necessary to produce rupture when acting in the direction of the axis, then

$$\frac{W'}{A} = T, \textbf{ the modulus of tenacity}. \quad . \quad . \quad (5)$$

Wood and iron are the two building materials most frequently exposed to this strain. The cohesive power of wood is greatest in the direction of the fibres, and in the tables showing the results of the experiments made on the strength of materials, the tensile strength there given is taken with reference to that direction, unless otherwise stated.

From eq. (5), we have $W' = TA$, from which knowing T and A, the force necessary to rupture the bar may be deduced.

159. The following table gives the tensile strength, per square inch, as obtained by experiment upon some of the materials frequently used in building:

Material.	Tensile Strength per sq. inch.			
Ash	10,803	lbs. to	24,033	lbs.
Chestnut	11,891	" "	13,066	"
Cedar	——	" "	10,300	"
Hickory	12,866	" "	40,067	"
Oak, white	12,300	" "	25,222	"
" live	——		15,800	"
Pine	11,400	" "	19,200	"
Fir	12,867	" "	16,833	"
Hemlock	——		16,533	"
Cast iron, common pig	——		15,000	"
" " good common iron	——		20,000	"
Bar iron	——		57,000	"
" " Swedish	——		72,000	"
Copper wire	——		60,000	"
Steel, cast	——		128,000	"
" shear	——		124,000	"
" puddled	——		105,000	"
Tin, cast	——		4,800	"
Lead, "	——		1,800	"
Zinc	——		7,500	"

The specimens of wood in the foregoing list were dry and seasoned. The time of seasoning varying from one to fifteen years. They were grown in different parts of the United States, extending from the extreme north to the farthest south, and from the Atlantic coast to the Pacific. The differences in the localities from whence they were brought and the times of seasoning, explain the differences observed in the tensile strength of specimens of the same wood.

The tensile strength of the metals is materially modified by the processes of manufacture and by the impurities they contain.

It is evident, from this table, and from what has been just stated, that it is not practicable to assume a value for the modulus of tenacity which will be safe and economical for a given material. Its value in any particular case should be determined by experiment; or before its value can be assumed, the quality of the material must in some way be known.

The work expended in the elongation of the bar.

160. The general formula from Anal. Mechanics is

$$Q = \int P ds,$$

in which P is the resistance, s the path of the point of application, and Q the quantity of work.

In this formula, substitute W for P, and l the elongation for s, and we have

$$Q = \int W dl.$$

Substituting for W its value from eq. (3), there obtains,

$$Q = \int EA \frac{l}{L} dl,$$

to represent the quantity of work.

Integrating between the limits $l = 0$ and $l = l'$, we have,

$$Q = \frac{1}{2} \frac{EA}{L} l'^2 = \frac{1}{2} \frac{EAl'}{L} l'.$$

From eq. (3) we have

$$\frac{EAl'}{L} = W',$$

W′ being the particular value of W producing the elongation, l'.

Substituting this value of W′ in the preceding equation, and we have

$$Q = \tfrac{1}{2} W'l'. \quad . \quad . \quad . \quad . \quad . \quad (6)$$

If, in the eq.

$$Q = \int W dl,$$

W were constant and equal to W′, then

$$Q = W' \int dl,$$

which integrated between the limits $l = 0$ and $l = l'$ will give

$$Q = W'l'.$$

This value of Q is twice that of Q in eq. (6); whence it follows that the work expended in producing the elongation l', by applying the force W′, at once, and keeping it constant, is twice the work which would be expended, if the force were applied by increments increasing gradually from zero to W′.

Combining eqs.

$$W' = \frac{EAl'}{L} \text{ and } Q = \tfrac{1}{2} W'l',$$

and eliminating l', we get

$$Q = \tfrac{1}{2} \frac{W^2 L}{E A},$$

whence it is seen that the work expended upon the elongation of the bar varies directly with the square of the force producing it, with the length of the bar, and inversely with the area of cross section and coefficient of elasticity.

Elongation of a bar, its weight considered.

161. To determine the elongation of a bar, under the same circumstances as the preceding case, *when its weight is taken into consideration.*

In eq. (2), the weight of the bar being very small compared with W, it was neglected.

To determine the elongation, considering the weight of the bar, represent (Fig. 13) by L, W, l, and A, the same quantities as before, by x, the distance from A of any section as C, by dx, the length of an elementary portion as C D, and by w, the weight of a unit of volume of the bar. The volume of the portion B C, will be expressed by $(L-x) A$; and its weight by $(L-x) Aw$.

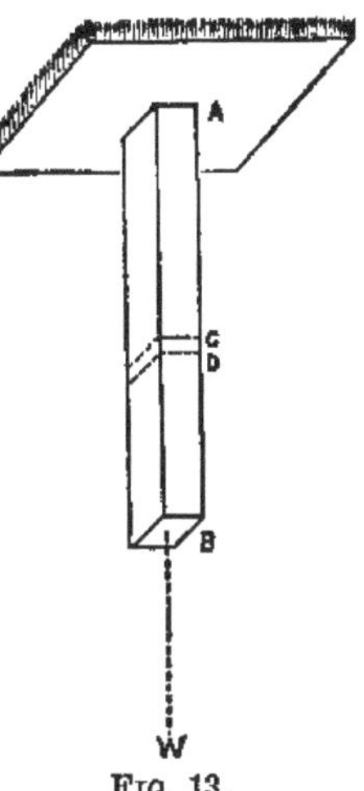

FIG. 13.

The total force acting to elongate the elementary portion C D, will be expressed by

$$W + (L-x) Aw.$$

Substituting this for W, and dx for L in eq. (2), we have

$$\text{elongation of } dx = \frac{W + (L-x) Aw}{EA} dx.$$

The total length of dx after elongation will, therefore, be

$$dx + \frac{W + (L-x) Aw}{EA} dx.$$

Integrating this between the limits $x = 0$ and $x = L$, there obtains,

$$L + l = L + \frac{WL}{EA} + \frac{\frac{1}{2}wL^2A}{EA} \quad . \quad . \quad . \quad . \quad (7)$$

for the total length of the bar after elongation

This may be written,

$$L + l = L\left(1 + \frac{W + \frac{1}{2}wAL}{EA}\right).$$

If, in this expression, we make $W = 0$, we have

$$l = \frac{\frac{1}{2}wAL}{EA}L.$$

In this, wAL is the weight of the bar; representing this weight by W' and substituting in last expressson, we have

$$l = \frac{\frac{1}{2}W'L}{EA},$$

or the elongation due to the weight of the bar, is one half of what it would be if a weight equal to that of the bar were concentrated at the lower end.

An examination of the expression, $W + (L - x)\,Aw$, shows that the strain on the different cross-sections varies with x, decreases as x increases, and is greatest for $x = 0$, or on the section at the top. Since the bar has a uniform cross-section, the strain on the unit of area is different in each section.

BAR OF UNIFORM STRENGTH TO RESIST ELONGATION.

162. *To determine the form a vertical bar should have, in order to be equally strong throughout, when strained only by a force acting in the direction of the axis of the bar*, the weight of the bar being considered.

Suppose the bar, fixed at one end and the applied force producing elongation to be a weight suspended from the other end. [Fig. 14.]

From the preceding article, it is seen that if the bar has a uniform cross-section, that the strain on each section is different. In order that the bar should be equally strong throughout, the strain on each unit of area of cross-section must be the same throughout the bar. This can only be effected by making the area of the cross-section proportional to the stress acting on it, or having the cross-sections variable in size.

Represent by

A, the area of the variable cross-section;

A', the area of cross-section at B, or the lower one;

A'', the area of cross-section at A, or the top section;

T_{1}, the strain allowed on the unit of area;

W, the force applied to the bar producing elongation;

x, the distance, B C, estimated upwards from B.

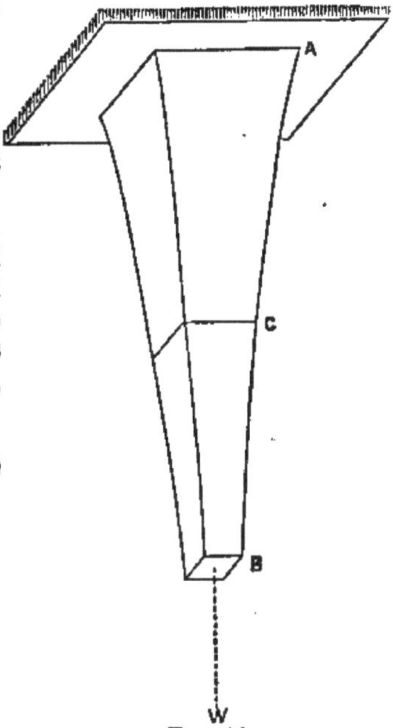

The total force acting on any section as C, to elongate it, is

$$W + w\int A\,dx,$$

w being the weight of the unit of volume of the bar.

Since T_1 is the strain allowed on the unit of area, $T_1 \times A$ will represent the total strain on the section at C, and will be equal to the force acting on this section to elongate it. Hence, we have

$$W + w\int A\,dx = T_1 A \quad (8)$$

Differentiating, we have

$$wA\,dx = T_1 dA,$$

which may be written

$$\frac{w\,dx}{T_1} = \frac{dA}{A}.$$

FIG. 14.

Integrating, we get

$$\frac{wx}{T_1} = \text{Nap. log } A + C. \quad . \quad . \quad . \quad (9)$$

Making $x = 0$, we have $A = A'$, whence

$$0 = \text{Nap. log } A' + C.$$

Substituting for C in eq. (9) its value obtained from the last equation, we get

$$\frac{w}{T_1}x = \text{Nap.log}\frac{A}{A'}$$

and passing to the equivalent numbers,

$$A = A'e^{\frac{wx}{T_1}}.$$

But

$$A' = \frac{W}{T_1},$$

which substituted above gives,

$$A = \frac{W}{T_1}e^{\frac{wx}{T_1}}.$$

Making $x = L$ and A becomes equal to A″, hence

$$A'' = \frac{W}{T_1} e^{\frac{wL}{T_1}}$$

the value for the area of the section at the upper end.

Form of bar when it has a circular cross-section.

163. No particular form has been assigned to the cross section of the bar in this discussion. Let it be a circle and represent the variable radius by r.

Then the area of any cross-section will be πr^2, which being substituted for A in eq. (8), gives

$$W + w\int \pi r^2 dx = T_1 \pi r^2.$$

Differentiating, there obtains

$$w\pi r^2 dx = T_1 2\,\pi r dr,$$

hence

$$\frac{dr}{r} = \frac{w}{2T_1} dx,$$

which integrated gives

$$\text{Nap. log. } r = \frac{w}{2T_1} x + C, \quad . \quad . \quad (10)$$

which shows the relation between x and r.

Eq. (10) is the equation of a line, which line being constructed will represent by its ordinates the law of variation of the different cross-sections of the bar. It also shows the kind of line cut from the bar by a meridian plane.

The most useful application of this problem is to determine the dimensions of pump-rods, to be used in deep shafts, like those of mines.

COMPRESSION.

164. The strains caused by pressure acting in the direction of the axis of the piece tend to compress the fibres and shorten the piece.

From the principle that all bodies are elastic, it follows that all building materials are compressible.

Within the limit of elasticity it is assumed that the resistances to compression are the same as tension. They are not really the same; but within the elastic limit the differences are so small, that for all practical purposes it is sufficiently exact to consider them equal.

The coefficient of elasticity of the material is assumed the same in both cases, and to distinguish it from the coefficients of elasticity when the fibres are displaced in other ways, it is sometimes called the **coefficient of longitudinal elasticity,** or resistance to direct lengthening or shortening.

To ascertain the force under which a given piece would be crushed, we first ascertain the weight necessary to crush a piece of the same material; and since experiment has shown that the resistances of different pieces of the same material to crushing are nearly proportional to their cross-sections, the required force can be easily determined.

Assuming that these resistances are directly proportional to the cross-sections, let W′ be the required force, A the area of cross-section of given piece, and C the force necessary to crush a piece of the same material whose cross-section is unity.

We have, $W' : C :: A : 1$, or

$$W' = AC, \quad . \quad . \quad . \quad . \quad . \quad . \quad . \quad (11)$$

hence

$$\frac{W'}{A} = C. \quad . \quad . \quad . \quad . \quad . \quad . \quad . \quad . \quad (12)$$

Many experiments have been made on different materials to find the value of C, and the results tabulated. If the experiments for finding C were not made on pieces whose cross-sections were unity, they were reduced to unity by means of eq. (12). The pieces used in the experiments were short, their lengths not being more than five times their diameter or least thickness.

This value of C, the **modulus of crushing,** is equal therefore to the pressure, upon the unit of surface, necessary to crush a piece whose length is less than *five* times its least thickness, the pressure being uniformly distributed over the cross section and acting in the direction of the length of the piece. Experiment shows that it requires a much less pressure to crush a piece when the force is applied *across the fibres,* than when it is applied in the direction of their length.

165. The following are the values of C for some of the ma-

terials in common use, and were obtained by crushing pieces of small size, and as a rule not longer than twice their diameter:

Material.	Crushing Forces per sq. inch, in lbs.		
Ash	4,475	to	8,783
Chestnut	——		5,000
Cedar	——		5,970
Hickory	5,492	"	11,213
Oak, white	5,800	"	10,058
Oak, live	——		6,530
Pine	5,017	"	8,947
Fir	6,644	"	9,217
Hemlock	——		6,817
Cast iron	56,000	"	105,000
Wrought iron	30,000	"	40,000
Cast steel	140,000	"	390,000
Brick	3,500	"	13,000
Granite	5,500	"	15,300

Rankine gives from 550 to 800 for common red brick, and 1,100 for strong red brick.

The remarks relative to the specimens of wood used to obtain the values of T in the table on page 83 apply equally to this case.

SHEARING STRAINS.

166. There are two kinds of simple **shear**; one in which the stress acts normally to all the fibres, like that developed in a rivet when the plates which it fastens are strained by tension or compression in the direction of their lengths; and one in which the stress acts in a plane parallel to the fibres, either in the direction of, or across, the fibre. The former is called a **transverse shear,** and the latter, **detrusion.**

The relations between the strains and the stresses developed by a shearing force may be expressed by equations analogous to those used for tension.

In describing the shearing strain, the section C D (Fig. 15) was supposed not to have rotated around any line in its plane, but to have had a motion of translation parallel to the plane A B, so that after the movement, any fibre, as *ab*, will have a new position, as *ab'*.

Suppose A B to remain fixed, and represent by
L, the original length of any fibre ab between the two consecutive planes A B and C D;

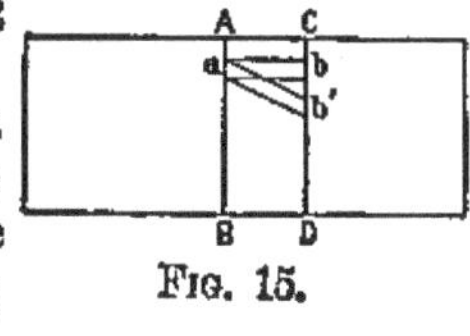

FIG. 15.

γ, the distance bb' which every point of the plane C D has moved in the direction of C D, relatively to the plane A B, owing to the force causing this displacement;

s, the amount of shearing stress in any fibre;

a, the area of the cross-section of the fibre;

E', a constant.

Then,

$\frac{s}{a}$ = the intensity of the shearing stress on a unit of area,

and $\frac{\gamma}{L}$ = the measure of displacement of the fibre per unit of length. Hence,

$$\frac{s}{a} = E' \frac{\gamma}{L} \quad . \quad . \quad . \quad . \quad . \quad . \quad (13)$$

from which we get

$$E' = \frac{\frac{s}{a}}{\frac{\gamma}{L}},$$

a value analogous to that obtained for E in equation (1).

This value of E' is constant within the limit of elasticity for each elementary fibre. If the material is homogeneous it has the same value for all the fibres, or is constant for the same material.

Represent by S_1 the total stress developed in the section C D; by A, the area of the section; and let the piece be of homogeneous material. Then,

$$S_1 = E' A \frac{\gamma}{L}. \quad . \quad . \quad . \quad . \quad (14)$$

which expresses the relation between the total stress developed in the section and the shearing strain.

The constant E' is the coefficient of elasticity corresponding to a transverse shearing strain, and is frequently called the **coefficient of lateral elasticity**, to distinguish it from the coefficient of **longitudinal elasticity**.

The shear is assumed to be distributed uniformly over the cross-section of the material. Suppose the shear to be increased until rupture takes place and let S′ represent the intensity of the total shearing stress on the cross-section.

Then,

$$\frac{S'}{A} = S,$$

in which S is the **modulus of shearing** for the material used.

167. The following are some of the values of S, obtained by experiment, for some of the building materials in use, viz.:

TRANSVERSE SHEARING.

Materials.	Value of S.
Ash	6,280 lbs.
Cedar	3,400 "
Hickory	6,500 "
Oak, White	4,000 "
Oak, Live	8,000 "
Pine, Yellow	4,500 "
Pine, White	2,500 "
Cast steel	92,400 "
Wrought iron	50,000 "
Cast iron	30,000 "
Copper	33,000 "

DETRUSION.

White pine	480 lbs.
Spruce	470 "
Fir	592 "
Hemlock	540 "
Oak	780 "

TRANSVERSE STRAIN.

168. Extraneous forces acting either perpendicularly or obliquely to the axis of a piece that is fixed, cause cross-strains and develop transverse stresses in the material.

In describing the nature of a cross-strain (Art. 150), it is assumed that a consecutive section of the piece, as C D (Fig. 16), could not take a position as C′ D′ unless the fibres on one side of the axis of rotation were lengthened and those on the other side shortened. Also, that the fibres farthest from this axis were elongated or shortened more than those nearest to it, and as a consequence the stresses in the fibres were variable in their intensities throughout the cross-section.

To determine the relations between the strains of the fibres caused by the bending forces and the corresponding stresses developed, a theory must be adopted relating to the strains produced, and a law assumed for the distribution of the stresses over the cross-section.

Suppose a piece of homogeneous material, in form of a bar or beam, to be placed in a horizontal position and fixed at one end, and suppose this piece to be acted upon by a system of extraneous forces, the resultant, W, of which is perpendicular to the axis and intersects it at the free end.

The action of this system of extraneous forces is to bend the piece, causing cross-strains and developing both transverse and shearing stresses throughout the piece.

Neglecting the shearing stress for the present, let it be required *to determine the relations between the cross-strains and the transverse stresses produced by the bending force*, W.

The cross-sections of the piece are assumed to be uniform, or to vary from each other by some law of continuity that is known; the forms of the cross-sections are similar, and for any two consecutive sections may be considered to be equal.

The common theory for the strains, deduced from observation and experiment, is as follows, viz.:

1. That the fibres on the convex side of the piece are extended, and those on the opposite side are compressed.

2. That the strains of the fibres caused by the bending force are either compressive or tensile.

3. That there is a surface between the compressed and extended fibres in which the fibres are neither compressed nor extended.

4. That the strains of the fibres are proportional to their distance from this surface, known as the **neutral** surface.

5. That the cross-sections of the piece normal to the fibres before bending will remain normal to them after bending.

6. That rupture will take place either by compression, or by extension, of the fibres on the surface of the piece when the stress is equal to the **modulus of rupture**.

The intersection of the neutral surface by the plane of cross-section is called the *neutral axis* of the section.

From this theory, it follows, that the intensities of the stresses of tension and compression in the fibres are also proportional to their distances from the neutral axis as long as the strain is within the elastic limit. The stress developed on a cross-section to resist the action of a bending force is, therefore, a uniformly varying one; being least, or zero, at the neutral axis, and greatest at the points farthest from this axis.

To find the stress in any fibre in terms of the strain, let A B and C D (Fig. 16) be the intersections of two consecutive cross-sections of the piece by the plane of the axis, E F, of the piece and the resultant, W, of the bending forces.

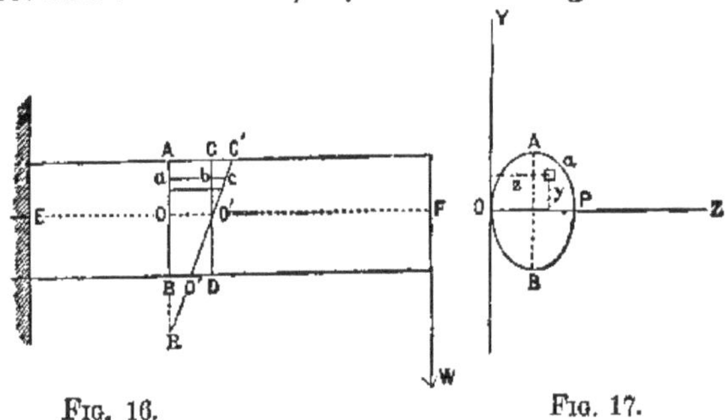

FIG. 16. FIG. 17.

Let O Y and O Z (Fig. 17) be two rectangular co-ordinate axes to which all points of the cross-section are referred.

Represent by

y and z, the co-ordinates of all points in the plane Y Z;
x, the distances measured on the line E F;
dx = O'O = the distance between the sections A B and C D;
$dydz = a$ = the cross-section of a fibre;
$\lambda = bc$ = the elongation of any fibre as ab;
ρ = O R, the radius of curvature.

Let the section A B remain fixed and the section C D take some position as C'D' under the action of the bending force; the strain being within the elastic limit.

Then, by hypothesis, the fibres above E F will be elongated, and the elongation bc of any one fibre, as ab, will be proportional to its distance, y, from the neutral axis.

From the similar triangles $bO'c$ and O R O' we have

$$bc : OO' :: bO' : OR,$$

or,

$$\lambda : dx :: y : \rho,$$

whence

$$\lambda = \frac{y}{\rho} dx \quad . \quad . \quad . \quad . \quad . \quad . \quad . \quad (15)$$

an expression for the amount of elongation of a fibre at the distance y from the neutral axis.

The expression for the intensity of the stress developed

in a bar to resist an elongation equal to l is (eq. 3) equal to $EA\frac{l}{L}$. In this expression substituting $dydz$ for A, the value of λ just obtained for l, and dx for L, we obtain

$$\frac{E}{\rho} ydydz \quad . \quad . \quad . \quad . \quad . \quad (16)$$

for the intensity of the stress developed in the fibre ab.

Since this expression is true for any fibre that is elongated, the total stress on the elongated fibres of this section will be expressed by

$$\frac{E}{\rho}\int\int ydydz.$$

In like manner the total stress on the compressed fibres will be expressed by

$$-\frac{E}{\rho}\int\int ydydz,$$

the negative sign being used to denote the contrary direction of the elastic resistance of the compressed fibres.

Since the strain is within the elastic limit the beam is strong enough to resist the action of the extraneous forces, and the moment of resistance at the cross-section is exactly equal and opposite to the moment (Wx) of the bending forces at the same cross-section.

The moment of resistance to elongation of a fibre, at the distance y from the neutral axis, is equal to the intensity of the stress in the fibre (eq. 16) multiplied by y, and, to compression, the same expression multiplied by $-y$.

The total moment of resistance at the cross-section will be

$$\frac{E}{\rho}\int\int y^2dydz + \frac{E}{\rho}\int\int y^2dydz, \quad . \quad (17)$$

which placed equal to Wx, gives an equation expressing the relation between the moments of the transverse stresses and those of the extraneous forces producing bending at any cross-section of the beam.

Let b be the greatest value of z, and $\frac{1}{2}d$ that of y (Fig. 17) and integrating expression (17) so as to include the whole cross-section, we may write this equation as follows:

$$\frac{E}{\rho}\int_{z=0}^{z=b}\int_{y=-\frac{d}{2}}^{y=\frac{d}{2}} y^2dydz = (Wx) \quad . \quad (18)$$

It will be seen that the quantity under the sign of integration when integrated twice will give the **moment of inertia** of the cross-section of the piece with respect to the neutral axis. Representing this by I and that of the extraneous force by M, we may write (eq. 18) as follows:

$$\frac{EI}{\rho} = M. \quad . \quad . \quad . \quad . \quad . \quad . \quad . \quad . \quad (19)$$

The first member is oftentimes called the **moment of elasticity**, sometimes the **moment of resistance**, and at others the **moment of flexure**, and the second member is called the **bending moment.**

169. This equation may be verified as follows:

We know that if all the elementary masses were concentrated at the principal centre of gyration, the moment of inertia would be unaltered; also, that the forces tending to produce rotation of the body might be concentrated at this point without thereby changing the conditions of equilibrium.

Suppose the resistances offered by the fibres to rotation concentrated at the principal centre of gyration, and equal to P′ acting with a lever arm, k. We have for equilibrium,

$$P'k = Wx = M.$$

From Mechanics, we have

$$k = \text{principal radius of gyration} = \sqrt{\frac{\Sigma mr^2}{A}},$$

in which m is the elementary mass, r its distance from the axis, and A the area of cross-section.

Substituting for Σ the sign of integration, and for m its value in terms of y and z (Fig. 17), we get,

$$k = \sqrt{\frac{\int\int y^2 dy dz}{A}}.$$

Squaring and dividing both members by k, we get

$$k = \frac{\int\int y^2 dy dz}{Ak}.$$

Hence,

$$P'k = \frac{P' \int\int y^2 dydz}{Ak} = M, \quad . \quad . \quad . \quad (20)$$

and

$$P' = \frac{M \times Ak}{\int\int y^2 dydz},$$

whence

$$\frac{P'}{A} = \frac{kM}{\int\int y^2 dydz},$$

which is the value the force would have on the unit of area at the principal centre of gyration, or the distance k from the neutral axis, under this hypothesis.

It has been assumed that the resistances are directly proportional to the distance from the neutral axis; hence, at the unit's distance, the force on the unit of area would be

$$\frac{P'}{Ak} = \frac{M}{\int\int y^2 dydz},$$

and at the distance, y, the force would be

$$\frac{P'y}{Ak} = \frac{My}{\int\int y^2 dydz}.$$

The strain on the unit of area at the distance, y, from the axis is shown by expression (16), to be equal to $\frac{E}{\rho} y$. Hence,

$$\frac{E}{\rho} y = \frac{My}{\int\int y^2 dydz},$$

or

$$\frac{E}{\rho} \int\int y^2 dydz = M,$$

which is the same result as that shown by eq. (18).

SHEARING STRAIN PRODUCED BY A FORCE ACTING TO BEND THE BAR.

170. No reference was made in the preceding article to the shearing strain produced in the bar by a bending force acting at one end, for the reason, that in prismatic bars of this kind it is rarely necessary in practice to consider this strain.

If in this bar (Fig. 16), the section A B had been taken consecutive to the section, at F, where the force was applied, the action of the force would not have been to turn this section F around a line in its plane, but to have sheared it off from its consecutive section. This action would have been resisted by the adhesion of the sections to each other. The force W is supposed to act uniformly over the entire section F, hence the resistance to shearing in the adjacent section will be uniformly distributed over its surface and equal to W. The resistance on the unit of surface would therefore be $\frac{W}{A}$.

The adhesion of these two sections prevents their separation by this force, hence the second section is drawn down by the force W, which tends to shear it from the third section, and so on.

In this particular case, the action of the force W to shear the sections off, is transmitted from section to section until the fixed end is reached, and the shearing strain of each section is the same and equal to W. And in general, the **shearing stress of any cross-section** of a bar or beam placed in a horizontal position is equal to the *sum of all the vertical forces transmitted through and acting at that section.*

CHANGES IN FORM OF THE BAR.

171. In a bar strained by a force acting in the direction of its axis, the lengthening and shortening of the bar have been the only changes of form considered. There is another change that invariably accompanies them. This is the contraction or enlargement of the area of cross-section, when the bar is extended or compressed. When the elongation or contraction is small, the change in cross-section is microscopically small; but when these strains are very great, this change is sensible in many materials.

In structures, the pieces are not subjected to strains of sufficient magnitude to allow this change of cross-section to be observed, and hence it is neglected.

It is well to keep this change in section in mind, as by it we are able to explain certain phenomena that are met with in experiments, when the strains to which the specimens are submitted pass the limits of elasticity.

STRAIN ON THE UNIT OF AREA PRODUCED BY A BENDING FORCE.

172. Expression (16) represents the stress of extension on the fibre whose cross-section is $dydz$. Dividing this expression by the area of cross-section of the fibre, we have

$$\frac{E}{\rho}y = P,$$

in which P represents the stress on the unit of area at the distance y from the neutral axis. Dividing through by y and multiplying both members by I, we have

$$\frac{EI}{\rho} = \frac{PI}{y} = M \quad . \quad . \quad . \quad . \quad . \quad (21)$$

whence

$$P = \frac{M}{I}y \quad . \quad . \quad . \quad . \quad . \quad . \quad (22)$$

which formula gives for a force of deflection, the stress on a unit of area at any point of the section.

When the bar has a uniform cross-section, I will be constant, and P will vary directly with y and M, and by giving to y its greatest value, we find the greatest strain in any assumed cross-section.

VALUES OF I.

173. In bars or pieces having a uniform cross-section, the moment of inertia for each section with reference to the neutral axis is the same, and hence I is constant for each piece, and is easily determined when the section is a known geometrical figure.

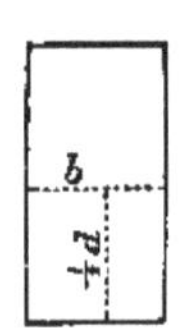

Fig. 18.

1. When the cross-section is a rectangle (Fig. 18) in which b is the breadth, and d the depth, the integral taken within the limits $z = 0$, and $z = b$, $y = \frac{1}{2}d$ and $y = -\frac{1}{2}d$, gives

$$I = \tfrac{1}{12}\, bd^3.$$

2. For a cross-section of a hollow girder, like that of (Fig. 19) in which b is the entire breadth, d the total depth, b' the breadth of the hollow interior, d' its depth, the integral gives

$$I = \tfrac{1}{12}\,(bd^3 - b'd'^3)$$

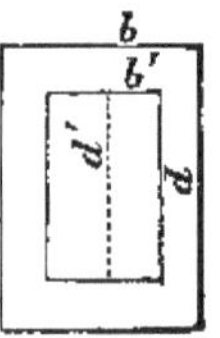

Fig. 19.

The expression will be of the same form in the case of the cross-section of the I-girder, (Fig. 20), in which b is the breadth of the flanges; b' the sum of breadths of the two shoulders; d the depth of the girder, and d' the depth between the flanges.

3. When the cross-section is a circle, and the axes of co-ordinates are taken through the centre, the limits of z will be $+r, -r$; and those of y will be $+$, and $-\sqrt{r^2 - z^2}$; and

$$I = \tfrac{1}{4}\,\pi r^4.$$

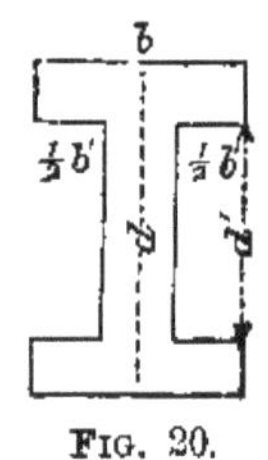

Fig. 20.

4. For a hollow cylinder, in which r is the exterior and r' the interior radius,

$$I = \tfrac{1}{4}\,\pi(r^4 - r'^4).$$

5. When the cross-section is an ellipse, and the neutral axis coincides with the conjugate axis, if the transverse axis be represented by d, and the conjugate by b, and the limits of z and y be taken in the same manner, as in the circle, then,

$$I = \tfrac{1}{64}\,\pi bd^3.$$

6. When the cross section is a rhombus or lozenge, in which b is the horizontal and d the vertical diagonal,

$$I = \tfrac{1}{48} bd^3.$$

FLEXURE.

174. In the preceding article on transverse strain, to simplify the investigation, without affecting the accuracy of the

results, the bar was placed horizontally, and no notice was taken of the change of position of the **mean fibre** after the application of the bending force.

The strain was within the limit of elasticity, and for this force the body was regarded as perfectly elastic.

The action of the force was to bend the bar, and hence to bend the mean fibre without lengthening or shortening it, making it assume a curved form.

When the bar is bent in this manner, the curve assumed by the mean fibre is called the **elastic curve or equilibrium curve.** Its equation is deduced by equating the moment of resistance and the bending moment, and proceeding through the usual steps.

All the external forces to the right, or to the left, of any assumed cross-section are held in equilibrium by the elastic resistances of the material in the section.

The general equation (19), $\frac{EI}{\rho} = M$, expresses the condition of equality between the moments of resistance and bending, and is the equation from which that of the curve assumed by the mean fibre after flexure may be deduced.

From the calculus, we have

$$\rho = \pm \frac{\left(1 + \frac{dy^2}{dx^2}\right)^{\frac{3}{2}}}{\frac{d^2y}{dx^2}},$$

which, substituted in eq. (19), gives

$$\frac{EI\frac{d^2y}{dx^2}}{\left(1 + \frac{dy^2}{dx^2}\right)^{\frac{3}{2}}} = M \quad . \quad . \quad . \quad . \quad . \quad (23)$$

When the deflection is very small, $\frac{dy^2}{dx^2}$ is very small compared with unity and may be omitted; and eq. (23) becomes for this supposition

$$EI\frac{d^2y}{dx^2} = M \quad . \quad . \quad . \quad . \quad . \quad . \quad . \quad (24)$$

which is the general equation expressing the relation between the moment of flexure and the bending moment of the extraneous forces for the mean fibre of any prismatic bar, when the deflection is small.

175. To find the equation of mean fibre of a bar placed horizontally, fixed at one end, and strained by a vertical force W at the other end.

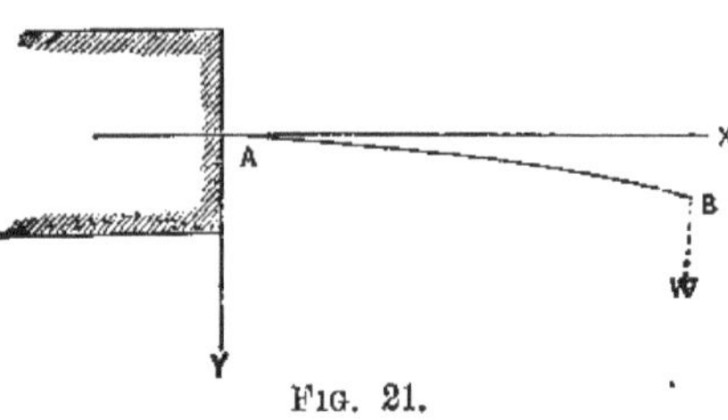

FIG. 21.

Denote by (Fig. 21) l, the length of the bar from the fixed end to the point of application of W, it will be equal to the length of the mean fibre, A B.

Let AX and AY be the co-ordinate axes and Y positive downwards. The bending moment of W for any point, x, will be W $(l - x)$, and substituting this for M in eq. (24), we have

$$\mathrm{EI}\frac{d^2y}{dx^2} = \mathrm{W}(l - x) \quad . \quad . \quad . \quad . \quad (25)$$

Integrating, we have

$$\mathrm{EI}\frac{dy}{dx} = \frac{\mathrm{W}}{2}(2lx - x^2) + \mathrm{C}. \quad . \quad . \quad (26)$$

If $x = 0$, by hypothesis $\frac{dy}{dx} = 0$, and hence $\mathrm{C} = 0$.

Integrating eq. (26) we have

$$\mathrm{EI}y = \frac{\mathrm{W}}{6}(3lx^2 - x^3) + \mathrm{C}' \quad . \quad . \quad (27)$$

Noting that for $x = 0$, $y = 0$, we have $\mathrm{C}' = 0$,

hence, $$y = \frac{\mathrm{W}}{6\mathrm{EI}}(3lx^2 - x^3) \quad . \quad . \quad . \quad (28)$$

which is the equation of the curve of mean fibre under these circumstances.

Inspection of eqs. (26 and 28) will show that the greatest slope of the curve and the greatest distance between any point of it and the axis of X will be at B. Eqs. (25) and (28) show that the curve is convex towards the axis of X.

Represent by f the maximum ordinate of the curve. Its value will be obtained by making $x = l$, hence

$$f = \frac{\mathrm{W}l^3}{3\mathrm{EI}} \quad . \quad . \quad . \quad . \quad . \quad (29)$$

If the bar had been loaded uniformly instead of by a weight acting at its extremity; representing by w the load on a unit of length, eq. (24) would have become for this case,

$$EI\frac{d^2y}{dx^2}=\frac{w}{2}(l-x)^2 \quad . \quad . \quad . \quad . \quad (30)$$

hence the equation of the curve of its mean fibre,

$$y=\frac{w}{24EI}(6l^2x^2-4lx^3+x^4). \quad . \quad . \quad (31)$$

The value of the maximum ordinate in this case would be

$$f=\frac{wl^4}{8EI} \quad . \quad . \quad . \quad . \quad . \quad (32)$$

Instead of W concentrated at the end as shown by eq. (28), suppose it to have been uniformly distributed over the bar, then $\frac{W}{l}$ would be the load on each unit of length in that case, and substituting this in eq. (32) for w, and calling the corresponding ordinate, f', we have,

$$f'=\frac{\frac{W}{l}\times l^4}{8EI}=\frac{Wl^3}{8EI} \quad . \quad . \quad . \quad (33)$$

Hence $f':f::\frac{1}{8}:\frac{1}{3}$, from which we see that concentrating the load at the end of the bar increases the deflection nearly three times that obtained when the load was uniformly distributed.

BEAMS OF UNIFORM CROSS-SECTION.

BEAMS RESTING ON TWO OR MORE SUPPORTS.

176. The term **bar** is used to designate a piece when the dimensions of its cross-section are not only small compared with the length of the piece, but are actually small in themselves. The term **beam** is used when the cross-section is of considerable size, consisting of several square inches.

A beam resting on three or more supports, or having its ends fixed so that they will not move is called a **continuous** beam. If it rests on two points of support only, and the ends are free to move, it is a non-continuous beam. If placed in a horizontal position, with one end fixed and the other free, it is known as a semi-girder or **cantilever**.

Beam Resting on two Points of Support.

177. Let it be required to determine the **bending moments, shearing stress, and equation of mean fibre of a straight beam resting in a horizontal position on two points of support.**

There are two cases: 1, when the beam is uniformly loaded; and, 2, when acted upon by a single force between the two points of support.

1st Case.—The external forces acting on the beam are the load uniformly distributed over it and the vertical reactions at the points of support.

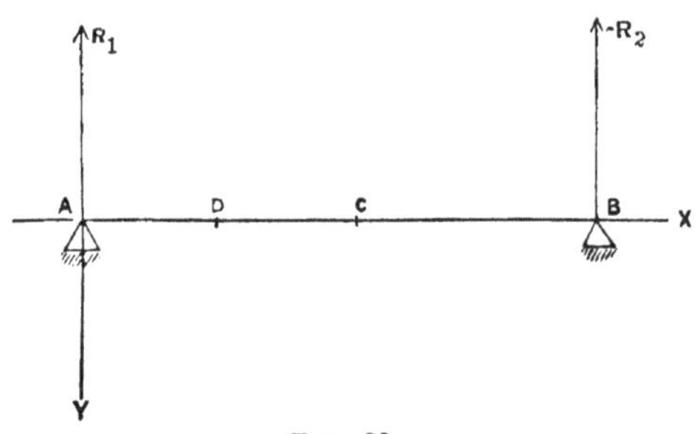

Fig. 22.

Let A B (Fig. 22) be the beam, A and B the points of support, and A the origin of co-ordinates. A X and A Y, the axes. Denote by $2l$ the distance between two points of support A B.

$w =$ weight on unit of length.

$x =$ abscissa of D, any section of the beam A B.

The total load on the beam is $2wl$ and the reactions at each point of support are respectively equal to $- wl$.

Bending moment.—Let D be any section of the beam made by a plane passed perpendicularly to the axis, through the point, whose abscissa is x, and let us consider all the forces acting on either side of D; in this case let it be on the side A D.

The forces acting on the beam from A to D are the weight on this portion of the beam, and the reaction at A. The algebraic sum of their moments will be the bending moment of the external forces acting on this segment. Let M be this moment and we have

$$M = wx \times \frac{x}{2} - wl \times x = \frac{wx^2}{2} - wlx \quad . \quad . \quad . \quad (34)$$

The second member of this equation is a function of a single variable, and may therefore be taken as the ordinate of a line of which x is the abscissa. Constructing the different values of the ordinate, the line may be traced. This line is a parabola, and shows the rate of increase or decrease in the bending moments.

The curve thus constructed may be called the **curve of the bending moments.**

Shearing strain.—The shearing stress in the beam at D is equal to the algebraic sum of all the vertical forces acting at this section, hence

$$S' = wx - wl. \quad . \quad . \quad . \quad . \quad . \quad (35)$$

The second member of this equation represents the ordinate of a right line. Constructing the line, the ordinates will show the rate of increase or decrease of the shearing strain for the different sections.

By comparing equations (34) and (35) it will be seen that

$$S' = \frac{dM}{dx} \quad . \quad . \quad . \quad . \quad . \quad . \quad (36)$$

which shows that **the shearing stress at any section** *is equal to the first differential coefficient of the bending moment of that section taken with respect to x.*

For convenience we used the segment A D, but the results would have been the same if we had taken B D. For, suppose we find the bending moment for this segment, we have for the moment of the weight, acting to turn it around D,

$$w(2l - x) \times \frac{2l - x}{2} = \frac{w}{2}(2l - x)^2.$$

And for the moment of reaction,

$$-\,wl(2l - x).$$

The algebraic sum of these moments will be

$$\frac{wx^2}{2} - wlx,$$

the same as (34), as it should be.

Equation of mean fibre.—Substituting the second member of eq. (34) for M in eq. (24), we have

$$EI\frac{d^2y}{dx^2} = \tfrac{1}{2}wx^2 - wlx. \quad . \quad . \quad (37)$$

Integrating, we get

$$EI\frac{dy}{dx} = \frac{w}{6}x^3 - \frac{w}{2}lx^2 + C.$$

For $x = l$, $\frac{dy}{dx} = 0$, and we have $C = \frac{1}{3}wl^3$.

Substituting this value of C, and integrating, we get

$$EIy = \frac{w}{24}x^4 - \frac{w}{6}lx^3 + \frac{1}{3}wl^3x + C'.$$

For $x = 0$, y is equal 0, and hence $C' = 0$, and we have

$$y = \frac{w}{24EI}(x^4 - 4lx^3 + 8l^3x) \quad . \quad . \quad . \quad (38)$$

which is the equation of the curve of mean fibre, and may be discussed as any other algebraic curve.

Deflection.—If we represent the maximum ordinate of the curve by f, we find

$$f = \frac{5}{24}\frac{wl^4}{EI},$$

the maximum deflection, which is at the middle point of the beam.

Equation (38) may be placed under the form,

$$y = \frac{w}{24EI}(2lx - x^2)\left[5l^2 - (x - l)^2\right] \quad . \quad (39)$$

For values of x, differing but slightly from l, the quantity $(x-l)^2$ may be omitted without materially affecting the value of the second member for these values. Omitting this quantity, and eq. (39) reduces to

$$y = \frac{5wl^2}{24EI}(2lx - x^2) \quad . \quad . \quad . \quad . \quad (40)$$

which is the equation of a parabola. Hence, a parabola may be constructed passing through the middle point of the curve of mean fibre and the points of support, which nearly coincides with the curve of mean fibre in the vicinity of its middle point.

The parabola whose equation is eq. (40) differs but slightly throughout from the curve given by eq. (38); for the greatest difference between the ordinates of the two lines for the same value of x, will be when $x = \frac{l}{2}(2 \pm \sqrt{2})$, which gives

$$y' = \frac{wl^4}{24EI} \times \frac{9}{4}, \; y'' = \frac{wl^4}{24EI} \times \frac{5}{2},$$

y', representing the ordinate of the curve for this value of x, and y'', the ordinate of the parabola for the same value of x

Whence, we get

$$\frac{y''-y'}{y'} = \frac{1}{5}$$

178. 2D CASE.—The external forces acting on the beam are the applied force, whatever it may be, and the vertical reactions at the points of support.

Let A B (Fig. 23) represent the beam resting on the supports, A and B, sustaining a weight, 2W, at any point, as P, between the points of support. Denote the reactions at A and B by R_1 and R_2, A B by $2l$, A P by l'.

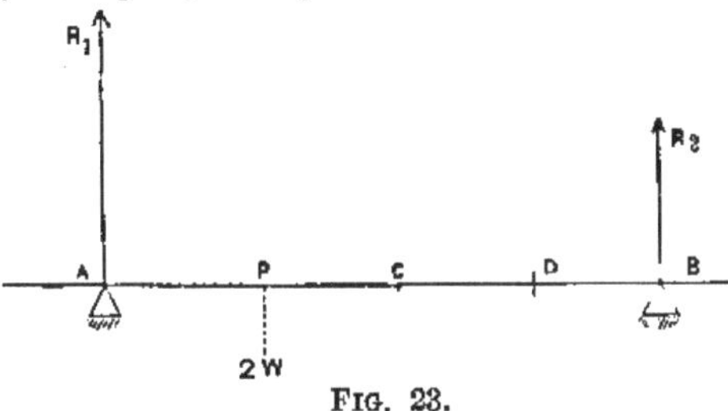

FIG. 23.

The reactions R_1 and R_2 will be proportional to the segments in which the beam is divided, and this sum, disregarding the weight of the beam, is equal to 2W. Hence,

$$R_1 : R_2 : 2W :: PB : AP : AB,$$

from which proportion we, knowing 2W and l', can determine the values of R_1 and R_2. Knowing these, we can obtain the bending moment and shearing strain of any section, and the deflection of the beam due to the force 2W.

179. The most important case of the single load is that in which the load is placed at the centre. Suppose 2W to act at the centre, then $R_1 = R_2 = -W$. Assume the origin of co-ordinates and the axis of X and Y to be the same as in the first case.

Bending moment.—For any section between A and C the bending moment will be $M = -Wx$.

Shearing strain.—The shearing stress on any section will be

$$S' = \pm W.$$

Equation of mean fibre.—Substituting in second member of eq. (24) the above value of M, we have

$$EI\frac{d^2y}{dx^2} = -Wx \quad . \quad . \quad . \quad . \quad . \quad (41)$$

Integrating, and substituting for C, its value, we get

$$EI\frac{dy}{dx} = \frac{W}{2}\,(l^2 - x^2) \quad . \quad . \quad . \quad . \quad (42)$$

Integrating again and substituting for C, its value, we get

$$y = \frac{W}{6EI}(3l^2x - x^3), \quad . \quad . \quad . \quad . \quad (43)$$

which is the equation of so much of the mean fibre as lies between the origin, A, and the middle point, C.

The right half of the mean fibre is a curve exactly similar in form. Assuming B as the origin and the abscissas as positive from B towards C, eq. (43) is also the equation of the right half of the curve.

Deflection.—The maximum deflection is at the centre, and is

$$f = \tfrac{1}{3}\frac{Wl^3}{EI}$$

Comparing this with the deflection at the centre in the previous case, it is seen that the *deflection produced by a load uniformly distributed over the beam is five-eighths of that produced by the same load concentrated and placed at the middle point.*

180. **Comparison of strains produced.**—The bending moment for any section, when the beam is uniformly loaded, is, eq. (34),

$$M = \frac{wx^2}{2} - wlx,$$

and when the beam is acted upon by a load at the middle point, is, eq. (41),

$$M = -Wx,$$

Both will have their maximum values for $x = l$.
Equating these values, we have

$$Wl = \tfrac{1}{2}wl^2,$$

whence

$$W = \frac{wl}{2},$$

which shows that the greatest strain on the unit of area of the fibres, when the load is uniformly distributed, is the same as that which would be caused by half the load concentrated and placed at the middle point of the beam.

Beam strained by a uniform load over its entire length and a load resting midway between the two points of support.

181. If a beam be uniformly loaded, and support also a load midway between the points of support, the corresponding

values for the strains can be obtained by adding algebraically the results determined for each case taken separately.

If the beam had other loads besides the one at C, we could in the same manner find the bending moments, shearing strains, and deflections due to their action. The algebraic sum of the moments, ordinates of deflection, etc., would give the results obtained by their simultaneous action.

Beam having its ends firmly held down on its supports.

182. In the preceding cases the beams are supposed to be resting on supports, and not in any way fastened to them. If the ends of the beams had been fastened firmly so that they could not move—as, for example, a beam having its ends firmly imbedded in any manner in two parallel walls—the results already deduced would have been materially modified.

Let it be required to determine the **strains** *and* **equation of curve of mean fibre** *in the case where the beam has its extremities horizontal, and firmly embedded so that they shall not move,* the beam being uniformly loaded.

If we suppose a bar fitted into a socket (Fig. 24) and acted upon by a force to bend it, it is evident, calling Q_1 the force of the couple developed at the points B and H, that the moment of the force W, whose lever arm is l, is opposed by the moment of resistance of the couple, B Q_1 and H Q_1 acting through the points H and B.

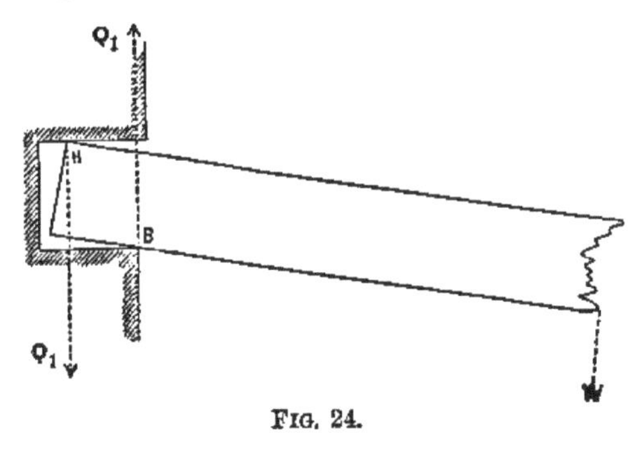

FIG. 24.

Hence, we have

$$Q_1 l' = l W,$$

l' being the lever arm of the couple.

We see that Q_1 increases proportionally to any decrease in l', and that these quantities themselves are unknown, although their product must be constant and equal to the bending moment of the beam at B.

To determine the bending moment at any section of a beam having its ends firmly held down; let A B (Fig. 25) be the beam before being loaded, and denote by

$2l$ = A B = the length;

w = the weight on unit of length;

x = the abscissa at any point, the origin of co-ordinates being at A, and A B coinciding with axis of X, as in preceding cases.

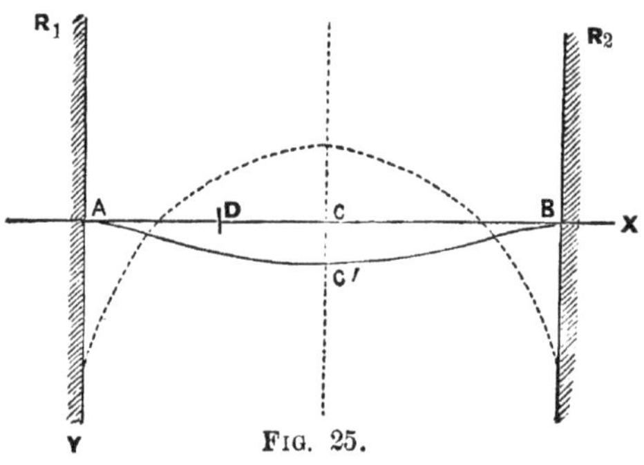

FIG. 25.

The total load on the beam will be $2wl$, and the reactions at the points of support are each equal to $-wl$.

The bending moment of any section D, is equal to the algebraic sum of the moments of vertical reaction at A, of the weight on A D, and of the unknown couple acting on the left of A.

Calling μ the moment of the unknown couple and substituting this algebraic sum in eq. (24), we have

$$EI\frac{d^2y}{dx^2} = -wlx + \frac{wx^2}{2} + \mu \quad . \quad . \quad . \quad (44)$$

Integrating and noting that for $x=0$, $\frac{dy}{dx}=0$, we have $C=0$, and

$$EI\frac{dy}{dx} = -\frac{wl}{2}x^2 + \frac{w}{6}x^3 + \mu x. \quad . \quad . \quad (45)$$

In this equation make $x = 2l$, for which $\frac{dy}{dx}=0$, and we find

$$\mu = \tfrac{1}{8}wl^2,$$

which is the value of the moment of the unknown couple acting at the left point of support. It is also the value of the one at the right point of support, B.

Writing this value for μ in equations (44) and (45), we have

$$EI\frac{d^2y}{dx^2}=-wlx+\frac{w}{2}x^2+\frac{w}{3}l^2 \quad . \quad . \quad . \quad (46)$$

$$EI\frac{dy}{dx}=-\frac{wl}{2}x^2+\frac{w}{6}x^3+\frac{wl^2}{3}x, \quad . \quad . \quad (47)$$

and then by integration,

$$EIy=-\frac{wl}{6}x^3+\frac{w}{24}x^4+\frac{wl^2}{6}x^2+C'.$$

We find $C'=0$, and substituting, etc., we get

$$y=\frac{wx^2}{24EI}(x-2l)^2 \quad . \quad . \quad . \quad . \quad . \quad (48)$$

which is the equation of the curve of mean fibre.

Deflection.—Denoting by f, the maximum value for y, and we have

$$f=\frac{w}{24EI}l^4.$$

The corresponding value obtained, from eq. (38), is

$$f'=\frac{5w}{24EI}l^4$$

A comparison of these values of f shows that by firmly fastening the ends of the beam to the points of support in a horizontal position, the deflection at the centre is *one-fifth* of what it was when they merely rested on the supports.

Bending moments.—The curve of the bending moments is given by the equation.

$$M=\frac{w}{2}x^2-wlx+\frac{w}{3}l^2,$$

which is that of a parabola.

The bending moments for $x=0$, and $2l$, are both equal to $\frac{w}{3}l^2$, and for $x=l$, $-\frac{wl^2}{6}$. The bending moment of the section at the middle point is therefore half that of the section at A or B. Assuming a scale, lay off $\frac{w}{3}l^2$, below the line A B, on perpendiculars passing through A and B. Lay off half this value on the opposite side of the line A B on a perpendicular through the middle point. This gives us three points of the curve of which one is the vertex. The perpendicular through the middle point is the axis of the parabola, and with the three points already found the curve may be constructed.

This curve of bending moments cuts the axis of X in two points, the abscissas of which are $l\ (1 \pm \sqrt{\frac{1}{3}})$, and at the sections corresponding to them the bending moments will be equal to 0.

These values substituted in eq. (46) for x, reduces $\frac{d^2y}{dx^2}$ to zero, and an examination of this equation shows that there is a change of sign in $\frac{d^2y}{dx^2}$ at these points. It therefore follows that the **curve of mean fibre** has a point of inflexion for each of these values of x, that is, the curve changes at these points from being concave to convex, or the reverse, towards the axis of X.

The greatest strains on the unit of area produced by the deflecting force, will be in the cross-sections at the ends and middle; the lower half of the cross-section at the middle being extended, and the lower halves of these at the points of the support being compressed.

Shearing strain.—The expression for the shearing force is

$$S' = \frac{dM}{dx} = wx - wl,$$

which is the same as eq. (35), and its values may be represented by the ordinates of a right line which passes through the middle point.

The uniform load concentrated and placed at the middle.

183. If instead of being uniformly loaded, the beam was only strained by a single load, 2W, at the middle point, the bending moment, disregarding the weight of the beam, would be for values of $x < l$.

$$M = -Wx + \mu$$

and by a process similar to that just followed, we would find

$$y = \frac{W}{12EI}(3lx^2 - 2x^3),$$

to be the equation of the mean fibre from A to C.

The maximum deflection will be

$$f = \frac{W}{12EI}l^3,$$

which is equal to *one-fourth* of that obtained, with a load at the centre, when the ends of the beam are free. It is also seen that the deflection caused by a concentrated load placed at the middle of the beam, is the same as that caused by double the load uniformly distributed over the whole length.

If the beam was loaded both uniformly and with a weight, 2W, the results would be a combination of these two cases.

Beam loaded uniformly, fixed at one end, and resting on a support at the other.

184. Let A B (Fig. 26) represent the beam in a horizontal position, fixed at the end, A, and resting on a support at the end B.

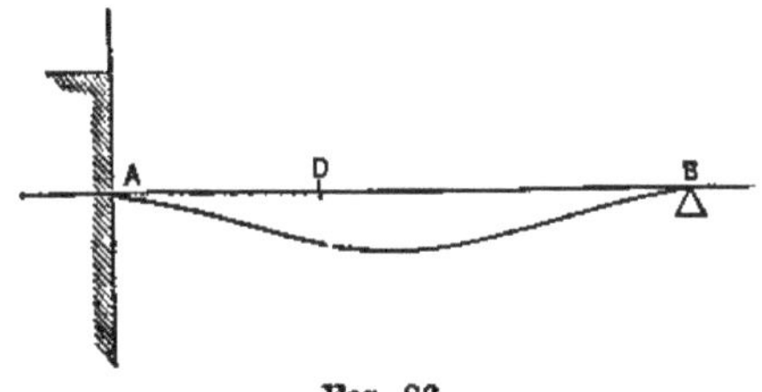

FIG. 26.

Adopting the notation used in previous case, we have $2wl$ for the total load on the beam.

The reactions at A and B are unequal. Represent by R_1 the reaction at A, and by μ the moment of the unknown couple at A. We have

$$EI\frac{d^2y}{dx^2} = -R_1x + \frac{wx^2}{2} + \mu \quad . \quad . \quad (49)$$

Hence by integration,

$$EI\frac{dy}{dx} = -\tfrac{1}{2}R_1x^2 + \frac{w}{6}x^3 + \mu x, \; C = 0 \quad (50)$$

$$EIy = -\tfrac{1}{6}R_1x^3 + \frac{w}{24}x^4 + \mu\frac{x^2}{2}, \; C' = 0 \quad (51)$$

The bending moment at B is equal to zero, hence for $x = 2l$, $\frac{d^2y}{dx^2}$ will be 0 and eqs. (49) and (51) reduce for this value of x to

$$0 = -R_1 2l + \frac{w}{2}(2l)^2 + \mu \quad . \quad . \quad . \quad . \quad (52)$$

$$0 = -\tfrac{1}{6}R_1(2l)^3 + \frac{w}{24}(2l)^4 + \mu\tfrac{1}{2}(2l)^2 \quad . \quad (53)$$

Combining these we find

$$R_1 = \tfrac{5}{8}\, w\,(2l) \text{ and } \mu = \frac{wl^2}{2}.$$

Hence the reaction at B is $\frac{3}{8}w\,(2l)$.

Substituting these values for R_1 and μ in eq. (49) the bending moment at any point, shearing strain, and curve of mean fibre can be fully determined. Placing the second member of eq. (49) equal to zero, and deducing the values of x, these will be the abscissas of the points of inflexion, and by placing the second member of eq. (50) equal to 0, the abscissa corresponding to the maximum ordinate of deflection may be obtained. The curve of bending moments, etc., may be determined as before.

Beam resting on three points of support in the same horizontal straight line.

185. Let it be required to determine the **bending moments, shearing strain, and equation of mean fibre** *of a single beam resting in a horizontal position on three points of support, each segment being uniformly loaded.*

Let A B C (Fig. 27) be the beam resting on the three points, A, B, and C.

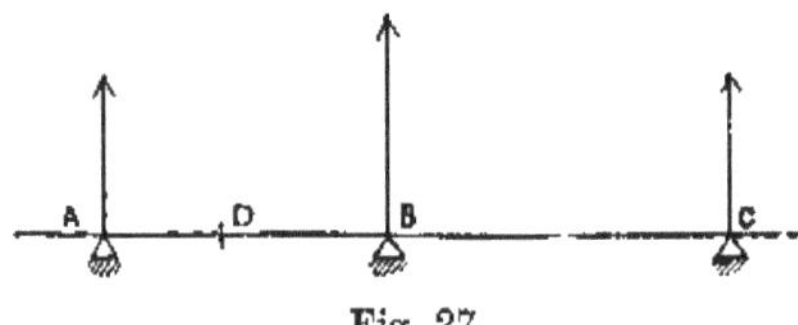

Fig. 27.

Let us consider the general case in which the segments are unequal in length and the load on the unit of length different for them.

Let l = A B, and w, the weight on each unit of its length,

l' = B C, and w' the weight on each unit of its length

R_1, R_2, R_3, the forces of reaction at the points of support, A, B, and C, respectively.

Take A B C as the axis of X and A the origin of coördinates with y positive downwards as in the other cases.

First, consider the segment A B, and let D be any section whose abscissa is x.

Since the reactions at the points of support are unknown, they must be determined.

We have

$$EI\frac{d^2y}{dx^2} = -R_1x + \frac{wx^2}{2} \quad . \quad . \quad . \quad (54)$$

Integrating, we get

$$EI\frac{dy}{dx} = -\tfrac{1}{2}R_1x^2 + \frac{wx^3}{6} + C. \quad . \quad . \quad (55)$$

Let ω represent the angle made by the curve of mean fibre with the axis of X at B, then for $x = l$ we have $\left(\frac{dy}{dx}\right)_{x=l} = \tan\omega$. whence

$$EI\tan\omega = -\tfrac{1}{2}R_1l^2 + \tfrac{1}{6}wl^3 + C. \quad . \quad . \quad (56)$$

Subtracting from preceding equation, member by member, we have

$$EI\left(\frac{dy}{dx} - \tan\omega\right) = -\tfrac{1}{2}R_1x^2 + \tfrac{1}{6}wx^3 + \tfrac{1}{2}R_1l^2 - \tfrac{1}{6}wl^3. \quad (57)$$

Integrating eq. (57) we get

$$EI(y - x\tan\omega) = -\tfrac{1}{6}R_1x^3 + \tfrac{1}{24}wx^4 + \tfrac{1}{2}R_1l^2x - \tfrac{1}{6}wl^3x. \quad (58)$$

the constant of integration in this case being equal to 0.

If in eq. (54) we make $x = l$, and denote the bending moment of the section at B by μ, we have

$$\mu = -R_1l + \frac{wl^2}{2} \quad . \quad . \quad . \quad . \quad (59)$$

In eq. (58) make $x = l$, hence $y = 0$, and we have

$$EI\tan\omega - \tfrac{1}{6}R_1l^2 + \tfrac{1}{24}wl^3 + \tfrac{1}{2}R_1l^2 - \tfrac{1}{6}wl^3 = 0 \quad . \quad (60)$$

by omitting common factor l. Combining this equation with the preceding one and eliminating R_1 and reducing, we get

$$EI\tan\omega = \tfrac{1}{3}l\mu - \tfrac{1}{24}wl^3 \quad . \quad . \quad (61)$$

which expresses the relation between the tan ω and μ.

Going to the other segment, taking C as the origin of coordinates and calling x positive towards B, we may deduce similar relations between the bending moment at B and the tangent of the angle made by the mean fibre at B with the axis of X. Since the beam is continuous, these curves are tangent to each other at the point B, and the angles made by both of them with the axis of X at that point are measured by a common tangent line through B. Therefore, the angles are supplements of each other and we may at once write the corresponding relation as follows,

$$-\mathrm{EI}\tan\omega = \tfrac{1}{3}l'\mu - \tfrac{1}{24}w'l'^3 \quad . \quad . \quad . \quad . \quad (62)$$

Since, for equilibrium, the algebraic sum of the extraneous forces must be equal to zero, we have

$$wl + w'l' - \mathrm{R}_1 - \mathrm{R}_2 - \mathrm{R}_3 = 0 \quad . \quad . \quad . \quad (63)$$

and since the algebraic sum of their moments with respect to any assumed section must be equal to zero, we have for the moments taken with respect to the section at B,

$$\mathrm{R}_1 \times l + w'l' \times \frac{l'}{2} = \mathrm{R}_3 \times l' + wl \times \frac{l}{2}. \quad . \quad . \quad (64)$$

These last four equations contain four unknown quantities, R_1, R_2, R_3, and $\tan\omega$.

By combining and eliminating, their values may be found. Combining equations (61) and (62), and eliminating $\tan\omega$, we have

$$\mu = \tfrac{1}{8}\,\frac{wl^3 + w'l'^3}{l + l'}. \quad . \quad . \quad . \quad . \quad (65)$$

The bending moment of any section, as D, is from equation (54)

$$-\mathrm{R}_1 x + w\frac{x^2}{2};$$

hence for $x = l$, we have M equal to the bending moment at B, which has been represented by μ, or eq. (59)

$$\mu = -\mathrm{R}_1\, l + \frac{w}{2}\, l^2,$$

from which we get

$$\mathrm{R}_1 = \frac{wl}{2} - \frac{\mu}{l} = \frac{wl}{2} - \tfrac{1}{8}\,\frac{wl^3 + w'l'^3}{l\,(l + l')}. \quad . \quad (66)$$

In a similar way, the value of R_3 may be found. These values of R_1 and R_3 substituted in eq. (63), will give the value of R_2.

The external forces, all being known, the bending moments, shearing strain, and equation of mean fibre may be determined as in previous examples.

186. *Example.*

The most common case of a beam resting on three points of support, is the one in which the beam is uniformly loaded throughout and the intermediate support is placed at the middle point.

In this case, $l = l'$ and $w = w'$. Substituting these values, in the expressions for μ and R_1, we have

$$\mu = \tfrac{1}{8} wl^2, \text{ and } R_1 = \tfrac{3}{8} wl.$$

The reaction at the middle point will therefore be

$$\tfrac{10}{8} wl \text{ or } \tfrac{5}{8} w(2l).$$

Substituting the value of R_1 in eq. (54) we obtain the bending moment for any section.

In the case of a beam resting on two supports, Fig. (22), and having a weight uniformly distributed along its length, it has been shown that each support bears one half of the distributed load; and that the deflection of the mean fibre at the middle point, represented by f, is the same as the beam would take were $\frac{5}{8}$ths of the load acting alone at the middle point. In the latter case the pressure upon a support, just in contact with the beam at its middle point, would be zero; and if the support were to be raised so as to bring the middle of the beam into the same right line with the extreme supports, the intermediate support would evidently sustain the total pressure at C to which the deflection was due, and which was $\frac{5}{8}$ths of the entire load; hence the reaction of the middle support will be equal to $\frac{5}{8}$ths. This conclusion agrees with the result determined by the previous analysis.

Each segment of the beam in this case might have been regarded as a beam having one end fixed and the other resting on a support; a case which has already been considered.

Theorem of Three Moments.

187. From the preceding, it is seen, that the reactions at the points of support can be determined whenever we know the bending moments at these points. These moments are readily found by the "**theorem of three moments.**"

This theorem has for its object to deduce a formula expressing the relation between the bending moments of a beam at any three consecutive points of support, by means of which the bending moments at these points may be obtained, without going through the tedious operations of combination and elimination practised in the last example.

Take any three consecutive points of support, as A, B, and

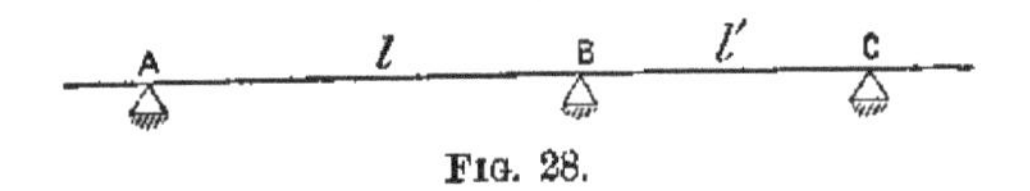

FIG. 28.

C, Fig. (28), of a beam resting on n supports. Denote by l and l', the lengths of the segments, A B and B C, w and w', the weights on each unit of length in each segment and M_1 M_2 M_3, the bending moments at these points, A, B, C.

The formula expressing the relation between these bending moments is

$$M_1 l + 2M_2(l + l') + M_3 l' = \tfrac{1}{4}wl^3 + \tfrac{1}{4}w'l'^3. \quad (67)$$

In every continuous beam, whose ends are not fixed, the bending moments at the end supports are each equal to zero. Hence, by the application of this formula, in any given case, as many independent equations can be formed as there are unknown moments, and from these equations the moments can be determined.

188. The demonstration of this theorem depends upon the principle, that the *bending moment at any point of support whatever*, and *the tangent of the angle made by the neutral fibre with the horizontal at that point*, may be expressed in functions of the first degree of the *bending moment at the preceding point of support, and the tangent of the angle made by the neutral fibre with the horizontal at that point.*

Let A B (Fig. 29) be any segment of a beam resting on n supports, A the origin, A X and A Y the axes of co-ordinates, and M_1 and M_2 the bending moments at A and B.

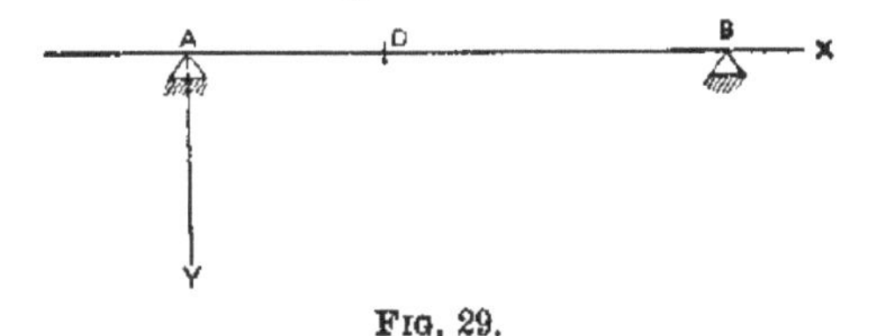

FIG. 29.

The applied forces acting on the beam and the reactions are taken vertical and in the plane of the mean fibre.

The external forces which act on the beam to the left of the support, A, may be considered as replaced by a resultant moment and a resultant shearing force, without disturbing the equilibrium. This resultant moment, represented by M_1, is equal and opposite to the moment of the internal forces at the section through the support A; the vertical force, which we represent by S_1, is equal and opposed to the shearing force at this section.

Represent by μ the algebraic sum of the moments of the external forces acting on the beam between A and any section as D, whose abscissa is x.

Then from eq. (24) we have

$$EI\frac{d^2y}{dx^2} = M_1 + \mu + S_1x \quad . \quad . \quad . \quad (68)$$

Denoting by ϕ the angle which the neutral fibre after deflection makes with the axis of X, at A, and integrating, we have

$$EI\left(\frac{dy}{dx} - \tan\ \phi\right) = M_1x + \int_0^x \mu dx + \tfrac{1}{2}S_1x^2. \quad (69)$$

Representing the quantity $\int_0^x \mu dx$ by M' and integrating, we have

$$EI\,(y - x\tan\phi) = \tfrac{1}{2}M_1x^2 + \int_0^x M'dx + \tfrac{1}{6}S_1x^3. \quad (70)$$

In these three equations, make $x = l$ and denote by N, Q, and K what μ, M', and $\int_0^x M'dx$ become for this value of x, and by ω the angle made by the curve of mean fibre with the axis of X at B; noting that for $x = l$, $EI\frac{d^2y}{dx^2} = M_2$, we have

$$\left.\begin{aligned} &M_2 = M_1 + N + S_1l, \\ &EI\ (\tan\omega - \tan\phi) = M_1l + Q + \tfrac{1}{2}S_1l^2, \\ &-EIl\tan\phi = \tfrac{1}{2}M_1l^2 + K + \tfrac{1}{6}S_1l^3. \end{aligned}\right\} (71)$$

Combining the first and third, and then the second and third of these equations and eliminating S_1, we have

$$\left.\begin{array}{l} \frac{1}{6}M_2 l^2 + EIl \tan\phi = -\frac{1}{3}M_1 l^2 + \frac{1}{6}Nl^2 - K, \\ \frac{1}{3}EIl \tan\omega + \frac{2}{3}EIl \tan\phi = -\frac{1}{6}M_1 l^2 + \frac{1}{3}Ql - K. \end{array}\right\} (72)$$

In these equations, N, Q, and K depend directly upon the applied forces, and are known when the latter are given. But M_1, M_2, $\tan\phi$ and $\tan\omega$ are unknown.

An examination of equations (72) shows that M_2 and $\tan\omega$ are functions of the first degree of M_1 and $\tan\phi$, whatever be the manner in which the external forces are applied.

Let us impose the condition that the system of forces acting on the beam shall be a load uniformly distributed over each segment, and denote by w the load on a unit of length of the segment A B.

For this case we have

$$\left.\begin{array}{r} \mu = \frac{1}{2}wx^2, \\ M' = \frac{1}{6}wx^3, \\ \int_0^x M\,dx = \frac{1}{24}wx^4, \end{array}\right\}$$

and in these, by making $x = l$, we have

$$N = \frac{1}{2}wl^2,$$
$$Q = \frac{1}{6}wl^3,$$
$$K = \frac{1}{24}wl^4.$$

Substituting in equations (72) these values for N, Q, and K, we have

$$\left.\begin{array}{l} M_2 = -2M_1 - \frac{6EI}{l}\tan\phi + \frac{1}{4}wl^2, \\ \tan\omega = -2\tan\phi - \frac{1}{2}\frac{l}{EI}M_1 + \frac{1}{24}\frac{wl^3}{EI}, \end{array}\right\} (73)$$

which agree with the principle already enunciated.

189. To deduce formula (67), let A, B, C (Fig. 28) be any three consecutive points of support of a beam resting on n supports.

From the first of equations (73) we may at once write

$$M_3 = -2M_2 - \frac{6EI}{l'} \tan\phi' + \tfrac{1}{4}w'l'^2,$$

and by considering x positive from B to A, and giving the proper sign to tan ϕ, we write

$$M_1 = -2M_2 + \frac{6EI}{l} \tan\phi' + \tfrac{1}{4}wl^2.$$

Multiplying these respectively by l' and by l, and adding them together, we have

$$M_1l + 2M_2(l + l') + M_3l' = \tfrac{1}{4}wl^3 + \tfrac{1}{4}w'l'^3,$$

which expresses the relation between the bending moments for any three consecutive points of support, and is the same as formula (67).

By a similar process we can find an equation expressing the relation between the tangents of the angles taken at the three points of support.

Applications of Formula (67).

190. 1st Case.—*Beam in a horizontal position, loaded uniformly, resting on three points of support, the segments being of equal length.*

In this case, we have $l' = l$, $w' = w$, and M_1 and M_3 each equal to zero. Substituting these values in eq. (67), we get

$$2(2l)M_2 = \tfrac{1}{4}(2wl^3),$$

whence

$$M_2 = \tfrac{1}{8}wl^2.$$

The bending moment of the section at B is, eq. (59),

$$-R_1l + \frac{wl^2}{2} = M_2 = \tfrac{1}{8}wl^2,$$

whence we get for the reaction at A,

$$R_1 = \tfrac{3}{8}wl,$$

which is the same value before found. The reaction at C is

the same, and that at B can now be easily determined, from the equation,

$$R_1 + R_2 + R_3 = 2wl.$$

Knowing all the external forces acting on the beam, the bending moment at any section, the shearing strain, etc., can be determined.

191. 2D CASE.—*Beam in a horizontal position resting on four points of support.*

Ordinarily a beam resting on four supports is divided into three unequal segments, the extreme or outside ones being equal to each other in length, and the middle one unequal to either.

If we suppose this to be the case, represent by A, B, C, and D the points of support in the order given. The bending moments at A and D are each equal to zero. To find those at B and C, take the general formula (67) and apply it first to the pair B C and B A, and then to the pair C B and C D, and determine the bending moments from the resulting equations. Having found them, the reactions are easily found; and knowing all the forces acting on the beam, the bending moments, shearing strains, and curve of mean fibre may be obtained.

192. 3D CASE.—*Beam in a horizontal position resting on five points of support*, the segments being equal in length.

When the number of supports is odd, the segments are generally equal in length, or if unequal, they are symmetrically disposed with respect to the middle point.

If the beam be uniformly loaded, it will only be necessary to find the bending moments at the points of support of either half of the beam, as those for corresponding points in the other half will be equal to them.

Suppose the case of five points of support.

Let A, B, C, D, and E be the points of support, C being the centre one. Represent by l the length of a segment, w the weight on a unit of length, M_2, M_3, M_4, the bending moments at B, C, and D, and the forces of reaction at A, B, and C, by R_1, R_2, R_3 respectively. From the conditions of the problem, M_2 is equal to M_4, and the reactions at A and B are equal to the reactions respectively at E and D.

Applying formula (67) to the first pair of segments, we have

$$4lM_2 + lM_3 = \tfrac{1}{2}wl^3,$$

and applying it to the second pair, BC and CD, we get

$$lM_2 + 4lM_3 + lM_4 = \tfrac{1}{2}wl^3.$$

In these equations, making M_4 equal to M_2 and combining the equations, we find

$$M_2 = \tfrac{3}{28}wl^2, \text{ and } M_3 = \tfrac{1}{14}wl^2.$$

The external forces acting on the first segment, AB, to turn it around the section at B, are $-R_1$ and wl. Hence we have

$$-R_1l + \tfrac{1}{2}wl^2 = M_2 = \tfrac{3}{28}wl^2,$$

whence

$$R_1 = \tfrac{11}{28}wl.$$

The external forces acting to turn the segment A C or half the beam around C are the reactions at A and B and the loads on the two segments A B and B C.

The algebraic sum of the moments for the section at C is,

$$-R_1 2l - R_2 l + \tfrac{3}{2}wl^2 + \tfrac{1}{2}wl^2 = M_3 = \tfrac{1}{14}wl^2.$$

Substituting in this the value just found for R_1 and solving with respect to R_2, we get

$$R_2 = \tfrac{32}{28}wl.$$

The sum of the reactions is equal to the algebraic sum of the applied forces, hence,

$$R_1 + R_2 + R_3 + R_4 + R_5 = 2R_1 + 2R_2 + R_3 = 4wl,$$

in which substituting for R_1 and R_2, their values, we find

$$R_3 = \tfrac{26}{28}wl.$$

The external forces acting on the beam are now all known, and hence the bending moments, shearing strain, etc., may be determined.

193. 4TH CASE.—*Beam in a horizontal position, resting on n points of support*, the segments being equal in length.

If the beam be uniformly loaded, it will, as in the last case, only be necessary to find the bending moments at the points of support of either half of the beam.

If n be even, the reaction of the $\frac{1}{2}n^{th}$ and $(\frac{1}{2}n+1)^{th}$ support will be equal; if n be odd, the $\frac{1}{2}(n+1)$ will be the middle support, and the reactions of the supports equidistant from the middle point will be equal.

The formula for the segments would become, n being even,

$$4M_2 + M_3 = \frac{1}{2}wl^2,$$

$$M_2 + 4M_3 + M_4 = \frac{1}{2}wl^2,$$

* * * * * * * *

$$M_{\frac{1}{2}n} + 4M_{\frac{1}{2}n+1} + M_{\frac{1}{2}n+2} = \frac{1}{2}wl^2.$$

In the last equation, $M_{\frac{1}{2}n+1}$ and $M_{\frac{1}{2}n+2}$ would be equal espectively to $M_{\frac{1}{2}n}$ and $M_{\frac{1}{2}n-1}$.

From these equations, $R_1, R_2, R_3, \ldots R_n$ could be obtained.

General Example.

194. 5TH CASE.—*Beam in a horizontal position resting on $n+1$ points of support,* segments unequal in length, and uniform load on unit of length being different for each segment.

Represent the points of support by $A_1\ A_2\ A_3 \ldots A_n\ A_{n+1}$, and the respective bending moments at these points of support by $M_1, M_2, M_3, \ldots . M_n, M_{n+1}$. Represent the length of the segments by $l_1, l_2, l_3, \ldots . l_n$ and the respective units of weight on the segments by $w_1, w_2, w_3, \ldots . w_n$.

The bending moments M_1, M_{n+1} being those at the extremities, are each equal to zero, and therefore there are only $n-1$ unknown moments to determine. Applying eq. (67) successively to each pair of segments, we obtain $n-1$ equations of the first degree with respect to these quantities, which by successive eliminations give us the values of the moments, $M_2, M_3, \ldots\ldots M_n$.

These equations will be of the following form:

$$2(l_1 + l_2)M_2 + l_2M_3 = \frac{1}{4}(w_1l_1^3 + w_2l_2^3)$$

$$l_2M_2 + 2(l_2 + l_3)M_3 + l_3M_4 = \frac{1}{4}(w_2l_2^3 + w_3l_3^3)$$

* * * * * * * *

$$l_{n-1}M_{n-1} + 2(l_{n-1} + l_n)M_n = \frac{1}{4}(w_{n-1}l^3_{n-1} + w_n l^3_n).$$

From these equations, the reactions at the points of support can be determined, and knowing all the external forces the strains on the beam may be calculated.

TORSION.

195. A beam strained by a system of extraneous forces, among which is a couple acting in a plane perpendicular to the axis of the piece, will be subjected to a stress of torsion in addition to the other stresses already described.

Suppose a beam fixed at one end (Fig. 30) and a couple applied to the free end, F, the axis of the couple intersecting the axis of the piece, and the plane of the couple perpendicular to the axis. The action of the couple will be to twist the beam around its axis, causing a twisting strain of the fibres and developing torsional stresses in the material.

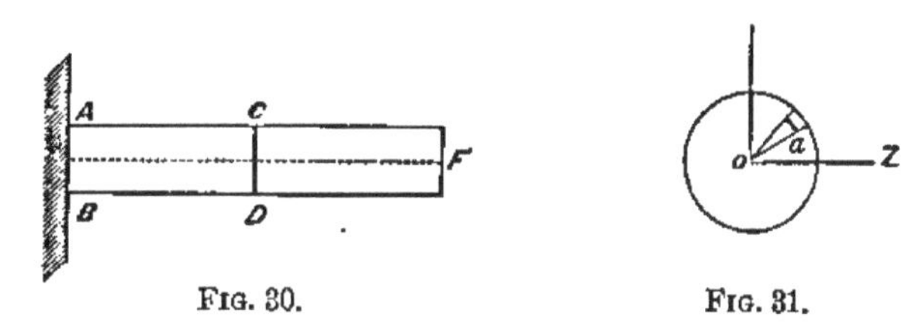

FIG. 30. FIG. 31.

To determine the stress of torsion at any cross-section as C D, let α be equal to the angular amount of torsion between any two cross-sections of the beam, and β the amount of angular change for a unit of length.

It is assumed that the total amount of angular change of any fibre between any two sections, or α, is directly proportional to the distance between the sections, and that the stress of torsion developed in the fibre is directly proportional to its distance from the axis of the piece.

Let $T_t' =$ the stress of torsion in any fibre, $a =$ the area of cross-section of the fibre, and $G =$ the coefficient of torsional elasticity; then

$$\frac{T_t'}{a} = G\beta, \quad \text{or} \quad T'_t = aG\beta.$$

Let O be taken as the pole (Fig. 31) O Z, the fixed line, and r and v the polar co-ordinates of points in the plane of cross-section C D. Then

$$a = rdr\,dv.$$

Since the stress is assumed to be directly proportional to the distance of the fibre from the axis, we get by substituting for a its value, and multiplying by r, the intensity of the stress in the fibre at the distance r from the axis to be

$$G \beta r^2 dr\, dv.$$

Suppose the section C D to be fixed. The twisting action of the couple at F is transmitted from section to section of the piece until it reaches C D, where it is opposed by the resistance developed in the section. The moment of resistance offered by the fibre at the distance r from the axis will be the intensity of the twisting stress in the fibre multiplied by its lever arm, r, or

$$G \beta r^3 dr\, dv.$$

The total moment of the resistance developed in the cross-section C D may be expressed as follows:

$$G \beta \int\int r^3\, dr\, dv, \quad . \quad . \quad . \quad . \quad (74)$$

Represent the moment of the couple acting at the section F by $F' \times \lambda$, and equating the moments, we have

$$G \beta \int\int r^3 dr\, dv = F'\lambda, \quad . \quad . \quad (75)$$

This expression $\int\int r^3 dr\, dv$ is called the **polar moment of inertia**; that is, the moment of inertia of a cross-section of the beam about an axis through its centre and perpendicular to the plane of cross-section.

Representing it by I_p, we have

$$G \beta I_p = F'\lambda, \quad . \quad . \quad . \quad . \quad (76)$$

Suppose the cross-section considered to be a circle, whose radius $= R$, and the section in which the resistance is considered is at the distance l from the plane of the twisting couple. Equation (76) would become for this case, by substituting $\frac{\alpha}{l}$ for β, and $\frac{1}{2}\pi R^4$ for I_p,

$$G = 2\frac{F'\lambda}{\pi R^4} \times \frac{l}{\alpha} \quad . \quad . \quad . \quad . \quad . \quad . \quad (77)$$

196. General Morin, in his work on Strength of Materials, gives the value for G for different materials.

The following are some of the values:

Wrought iron....	G =	8,533,700 lbs.
Cast-iron	G =	2,845,000 lbs.
Cast-steel..................	G =	14,223,000 lbs.
Copper....................	G =	6,210,000 lbs.
Oak........................	G =	569,000 lbs.
Pine.......................	G =	616,000 lbs.

Rupture by Twisting.

197. It is assumed that the torsional stress developed in the fibres of a piece varies directly with the distance of the fibre from the axis of torsion, and is greatest in the fibres farthest from this axis. If the strain be increased until rupture takes place, those fibres farthest from the axis will be the ones to give way first.

The intensity of the torsional stress for any cross-section developed in a fibre at the distance r from the axis is

$$G\,\beta\,r^2 dr\,dv.$$

This expression divided by the area of cross-section of the fibre, $r\,dr\,dv$, gives $G\,\beta\,r$ as the intensity of the torsional stress on the unit of surface at the distance r from the axis. Represent this intensity by T', and we have

$$T' = G\,\beta\,r.$$

Multiplying both members of this equation by I_p, and dividing by r, we get

$$\frac{T'}{r} I_p = G\,\beta\,I_p,$$

in which the second member is the same as the first member of equation (76). Hence,

$$\frac{T'}{r} I_p = F'\lambda$$

from which we get

$$T' = \frac{F'\lambda}{I_p}r, \quad . \quad . \quad . \quad (78)$$

or, an expression for the torsional stress on any unit of cross-section of a piece strained by a twisting force.

Let d = the greatest value that r can have for any cross-section. If d be substituted for r in equation (78) the resulting value of T′ will be the stress on the unit farthest from the axis for the cross-section considered.

Suppose $F'\lambda$ to be increased until rupture is produced, then T′ for this value of $r = d$, in the section where rupture begins, will be T_t, the **modulus of torsion,** or

$$T_t = F'\lambda \times \frac{d}{I_p} \quad . \quad . \quad . \quad . \quad . \quad . \quad (79)$$

from which the values of the modulus of torsion may be deduced.

INFLUENCE OF TEMPERATURE.

198. The influence of changes in temperature, especially in the metals, forms an important element to be considered in determining the amount of strain on a beam.

If the beam is free to move at both ends, there will be no strain in the beam arising from the changes of temperature; if the ends are fixed, there will be, and these strains must be determined.

The elongation or contraction produced by the changes of temperature is known for the different metals. The amount of strain upon the unit of area will be the same as that produced by a force elongating or contracting the beam an amount equal to that resulting from the change of temperature under consideration.

CHAPTER VII.

STRENGTH OF BEAMS.

PROBLEMS.

199. The object of the previous discussions has been to find the strains to which a beam is subjected by certain known forces applied to it.

The problems which now follow are:

Knowing all the external forces acting on a beam, to determine the form and dimensions of its cross-section, so that the strain on the unit of surface shall at no point be greater than the limit allowed; and knowing the form and dimensions of the cross-section of a beam, to determine the load which it will safely bear.

There are two cases; one is where the cross-section is constant throughout the beam; and the other is where it varies from one point to another.

1st CASE.—BEAMS OF UNIFORM CROSS-SECTION.

200. Strength of beam strained by a tensile force. Let W be the resultant force whose line of direction is in the axis of the beam and whose action is to elongate it.

From the equation preceding eq. (5), we have

$$\frac{W}{A} = \text{the stress on a unit of cross-section.}$$

Knowing the value of T for different materials, a value less than T for the given material is assumed for the stress to be allowed on the unit of cross-section. Assuming this value of the stress and calling it T_1, we have

$$A = \frac{W}{T_1}. \quad . \quad . \quad . \quad . \quad . \quad . \quad . \quad (80)$$

From which, knowing the form of cross-section and its area, the problem can be solved.

Suppose the form to be rectangular, and let b be the breadth and d the depth. Then

$$A = b \times d, \text{ or } bd = \frac{W}{T_1};$$

in which, if b be assumed, d can be determined, and the converse.

The solution of the reverse problem is evident. Knowing A and T_1, the value of W, or the load which will not produce a stress greater than T_1 on the unit of area, is easily determined.

201. Strength when strained by a compressive force. For all practical purposes, it is assumed sufficiently exact for short pieces to apply the methods just given for tension, substituting C_1 for T_1; the former being the assumed limit of compressive stress on the unit of area. When the pieces are longer than five times their diameter, they bend under the crushing load and break by bending, or by bending and by crushing.

Rankine gives the following limits of proportion between length and diameter, within which failure by crushing alone will take place, and beyond which there is a sensible tendency to give way by bending sideways.

Pillars, rods, and struts of **cast iron,** in which the length is not more than five times the diameter.

The same of **wrought iron,** not more than ten times the diameter.

The same of **dry timber,** not more than twenty times the diameter.

202. **Formulas for obtaining the strength of columns or pillars whose lengths are greater than five times the diameter of cross-section, when subjected to a compressive strain.**

The formulas deduced by Mr. Hodgkinson, from a long series of experiments made upon pillars of wood, wrought iron, and cast iron are much used in calculating the strength of pillars or columns strained by a force of compression.

Hodgkinson's Formulas.

Table for finding the strength of pillars, in which
W = the breaking weight, in tons of 2,000 pounds;
L = the length of the column in feet;
D = the diameter of exterior in inches;
d = the diameter of interior in inches.

Nature of column.	Both ends being rounded, length of column exceeding 15 times its diameter.	Both ends being flat, the length of column exceeding 30 times its diameter.
Solid square pillar of red cedar (dry). . . .	"	$W = 8.7\frac{D^4}{L^2}$
Same of oak (Dantzic) dry	"	$W = 12.2\frac{D^4}{L^2}$
Solid cylindrical col. of wrought iron.	$W = 47.9\frac{D^{3.76}}{L^2}$	$W = 149.7\frac{D^{3.55}}{L^2}$
Solid cylindrical col. of cast iron	$W = 16.6\frac{D^{3.76}}{L^{1.7}}$	$W = 49.4\frac{D^{3.55}}{L^{1.7}}$
Hollow cylindrical col. of cast iron	$W = 14.5\frac{D^{3.76}-d^{3.76}}{L^{1.7}}$	$W = 49.6\frac{D^{3.55}-d^{3.55}}{L^{1.7}}$

If the column be shorter than that given in the table, and more than five times its diameter, the strength may be determined by the following formula:

$$W = \frac{W'AC}{W' + \frac{3}{4}AC}. \quad . \quad . \quad . \quad . \quad (81)$$

in which W′ = the breaking weight, computed from the formulas in the above table;
C = the modulus of crushing in tons;
A = the cross-section in square inches; and
W = the strength of the column in tons.

Gordon's Formulas.

These are deduced from the same experiments, and are as follows:

Solid Pillars.

Cross-section a square.

$$\left.\begin{array}{ll} \text{Of cast iron} \quad . \quad . \quad . \quad . & W = \dfrac{80{,}000\ A}{1 + \dfrac{l^2}{266\ b^2}} \\[2em] \text{Of wrought iron} \quad . \quad . & W = \dfrac{36{,}000\ A}{1 + \dfrac{l^2}{3{,}000\ b^2}} \end{array}\right\} \quad . \quad . \quad . \quad (82)$$

Hollow Pillars.

Circular in cross-section.

$$\left.\begin{array}{ll} \text{Of cast iron} \quad . \quad . \quad . \quad . & W = \dfrac{80{,}000\ A}{1 + \dfrac{l^2}{400\ d^2}} \\[2em] \text{Of wrought iron} \quad . \quad . & W = \dfrac{36{,}000\ A}{1 + \dfrac{l^2}{3{,}000\ d^2}} \end{array}\right\} \quad . \quad . \quad . \quad (83)$$

Cross-section a square.

$$\left.\begin{aligned} &\text{Of cast iron} \quad . \quad . \quad . \quad . \quad W = \frac{80{,}000 \text{ A}}{1 + \dfrac{l^2}{533\ b^2}} \\ &\text{Of wrought iron} \quad . \quad . \quad W' = \frac{36{,}000 \text{ A}}{1 + \dfrac{l^2}{6{,}000\ b^2}} \end{aligned}\right\} \quad . \quad . \quad (84)$$

in which,

W = the breaking load in pounds;
A = the area of cross-section in square inches;
l = the length of the pillar in inches;
b = the length of one side of the cross-section; and
d = the diameter of the outer circumference of the base.

These formulas apply to *pillars with flat ends, the ends being secured so that they cannot move laterally and the load uniformly distributed over the end surface.* In the hollow columns, the thickness of the metal must not exceed $\frac{1}{8}$ of the outer diameter.

Mr. C. Shaler Smith's Formula.

This formula is deduced from experiments made by Mr. Smith on pillars of both white and yellow pine, and is

$$W' = \frac{5{,}000}{1 + \dfrac{.004\ l^2}{b^2}}, \quad . \quad . \quad . \quad . \quad (85)$$

in which b and l are in inches, and represent the same quantities as in the preceding formulas. W′ is the breaking load on the square inch of cross-section in pounds.

203. Mr. Hodgkinson, in summing up his conclusions derived from the experiments made by him on the strength of pillars, stated that:

"1st. In all long pillars of the same dimensions, the resistance to crushing by flexure is about three times greater when the ends of the pillars are flat than when they are rounded.

"2d. The strength of a pillar, with one end rounded and the other flat, is the arithmetical mean between that of a pillar of the same dimensions with both ends round, and one with both ends flat. Thus of three cylindrical pillars, all of the same length and diameter, the first having both its ends rounded, the second with one end rounded and one flat, and the third with both ends flat, the strengths are as 1, 2, 3, nearly.

"3d. A long, uniform, cast-iron pillar, with its ends firmly fixed, whether by means of disks or otherwise, has the same power to resist breaking as a pillar of the same diameter, and half the length, with the ends rounded or turned so that the force would pass through the axis.

"4th. The experiments show that some additional strength is given to a pillar by enlarging its diameter in the middle part; this increase does not, however, appear to be more than one-seventh or one-eighth of the breaking weight."

Similar pillars.—"In similar pillars, or those whose length is to the diameter in a constant proportion, the strength is nearly as the square of the diameter, or of any other linear dimension; or, in other words, the strength is nearly as the area of the transverse section.

"In hollow pillars, of greater diameter at one end than the other, or in the middle than at the ends, it was not found that any additional strength was obtained over that of cylindrical pillars.

"The strength of a pillar, in the form of the connecting rod of a steam-engine" (that is, the transverse section presenting the figure of a cross +), "was found to be very small, perhaps not half the strength that the same metal would have given if cast in the form of a uniform hollow cylinder.

"A pillar irregularly fixed, so that the pressure would be in the direction of the diagonal, is reduced to one-third of its strength. Pillars fixed at one end and movable at the other, as in those flat at one end and rounded at the other, break at one-third the length from the movable end; therefore, to economize the metal, they should be rendered stronger there than in other parts.

"Of rectangular pillars of timber, it was proved experimentally that the pillar of greatest strength of the same material is a square."

Long-continued pressure on pillars.—"To determine the effect of a load lying constantly on a pillar, Mr. Fairbairn had, at the writer's suggestion, four pillars cast, all of the same length and diameter. The first was loaded with 4 cwt., the second with 7 cwt., the third with 10 cwt., and the fourth with 13 cwt.; this last load was $\frac{97}{100}$ of what had previously broken a pillar of the same dimensions, when the weight was carefully laid on without loss of time. The pillar loaded with 13 cwt. bore the weight between five and six months, and then broke."

STRENGTH OF BEAM TO RESIST A SHEARING FORCE.

204. It has been shown that the transverse shearing stress varies directly with the area of cross-section, and that we have

$$S' = AS,$$

in which S is the modulus of shearing. Assuming a value which we represent by S_1 less than S for the given material, and we have

$$W = AS_1,$$

in which W is the force producing shearing strain and S_1 the limit of the shearing stress allowed on the unit of surface. Knowing the form, the dimensions to give the cross-section for any assumed stress are easily obtained.

TRANSVERSE STRENGTH OF BEAMS.

205. The stress on the unit of area of the fibres of a beam at the distance y from the neutral axis, in the case of transverse strain, is obtained from eq. (21),

$$\frac{PI}{y} = M.$$

As previously stated, the hypothesis is that the stress on the unit of area increases as y increases, and will be greatest in any section when y has its greatest value. That unit of area in the section farthest from the neutral axis will therefore be the one that has the greatest stress upon it. Now suppose M to be increased gradually and continually. It will at length become so great as to overcome the resistance of the fibres and produce rupture. Since the material is homogeneous, and supposed to resist equally well both tension and compression, the stresses on the unit of area at the same distance on opposite sides of the neutral surface are considered equal.

Representing by R the stress on the unit of area farthest from the neutral surface in the section where rupture takes place, and the corresponding value of y by y', we have

$$\frac{RI}{y'} = M', \quad . \quad . \quad . \quad . \quad (86)$$

in which M' is the bending moment necessary to produce rupture at this section.

When the cross-section is a rectangle, in which b is the breadth and d the depth, I is equal to $\frac{1}{12}bd^3$, and the greatest value of y' is $\frac{d}{2}$; substituting these values in eq. (86) we have for a beam with rectangular cross-section,

$$R \times \tfrac{1}{6}bd^2 = M'. \quad . \quad . \quad . \quad . \quad . \quad (87)$$

The first member is called the **moment of rupture** and its value for different materials has been determined by experiment.

These experiments have been made by taking beams of known dimensions, resting on two points of support, and breaking them by placing weights at the middle point.

From equation (87) we have

$$R = \frac{M'}{\tfrac{1}{6}bd^2} \quad . \quad . \quad . \quad . \quad . \quad . \quad (88)$$

in which, substituting the known quantities from the experiment, the value of R, called the **modulus of rupture**, is obtained.

These values, thus obtained, are especially applicable to all beams with a rectangular cross-section, and with sections that do not differ materially from a rectangle. Where other cross sections are used, special experiments must be made.

206. In a beam of uniform cross-section the stresses on the different sections vary, and that particular section at which the moment of the external forces is the greatest is the one where rupture begins, if the beam break. This section most liable to break may be called the **dangerous section**.

In rectangular beams the dangerous section will be where the moments of the straining forces are the greatest.

Let W denote the total load on a beam, and l its length, we have for the greatest moments in the following cases:

$M = Wl$, when the load is placed at one end of a beam, and the other end fixed.

$M = \frac{wl}{2} \times l = \frac{1}{2}Wl$, for the same beam uniformly loaded.

$M = \frac{W}{2} \times \frac{l}{2} = \frac{1}{4}Wl$, when the load is placed at the middle point of a beam resting its extremities on supports.

$M = \frac{wl}{2} \times \frac{1}{2}\frac{l}{2} = \frac{1}{8}Wl$, for the same beam uniformly loaded.

If a less value than that necessary to break the beam be

substituted in eq. (88) for M′, the corresponding value for R, will not be that for the modulus of rupture, but will merely be the stress on the unit of area farthest from the neutral axis in the dangerous section. Suppose a beam strained by a force less than that which will produce rupture and find for M the corresponding maximum value for each case. Substituting these in eq. (87), we have

$$\left.\begin{array}{l} R' \times \frac{1}{6}bd^2 = Wl \\ R' \times \frac{1}{6}bd^2 = \frac{1}{2}Wl \\ R' \times \frac{1}{6}bd^2 = \frac{1}{4}Wl \\ R' \times \frac{1}{6}bd^2 = \frac{1}{8}Wl \end{array}\right\} \quad . \quad . \quad . \quad . \quad (89)$$

in which R′ is the greatest stress on the unit of area in the dangerous section for the corresponding case of rectangular beams, whose moments are given above.

The value of R for a material may be determined by finding the force that will break a piece of the same material, of a **similar form**, and substituting the moment of this force for M′ in eq. (86), and deducing the value of R.

Some of the values of R for pieces of rectangular cross-section are as follows:

Material.	Value of R.
Ash	12,156 lbs.
Chestnut	10,660 "
Oak	10,590 "
Pine	8,946 "
Fir	6,600 "
Cast-iron	33,000 "

The value of R is also taken as equal to eighteen times the force required to break a piece of one inch cross-section, resting on two supports one foot apart, and loaded at the middle.

207. From the definition for R, it would seem, as before stated, that it should be equal either to C or to T, depending upon whether the beam broke by crushing or tearing of the fibres. In fact, it is equal to neither, being generally greater than the smaller and less than the greater; as shown in the case for cast iron, for which

The mean value of C = 96,000 pounds;
The mean value of T = 16,000 pounds; and
The mean value of R = 36,000 pounds.

If, then, instead of taking R from the tables, the value of T or C be used, taking the smaller value of the two, the calculated strength of the beam will be on the *safe side*. That is, the strength of the beam will be greater than that found by calculation.

Experiments should be made upon the materials to be used in any important structure, to find the proper value for R.

In determining the safe load to be placed on a beam, the following values for R′ may be taken as a fair average:

For seasoned timber, R′ = 850 to 1,200 pounds;
For cast iron, R′ = 6,000 to 8,000 pounds;
For wrought iron, R′ = 10,000 to 15,000 pounds.

INFLUENCE OF THE FORM OF CROSS SECTION ON THE STRENGTH OF BEAMS.

208. The resistance to shearing and tensile strains in any section of a beam is the same for each unit of surface throughout the section. The same has been assumed for the resistance to compressive strains within certain limits. Hence so long as the area of cross-section contains the same number of superficial units, the form has no influence on the resistance offered to these strains.

This is different in the case of a transverse strain.

We may write equation (21) under this form,

$$P = M\frac{y}{I}.$$

In this, if we suppose M to have a constant value, P will then vary directly with the factor $\frac{y}{I}$; that is, as this factor increases or decreases, there will be a corresponding increase or decrease in P.

Represent by d the depth of the beam, $\frac{1}{2}d$ will be the greatest value that y can have. It is readily seen, that for any increase of $\frac{1}{2}d$, I will increase in such a proportion as to decrease the value of $\frac{\frac{1}{2}d}{I}$, and hence decrease the amount of stress on the unit of area farthest from the neutral axis. Therefore we conclude that for two sections having the same area, the stress on the unit of surface farthest from the neutral axis is less for the one in which $\frac{d}{2}$ is the greater.

This principle affords a means of comparing the relative resistances offered to a transverse strain by beams whose cross-sections are different in form but equivalent in area.

For example, compare the resistances offered to a transverse strain by rectangular, elliptical, and I-girders, with equivalent cross-sections.

The values of I for the rectangle, ellipse, and I-section are respectively,

$$I = \tfrac{1}{12}bd^3, I = \tfrac{1}{64}\pi bd^3, \text{ and } I = \tfrac{1}{12}(bd^3 - b'd'^3).$$

Represent the equivalent cross-section by A, and we will have $A = bd$ for the rectangle, $A = \frac{1}{4}\pi bd$ for the ellipse, and $A = b(d-d')$ for the I-section. The latter is obtained by neglecting the breadth of the rib joining the two flanges, its area being small compared with the total area, and by regarding $d^2 = dd' = d'^2$, $d - d'$ being small compared with d.

Substituting these values of A in the factor $\frac{\frac{1}{2}d}{I}$, we get for the rectangle, $\frac{6}{Ad}$; for the ellipse, $\frac{8}{Ad}$; for the I-section $\frac{2}{Ad}$. Hence we see that $\frac{\frac{1}{2}d}{I}$ is least for the third, and greatest for the second, and therefore conclude that the stress on the unit of surface farthest from the neutral axis is the least for the I-girder, and its resistance to a transverse strain is greater than either of the other two forms.

Since the quantity A contains b and d, by decreasing b and increasing d, within limits, the resistance of any particular form will be increased. And hence, in general, the mass of fibres should be thrown as far from the neutral axis as the limits of practice will allow.

The strongest Beam that can be cut out of a Cylindrical Piece.

209. It is oftentimes required to cut a rectangular beam out of a piece of round timber. The problem is to obtain the one of greatest strength.

Denote by D the diameter of the log, by b the breadth, and d the depth of the required beam.

From the value

$$R' = \frac{M}{\frac{1}{6}bd^2},$$

it is evident that the strongest beam is the one in which bd^2 has its maximum value.

Representing the cross-section of the beam and of the log by a rectangle inscribed in a circle, we have

$$d^2 = D^2 - b^2,$$

D being the diameter of the circle. Multiplying by b, gives

$$bd^2 = bD^2 - b^3.$$

In order to have bd^2 a maximum, $D^2 - 3b^2$ must be equal to zero, which gives

$$b = D\sqrt{\tfrac{1}{3}},$$

and this, substituted in the expression for d^2, gives

$$d = D\sqrt{\tfrac{2}{3}}.$$

To construct this value of b, draw a diameter of the circle, and from either extremity lay off a distance equal to one-third of its length. At this point erect a perpendicular to the diameter, and from the point where it intersects the circumference draw the chords joining it with the ends of the diameter. These chords will be the sides of the rectangle.

2d CASE.—BEAMS OF VARIABLE CROSS-SECTION.

210. Beams of uniform strength.—Beams which vary in size so that the greatest stress on the unit of area in each section shall be constant throughout the beam, form the principal class of this second case.

In the previous discussions and problems the bar or beam has, with but one exception, been considered as having a uniform cross-section throughout, and in these discussions the moment of inertia, I, has been treated as a constant quantity.

Since the beams had a uniform cross-section it is evident that the greatest stress was where the moment of the external forces was the greatest.

Finding this greatest moment of the external forces, we determined the greatest stress and the section at which it acted. If this section was strong enough to resist this action, it follows that all other sections were strained less and were larger than necessary to resist the stresses to which they were exposed; in other words, there was a waste of material.

The greatest stress on a unit of surface of cross-section being known or assumed, let us impose the condition that it shall be the same for every section of the beam. This will necessitate variations in the cross-sections, hence I will vary and must be determined for each particular case.

A beam is called a "**solid of equal resistance**" when so proportioned that, acted on by a given system of external forces, the greatest stresses on the unit of area are equal for every section.

This subject was partly discussed under the head of tension in determining the form of a bar of uniform strength to resist elongation. The method there used could be applied to the case of a beam to resist compression.

Beams of Uniform Strength to resist a Transverse Strain.

211. Suppose the beam to be acted upon by a force producing a transverse strain, and let the cross-section be rectangular.

Let b and d represent the breadth and depth of the beam, and we have

$$I = \tfrac{1}{12}bd^3.$$

Substituting in eq. (21) this value of I, and giving to y its greatest value, which is $\frac{1}{2}d$, we have

$$P' \times \tfrac{1}{6}bd^2 = M,$$

or

$$P' = \frac{M}{\frac{1}{6}bd^2} \quad . \quad . \quad . \quad . \quad . \quad . \quad . \quad (90)$$

for the stress on a unit of surface at the distance $\frac{1}{2}d$ from the neutral axis in the cross-section under consideration.

The greatest stress will be found in that section for which M is the greatest. Representing this moment by M″ and the corresponding value of P′ for this section by P″, we have

$$P'' = \frac{M''}{\frac{1}{6}bd^2}. \quad . \quad . \quad . \quad . \quad . \quad . \quad (91)$$

This value of P″ is then the greatest value of the stress, upon the unit of surface, produced by the deflecting forces acting to bend the beam.

From the conditions of the problem, the greatest stress on the unit of surface must be the same for every cross-section. Eq. (90) gives the greatest stress on the unit of surface in any cross-section. It therefore follows that for a rectangular beam of uniform strength to resist a cross-strain, we must have

$$P'' = \frac{M}{\frac{1}{6}bd^2}. \quad . \quad . \quad . \quad . \quad . \quad . \quad (92)$$

Since P″ is constant, b or d, or both of them, must vary as M varies, to make the equation a true one; that is, the area of cross-section must vary as M varies.

We may assume b constant for a given case, and giving different values to M, deduce the corresponding ones for d; or, assuming d constant, do the same for b; or we may assume that their ratio shall be constant.

For the first case, b, the breadth constant, we have

$$d = \pm \sqrt{\frac{M}{\frac{1}{6}P''b}}. \quad . \quad . \quad . \quad . \quad (93)$$

For the second case, d, the depth constant, we have

$$b = \frac{M}{\frac{1}{6}P''d^2}, \quad . \quad . \quad . \quad . \quad . \quad (94)$$

and for the third, their ratio constant, $b = rd$, we have

$$d^3 = \frac{M}{\frac{1}{6}P''r}. \quad . \quad . \quad . \quad . \quad . \quad (95)$$

The assumed values of M with the deduced values of d, from eq. (93), will show the kind of line cut out of the beam by a vertical section through the axis, when the breadth is constant; and the deduced values of b, from eq. (94), will show the kind of line cut out of the beam by a horizontal section through the axis when the depth is constant. These lines will show the law by which the sections vary from one point to another throughout the beam.

As examples take the following cases:

212. CASE 1ST.—*A horizontal beam firmly fastened at one*

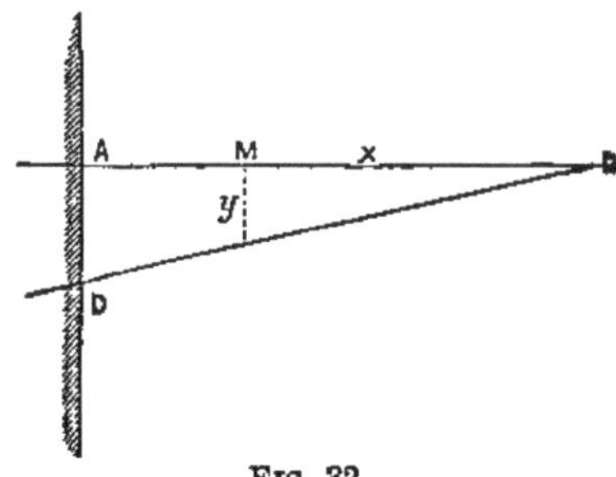

FIG. 32.

end (Fig. 32), *and the other end free to move, strained by a load uniformly distributed along the line, A B.*

Take B as the origin of co-ordinates, B A the axis of X, y positive downwards, the axis of Z horizontal, and w the weight on a unit of length.

The moment of the weight acting at any section as D is $\frac{wx^2}{2}$, substituting which for M in the expression (93) for d, we have

$$d = y = \pm x \sqrt{\frac{3w}{P''b}},$$

which is the equation of a right line as B D, passing through the origin of co-ordinates.

If the depth be constant, the breadth will vary from point to point, and the different values of the ordinate may be obtained by substituting this moment for M in expression (94), and we have

$$b = z = \frac{3w}{P''d^2} x^2,$$

which is the equation of a parabola having its vertex at B, as in Fig. 33.

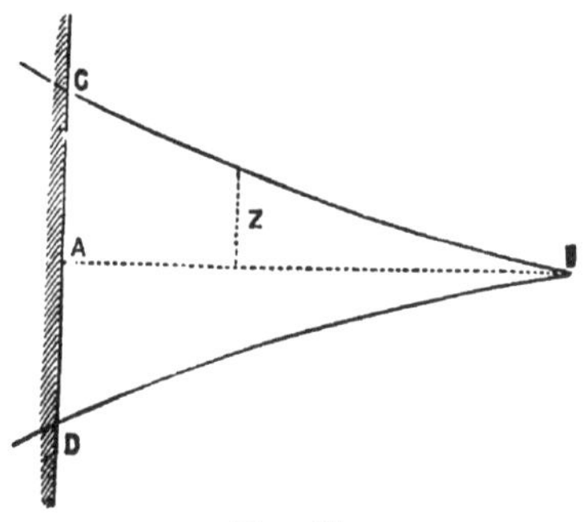

FIG. 33.

213. CASE 2D.—*A beam as in preceding case strained by a load, W, concentrated and acting at B, the weight of the beam disregarded.*

The breadth being constant, we have

$$d = y = \pm \sqrt{\frac{6W}{P''b}\, x},$$

or

$$y^2 = \frac{6W}{P''b} x,$$

which is the equation of a parabola, the vertex of which is at B. (Fig. 34.)

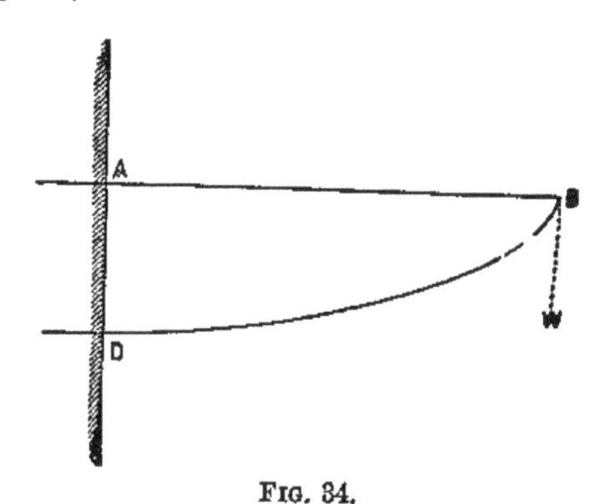

FIG. 34.

Suppose the depth constant; in this case we have

$$b = z = \frac{6W}{P''d^2} x,$$

which is the equation of a right line, and shows that the plan of the beam is triangular.

214. CASE 3D.—*Suppose the beam resting on two supports at its ends and uniformly loaded.*

Represent by $2l$ the distance between the supports, by w the load on a unit of length, and take A (Fig. 22) as the origin of co-ordinates.

The moment of the external forces at any section, as D, will be $\frac{1}{2}wx^2 - wlx$, which substituted in eq. (93), gives

$$d^2 = \frac{3w}{P''b} x^2 - \frac{6wlx}{P''b},$$

which is the equation of an ellipse.

This moment substituted in eq. (94), gives

$$b = \frac{3w}{P''d^2} x^2 - \frac{6wlx}{P''d^2}$$

which is the equation of a parabola.

215. In a similar way we may determine the forms of beams of rectangular cross-section, when other conditions are imposed.

If we had supposed the sections circular, then $I = \frac{1}{4}\pi r^4$, and this being substituted for I in the general expression for the stress on a unit of surface farthest from the neutral axis a similar process would enable us to determine the form of the beam.

Hence, knowing the strains to which any piece of a structure is to be subjected, we may determine its form and dimensions such that with the least amount of material it will successfully resist these strains.

RELATION BETWEEN STRESS AND DEFLECTION PRODUCED BY A BENDING FORCE.

216. Within the elastic limit, the relation between the greatest stress in the fibres and the maximum deflection of the beam produced by a bending force, may be easily determined.

Take a rectangular beam, supported at the ends and loaded at its middle point.

The third of equations (89) gives for this case

$$R' \times \tfrac{1}{6}bd^2 = \tfrac{1}{4}Wl,$$

and solving with respect to W, we have

$$W = \tfrac{2}{3}\frac{R'bd^2}{l},$$

in which W is the load on the middle point of the beam.

The maximum deflection produced by a load, 2W, in this case has been found, the length of beam being $2l$, to be

$$f = \tfrac{1}{3}\frac{Wl^3}{EI}.$$

Substituting for I, W, and l, the proper values, we have

$$f = \tfrac{1}{4}\frac{W}{E}\frac{l^3}{bd^3}.$$

Solving with respect to W, and placing it equal to the value of W obtained from eq. (89), we have

$$\tfrac{2}{3}R'\frac{bd^2}{l} = 4E\frac{bd^3}{l^3}f,$$

from which we get

$$R' = \frac{6Ed}{l^2}f. \quad . \quad . \quad . \quad . \quad . \quad (96)$$

Hence, knowing the deflection and the coefficient of elasticity, the greatest stress on the unit can be obtained and the converse.

FORCES ACTING OBLIQUELY.

217. The forces acting on the beam have been supposed to be in the plane of, and perpendicular to, the mean fibre.

The formulas deduced for this supposition are equally applicable if the forces act obliquely to the mean fibre.

Suppose a force acting obliquely in the plane of the mean fibre, it can be resolved into two components, one, P, perpendicular, and the other, Q, parallel to the fibre. The component P will produce deflection, and the component Q, extension or compression depending on the angle, whether obtuse or acute, made by the force with the piece.

The strains caused by each of the components can be determined as in previous cases.

For suppose the force applied in the plane of the axis of a beam, at F (Fig. 35), and let x be the distance to any section, as K, measured on the axis of the beam E F.

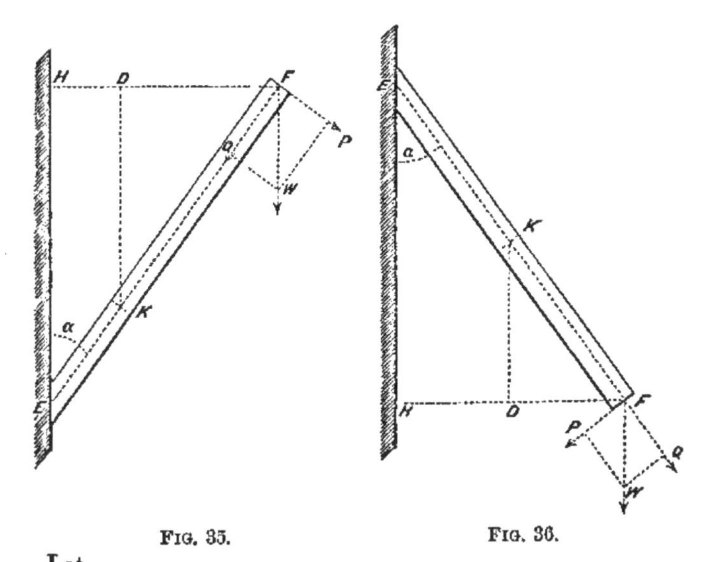

FIG. 35. FIG. 36.

Let

$l =$ E F, the length of the beam, and $a =$ the angle made by the axis E F with the vertical.

The bending moment at any section, as K, is equal to

$$Wx \sin a,$$

and its value for the dangerous section will be $Wl \sin \alpha$, l being the greatest lever arm of W.

The greatest stress caused by P on the unit, at the dangerous section of a rectangular beam, b and d being the dimensions of cross-section, will be

$$6 \frac{Wl \sin \alpha}{bd^2}.$$

The stress caused by Q on the unit will be either compressive, as Fig. (35), or tensile, as Fig. (36), and its intensity will be

$$\frac{W \cos \alpha}{bd}.$$

The total stress on the unit subjected to the greatest strain will therefore be

$$6 \frac{Wl \sin \alpha}{bd^2} + \frac{W \cos \alpha}{bd}.$$

If a value, as R', be assumed as the limit of the stress on the unit of material, it will be necessary to deduct from R' the intensity of the stress caused by Q, so as to avoid developing a greater stress on the unit than that assumed, or, we must have at the dangerous section for a rectangular beam,

$$W\,l \sin \alpha = \tfrac{1}{6}\left(R' - \frac{W \cos \alpha}{bd}\right)bd^2. \quad . \ (97)$$

and in general,

$$\frac{Px}{I} \times \tfrac{1}{2}d + \frac{Q}{A} < R'.$$

STRENGTH OF BEAMS TO RESIST TWISTING.

218. Strains of torsion are not common in structures and are prevented by distributing the loads symmetrically over the pieces, making the resultants of the straining forces intersect the axes of the pieces.

Whenever such a strain does exist, the intensity of the stress may be determined by the use of formula (79). In determining the value of T_t by this formula, the experiment must, as in the case of transverse strain, be made upon a piece similar in form to that for which the stress is to be found.

If the piece be circular in cross section, formula (79) may be placed under the form,

$$F' = \frac{T_t \pi R^3}{2\lambda},$$

which gives the force necessary to produce rupture by twisting.

It will be seen that the modulus of torsion is independent of the length of the piece, being dependent upon the moment of the twisting couple and upon the form and dimensions of the cross-section.

The length of the piece affects the value of the angle of torsion, α; the total angle being greater as l is greater. In using formula (77) a limit should be assumed for α such that the limit of torsional elasticity shall not be passed.

ROLLING LOADS.

219. Systems of forces, the points of application of which, like those of stationary loads, do not move, have been the only kinds considered in the previous discussions.

Many structures, such as bridges for example, are built to sustain loads in motion, the load coming upon the structure in one direction and moving off in another. A load of this kind is called a **moving**, a **rolling**, or **live** load, to distinguish it from the stationary kind usually called a **dead** load.

220. In determining the strength of a beam to resist the stresses developed by a live load, it is necessary *to determine the positions the load should have that will cause the greatest bending moment and the greatest transverse shearing strain at any section of the beam.*

Let the beam (Fig. 37) be horizontal, uniformly loaded, and strained by a uniformly distributed live load that gradually covers the entire beam. Let

$2l$ = A B, the length of the beam;

w = the weight of the uniform stationary load on the unit of length;

w' = the weight of the rolling load per unit;

m = the length of the rolling load in any one position;

n = the length of beam not covered by the rolling load;

R_1, R_2, the reactions at the points of support.

Take the origin of co-ordinates at A, the axes X and Y as in the previous cases, and suppose the live load to have come on at the end A, and to occupy, in one its positions, the space from A to D.

The reactions at the points of support, due to the uniform load on the beam and the live load from A to D, are

$$R_1 = wl + w'm \frac{2l - \frac{m}{2}}{2l}, \text{ and } R_2 = wl + w'm \frac{m}{4l}.$$

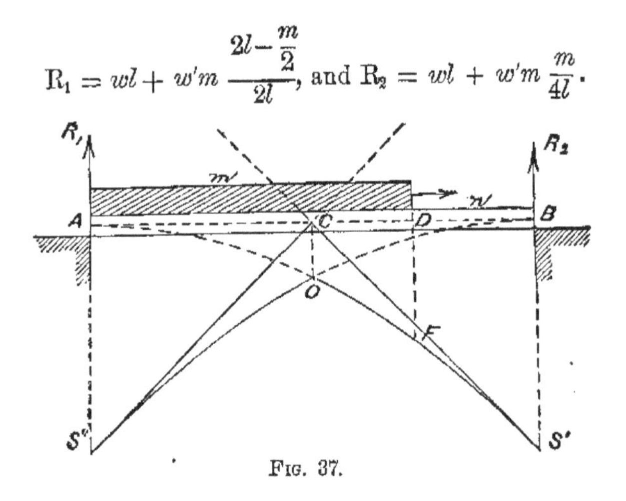

FIG. 37.

The bending moment at any section whose abscissa is x, and which lies between A and D, for this position of the load, is

$$M = -R_1x + (w + w')\frac{x^2}{2}, \quad . \quad . \quad (99)$$

and for any section between D and B, the abscissa being x,

$$M = -R_1x + \frac{wx^2}{2} + \frac{w'm}{2}(2x - m). \quad . \quad (100)$$

and, as seen, increases as m increases. The bending moment will, therefore, be greatest when m is greatest, or when $m = 2l$. Hence,

$$M = (w + w')\left(\frac{x^2}{2} - lx\right) \quad . \quad . \quad (101)$$

is the expression for the greatest bending moment at any section of the beam, and it exists *when the rolling load covers the whole beam.*

The shearing stress at any section between A and D is

$$S' = (w + w')x - R_1 \quad . \quad . \quad . \quad (102)$$

and for any section between D and B is

$$S' = (wx + w'm) - R_1 \quad . \quad . \quad . \quad (103)$$

in which substituting for R_1 its value, we have

$$S' = w(x - l) - w'\left(m - \frac{m^2}{4l} - x\right) \quad . \quad (104)$$

and

$$S' = w(x - l) + w'\frac{m^2}{4l} \quad . \quad . \quad . \quad . \quad (105)$$

from which the shearing stress at any section is obtained.

Let x' be the abscissa of the section, D, at the end of the live load in any one of its positions as its moves from A toward B. Substituting x' for x in eq. (105) we have

$$S'' = w(x' - l) + w'\frac{m^2}{4l} \quad . \quad . \quad . \quad (106)$$

for the shearing stress at this section when the live load extends to D.

If the rolling load extends entirely over the beam, the shearing stress at any section is

$$S' = (w + w')(x - l) \quad . \quad . \quad . \quad (107)$$

and for the section D,

$$S'' = (w + w')(x' - l)$$

which may be written

$$S'' = w(x' - l) + w'(x' - l) \quad . \quad . \quad (108)$$

The values of S'' at the section D, for the positions of these two loads, one extending to D and the other entirely over the beam, only differ from each other in the terms, $w'(x' - l)$ and $w'\frac{m^2}{4l}$.

Since $2l = m + n$, we may write

$$w'(x' - l) = \frac{w'}{2}(m - n), \text{ and } w'\frac{m^2}{4l} = \frac{w'}{2}\frac{m^2}{m + n}.$$

By comparing $\frac{m^2}{m + n}$ with $m - n$, it is seen that the term $w'(x' - l)$ is less than $w'\frac{m^2}{4l}$ whenever $m > l$, or at any section of a beam the *greatest shearing stress occurs when the moving load covers the longer of the two segments into which the section divides the beam.*

When the rolling load covers the longer segment, the shearing stress is said to be a **main** shear; when it covers only the shorter segment, it is called a **counter** shear.

The difference in the intensity of the shearing stress, at a given section, caused by a partial rolling load and by one that covers the beam can be shown graphically.

The term, $w(x - l)$, in equation (105), expresses the intensity of the shearing stress at any section caused by the dead load; the term, $w' \frac{m^2}{4l}$ expresses the shearing stress at the sections between D and B caused by the live load. If we place $y' = w(x - l)$ and $y' = \frac{w'}{4l} m^2$, two equations will be formed, one that of a right line, the other a parabola, in which the ordinates represent the shearing stresses caused by these loads. Construct the parabola, and let A O S′ be the arc determined. The ordinate D F of this arc will represent the shearing stress at the end section D, and at all sections between D and B produced by the live load, A D, in this position.

When the live load covers the beam, the total shearing stress at any section is given by equation (108). That part of the stress produced by the live load is expressed by the term $w'(x' - l)$, which is the ordinate of a right line passing through C and S′. No ordinate of this line between C and B is equal in length to the corresponding ordinate of O F S′; hence, the shearing stress in any section between C and D is greater when the live load extends from A to the section considered, than when it extends entirely over the beam.

Let m and x have simultaneous and equal values in equation (105) and the equation will be that of a parabola, the ordinates of which will express the intensity of the shearing stress in that section coinciding with the end of the moving load in all of its positions.

It will be seen that this parabola intersects the axis of X between A and C, which shows that there is a section of the beam at which there is no shearing stress when the end of the rolling load reaches it. The expression for the distance from the origin to this section may be obtained by placing the second member of eq. (105) equal to zero, and solving it with respect to x; there results

$$x_{(S' = 0)} = 2l \frac{w}{w'} \left(-1 \pm \sqrt{1 + \frac{w'}{w}} \right) . \quad (109)$$

An equal moving load coming on the beam from B produces a similar effect to that of the one coming from A.

It therefore follows that there is a point of "no shearing" bewcen B and C, and that this point, in this case, coincides with the section at the rear end of the rolling load coming from A as it rolls off the beam. These points of "no shear" are of interest in "built" beams or beams composed of several pieces.

LIMITS OF PRACTICE.

221. Until quite recently, comparatively speaking, it was the custom of most builders, in planning and erecting a structure, to fix the dimensions of its various parts from precedent, that is, by copying from structures already built.

So long as the structure resembled those already existing that had stood the test of time, this method served its purpose. But when circumstances forced the builder to erect structures different from any in existence or previously known, and to use materials in a way in which they had never before been applied, the experience of the past could no longer be his guide. Practical sagacity, a most excellent and useful qualification, was not sufficient for the emergency. Hence arose the necessity that the builder should acquire a thorough knowledge of the theory of strains, the strength of materials, and their general properties.

The principal object of "**strength of materials**" is to determine the stresses developed in the different parts of a structure, and to ascertain if the stresses are within the adopted limits. And as a consequent, knowing the strains, to determine the forms and dimensions of the different parts, so that with the least amount of material they shall successfully resist these strains.

The limits adopted vary with the materials and the character of the strain. The essential point is that the limit of elasticity of the material should not be passed, even when by some unforeseen accident the structure is subjected to an unusual stress. The adopted limit to be assigned is easily selected if the limit of elasticity be known; but as the latter is obtained with some difficulty, certain **limits of practice** have been adopted.

In many cases this practice is to arbitrarily assume some given weight as the greatest load per square inch on a given material, and to use this weight for all pieces of the same material. From the varying qualities of the same material it is easily seen that this method of practice differs but little from a "mere rule of thumb."

The most usual practice, especially for structures of importance, as bridges, is to determine the breaking weights or

ultimate strength of the different parts, and take a fractional part of this strength as the limit to be used. The reciprocal of this fraction is called the **factor of safety**.

A more accurate method would be to calculate the dimensions of the pieces necessary to resist the strains produced by the maximum load, and then enlarge the parts sufficiently to give the strength determined by the factor of safety.

When the structure is one of great importance, actual experiments should be made on each kind of material used in its construction, so that the values deduced for the ultimate strength shall be as nearly correct as possible.

222. These factors of safety are arbitrarily assumed, being generally about as follows:

Material.	Factor of safety.
Steel and wrought iron	3
Cast iron	6
Timber	6
Stone and brick	8 to 10.

These are for loads carefully put on the structure.

If the materials and workmanship were perfect, these factors could be materially reduced.

It has been shown (Art. 160) that the work expended by the sudden application of a given force, W, is equal to that expended by 2W if applied gradually at a uniform rate from zero to 2W. Hence a force, W, applied **suddenly** to a beam will produce the same strain on the beam as 2W applied gradually.

A rolling load moving swiftly on a structure approximates nearly to the case of a force suddenly applied.

Hence, for rolling loads, the factors of safety should be doubled.

CURVED BEAMS.

223. A beam which before it is strained has a curvilinear shape in the direction of its length is called a **curved beam.** The curve given to the mean fibre is usually that of a circular or a parabolic arc.

For the purposes of discussing the strains on beams of this class, it is supposed that:

1. The beam has a uniform cross-section;

2. That its cross-section is a plane figure, which if moved along the mean fibre of the beam and normal to it, keeping

the centre of gravity of the plane figure on the mean fibre, would generate the solid; and

3. That the dimensions of the cross-section in the direction of the radius of curvature of the mean fibre are very small compared with the length of this radius.

If the beam be intersected by consecutive planes of cross-section, the hypotheses adopted for a straight beam subjected to a cross strain are assumed as applicable to this case.

224. **General equations.**—Suppose the applied forces to act in the plane of mean fibre, let it be required to **determine the relations between the moment of resistance at any section and the moment of the external forces acting on the beam.**

Let E F (Fig. 38) be a curved beam; the ends E and F so arranged that the horizontal distance between them shall remain constant.

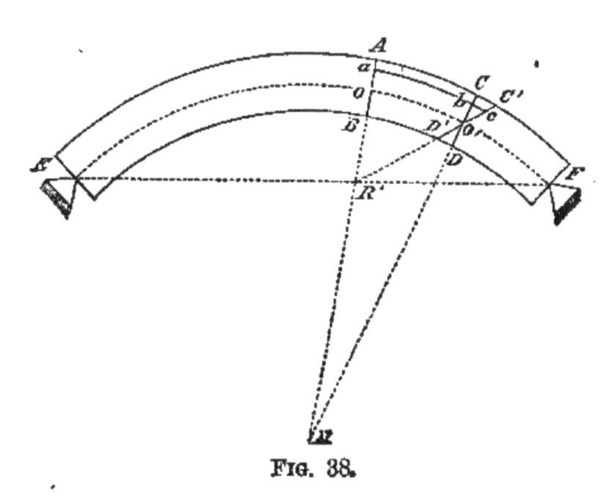

FIG. 38.

Let A B be any cross-section. The external forces acting on either side of this section are held in equilibrium by the resistances developed in this section. Suppose A B to be fixed, and let C′D′ be the position assumed by the consecutive section under the action of the external forces, on the right of A B. The resultant of these external forces may be resolved into two components, one normal and the other parallel to the tangent, to the curve of the mean fibre at O. Represent the former by F, the latter by P, and by M, the sum of the moments of the external forces around the neutral axis in the section A B.

The fibre ab is elongated by an amount bc, proportional to its distance from the neutral axis.

The force producing this elongation is

$$\frac{Ea \times bc}{ab},$$

or since ab may be considered equal to O O′,

$$\frac{Ea \times bc}{O\,O'},$$

in which E is the co-efficient of elasticity and a the area of cross-section of the fibre, ab.

Hence, there obtains to express the conditions of equilibrium,

$$\Sigma\frac{Ea \times bc}{O\,O'} = 0, \text{ and } \Sigma\frac{Ea \times bc}{O\,O'} \times bO' = M. \quad (110)$$

Represent by ρ and ρ', the radii of curvature, R O′ and R′O′.

The triangle, aRb, has its three sides cut by the right line, R′C′. Hence the product of the segments, R O′, bc, and aR′ is equal to the product of the three segments, R R′, bO′, and ac.

Substituting ρ for R O′, $\rho - \rho'$ for R′R, and ρ' for aR′, since O′b is very small in comparison with ρ', and we have

$$\rho \times bc \times \rho' = (\rho - \rho') \times bO' \times ac.$$

From which we get

$$\frac{bc}{ac} = \frac{\rho - \rho'}{\rho\rho'} \times bO' = \left(\frac{1}{\rho'} - \frac{1}{\rho}\right) \times bO'.$$

Since ac differs from O O′ by an infinitely small quantity, the expression obtained for $\frac{bc}{ac}$ may be taken as the value of $\frac{bc}{O\,O'}$. Substituting this value for $\frac{bc}{O\,O'}$, in the second of equations (110), we get.

$$E \times \left(\frac{1}{\rho'} - \frac{1}{\rho}\right) \times \Sigma(a \times \overline{bO'}^2) = M. \quad (111)$$

This sum, $\Sigma(a \times \overline{bO'}^2)$, is the moment of inertia of the cross-section taken with respect to the neutral axis passing through the centre of gravity of the section. Representing this by I, equation (111) may be written

$$EI\left(\frac{1}{\rho'} - \frac{1}{\rho}\right) = M, \quad . \quad . \quad . \quad . \quad (112)$$

which is the general equation, showing the relation existing between the moments of resistances of any section and the moments of the external forces acting on that section.

225 **Displacement of any point of the curve of mean fibre.**—Let A B (Fig. 39) be the curve of mean fibre before the external forces are applied to the beam.

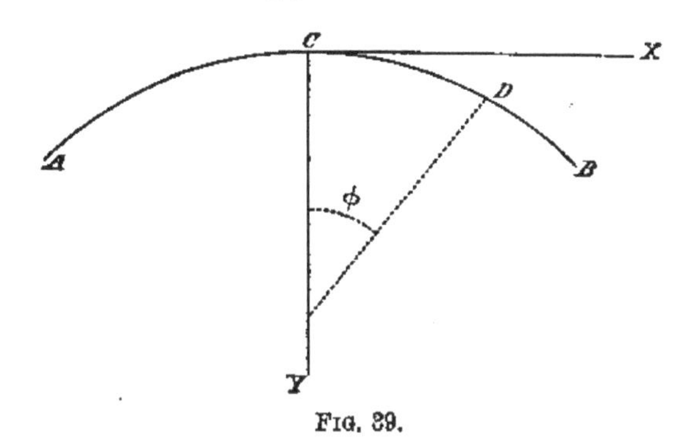

FIG. 39.

Take the origin of co-ordinates at the highest point, C, and the axes X and Y as shown in the figure.

Let D be any point whose co-ordinates are x and y, and represent by ϕ the angle made by the plane of cross-section at D with the axis of Y.

Suppose the external forces applied, and denote by x' and y' the co-ordinates of D in its new position, and by ϕ' the new angle made by the plane of cross-section with the axis of Y.

It is supposed that the displacement of the point, D, is so slight that M remains unchanged.

From the calculus we have

$$\rho = \frac{dz}{d\phi}, \text{ and } \rho' = \frac{dz'}{d\phi'},$$

in which dz and dz' are the lengths of the elementary prism before and after the strain measured along the mean fibre. Since they differ by an infinitely small quantity from each other, by making $dz = dz'$ and substituting in equation (112) we get

$$\mathrm{EI}\left(\frac{d\phi' - d\phi}{dz}\right) = \mathrm{M}.$$

Integrating we obtain

$$\phi' - \phi = \int \frac{\mathrm{M}}{\mathrm{EI}}\, dz + \mathrm{C}. \quad . \quad . \quad (113)$$

The component force, parallel to the tangent at D, acts in the direction of the length of the fibre. Since the points E and F are fixed, this force produces a strain of compression on the fibre. The length of this fibre, after compression between the two consecutive planes, is represented by dz', and is

$$dz' = dz - \frac{Pdz}{EA} = dz\left(1 - \frac{P}{EA}\right).$$

The values of $\cos \phi$, $\sin \phi$, $\cos \phi'$, and $\sin \phi'$ may be written as follows:

$$\cos \phi = \frac{dx}{dz} \qquad \sin \phi = \frac{dy}{dz}$$

$$\cos \phi' = \frac{dx'}{dz'}, \qquad \sin \phi' = \frac{dy'}{dz'}.$$

Substituting, in the last two of these, the value just found for dz', we get

$$\cos \phi' = \frac{dx'}{dz\left(1 - \frac{P}{EA}\right)}, \text{ and } \sin \phi' = \frac{dy'}{dz\left(1 - \frac{P}{EA}\right)}.$$

If $\phi' - \phi$ is very small, we may write

$$\cos \phi' = \cos \phi - (\phi' - \phi) \sin \phi, \text{ and}$$
$$\sin \phi' = \sin \phi + (\phi' - \phi) \cos \phi.$$

Substituting these values of $\cos \phi'$ and $\sin \phi'$, in the expressions above, and solving with respect to dx' and dy', we get

$$dx' = dz\left(1 - \frac{P}{EA}\right)(\cos \phi - (\phi' - \phi) \sin \phi),$$

$$dy' = dz\left(1 - \frac{P}{EA}\right)(\sin \phi + (\phi' - \phi) \cos \phi).$$

Substituting in these for $\sin \phi$ and $\cos \phi$, their values in terms of dz, dy, and dx, we get

$$dx' = \left(1 - \frac{P}{EA}\right)(dx - (\phi' - \phi)\, dy),$$

$$dy' = \left(1 - \frac{P}{EA}\right)(dy + (\phi' - \phi)\, dx),$$

whence, by omitting the products of the second terms, we get

$$dx' - dx = -\frac{P}{EA}dx - (\phi' - \phi)\,dy,$$

$$dy' - dy = -\frac{P}{EA}dy + (\phi' - \phi)\,dx.$$

Integrating, there obtains

$$\left.\begin{aligned} x' - x &= -\int \frac{P}{EA}dx - \int(\phi' - \phi)\,dy, \\ y' - y &= -\int \frac{P}{EA}dy + \int(\phi' - \phi)\,dx. \end{aligned}\right\} \quad (114)$$

The constants of integration reduce to zero for both equations, since from hypothesis there is no displacement of the points at the ends of the curve of mean fibre.

If the beam is metal, the effect of temperature must be included in these expressions for the displacement.

The constant of integration which enters the expression for $\phi' - \phi$, also enters in the last two equations for the displacement. The value of this constant must be known in order to determine the displacement. Besides the constant, there is also an unknown moment in M which must be determined.

The applied forces acting on the beam are fully given, and are taken, as before stated, in the plane of mean fibre. The reactions at the points of support are not known, and must be determined.

Let X_1 represent the algebraic sum of all the components of the applied forces parallel to the axis of X; Y_1 the sum of the components parallel to the axis of Y; R_1 and R_2 the vertical components of the reactions at A and B, respectively; and Q_1 and Q_2 the horizontal components of these reactions.

For equilibrium, there obtains,

$$\left.\begin{aligned} X_1 + Q_1 - Q_2 &= 0, \\ Y_1 - R_1 - R_2 &= 0, \\ \mu_1 - R_2 l_1 + Q_2 l_2 &= 0. \end{aligned}\right\} \quad . \quad . \quad . \quad (115)$$

In the last equation, μ_1 represents the sum of the moments of the known applied forces taken with respect to the point of support, A, l_1, and l_2, the lever arms of R_2 and Q_2, with respect to the same point.

We have three equations and four unknown quantities. By introducing the condition that the point B, shall occupy the same position after the application of the forces as it had before, that is, be *fixed*, a fourth equation may be obtained, and the problem made determinate.

To express this last condition, let x_1 and y_1 be the co-ordinates of the extremity B (Fig. 40), x and y the co-ordinates

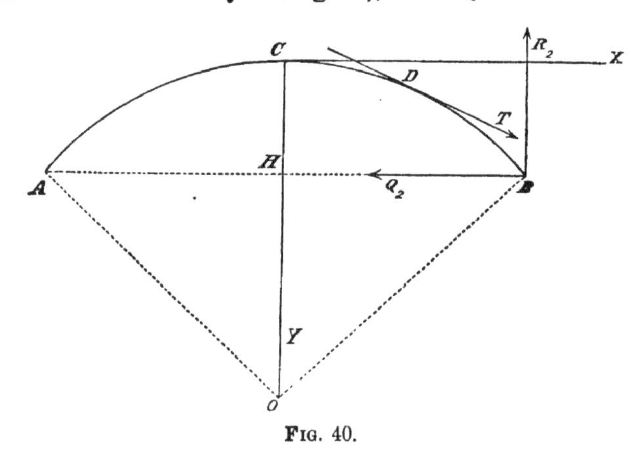

FIG. 40.

of any point as D, and ϕ the angle made by the tangent line at D with the axis of X. Represent by T_1 the sum of the components of the applied forces parallel to the tangent DT, and by μ the sum of the moments of the applied forces with respect to to the section at D.

The bending moment at D will be

$$M = \mu + Q_2 (y_1 - y) - R_2 (x_1 - x) \quad . \quad . \quad (116)$$

and for the force acting in the direction of the tangent DT,

$$P = T_1 + Q_2 \cos \phi + R_2 \sin \phi. \quad . \quad . \quad . \quad (117)$$

In these two equations, whenever the applied forces are given, μ, T_1, $y_1 - y$, and $x_1 - x$, are known functions; but R_2 and Q_2 are unknown constants.

But from the third of equations (115) we have

$$R_2 = \frac{\mu_1 + Q_2 l_2}{l_1},$$

which substituted in the expressions just obtained for M and P give them in terms of one unknown constant and known functions.

We are now able to find the values of the constant of integration before referred to, and the component Q_2. Knowing the latter, those of Q_1, R_1, and R_2 are easily found.

226. Having found all the external forces acting on the beam, the intensity of the stress on any cross-section may be determined.

The stress on the unit of area of any cross-section, at the distance y from the neutral axis, is

$$R' = \frac{P}{A} + \frac{My}{I} \text{ and } S' = \frac{F}{A},$$

in which P and F are the components of the external forces, perpendicular and parallel to the plane of cross-section; A, the area of the cross-section; I, its moment of inertia; and M the bending moment of the external forces with respect to the neutral axis of the cross-section.

227. In chapters IV. and V. of his "Cours de Mécanique Appliquée," M. Bresse has given a complete discussion of the strains in curved beams resting on two points of support, produced by external forces acting in the plane of mean fibre; the cross-section of the beam being uniform and the curve of mean fibre a circular arc.

He has deduced exact formulas for the *horizontal thrust*, Q_2, and reduced these formulas to forms of easy application for the cases most commonly used. He has besides constructed tables containing the values of the quantities found in the formulas, under the different suppositions usually made.

If the beam has its ends in the same horizontal plane and is loaded symmetrically with reference to its middle point, or strained by vertical loads only, Q_1 and Q_2 are equal.

The following formula for a load uniformly distributed over the beam, along the mean fibre, when the rise, H C, is small compared with the span, A B, is given by him:

$$Q_1 = Q_2 = w\rho\phi\frac{l}{2f}\left(\frac{1 - \frac{3}{7}\frac{f^2}{l^2}}{1 + \frac{15}{18}\frac{k^2}{f^2}}\right), \quad . \quad . \quad (118)$$

in which w is the load on the unit of length of the curve; ρ, the radius of the curve of mean fibre; ϕ, the half of the angle, A O B, included between the radii drawn to the extremities A and B; $2l$, the length of the chord, A B; f, the rise, H C; and k, the radius of gyration of the cross-section of the beam.

And under the same circumstances, the load being distributed on the beam uniformly over the chord A B or a horizontal tangent at C, he gives the following formula:

$$Q_1 = Q_2 = \frac{wl^2}{2f}\left(\frac{1-\frac{1}{7}\frac{f^2}{l^2}}{1+\frac{15}{18}\frac{l^2}{f^2}}\right). \quad . \quad . \quad (119)$$

228. Suppose a curved beam, the mean fibre being a parabolic arc, to be strained by a uniform load distributed so as to be directly proportional to its horizontal length. An **approximate method** of determining the stresses is as follows:

Let A V (Fig. 41) be half of the curve of mean fibre; V, the origin of coördinates; the tangent V X to be the axis of X, and V Y

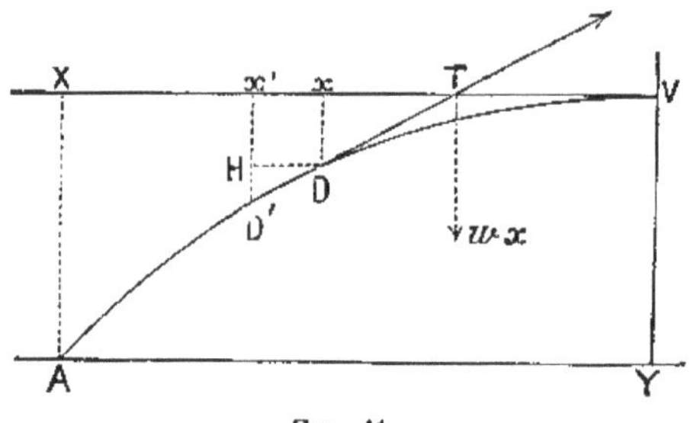

FIG. 41.

the axis of Y. Let D and D′ be any two consecutive points whose abscissas are x and x'. Denote by l the half-span A Y, by f the rise V Y, and by w the weight on the unit of length measured on V X.

Assuming the bending moment at V to be zero, suppose the right half of the beam to be removed. The equilibrium among the external forces acting on the remaining half may be preserved by the substitution of a horizontal force, H, acting at V. The external forces acting on the beam between V and any section as D, will therefore be the force H, the weight wx, and the reaction at D, which denote by P.

Since there is an equilibrium in the system of forces acting on the arc D V, the intensities of these forces H, P, and wx must be proportional to the sides of the triangle DxT. Since D H and D′H are respectively parallel to Tx and Dx, we have

$$Tx : Dx :: HD : D'H,$$

or $$H : wx :: dx : dy,$$

whence $$dy = \frac{w}{H} x dx.$$

Integrating, we obtain

$$y = \frac{w}{2H} x^2 + C. \quad . \quad . \quad . \quad . \quad (120)$$

Taking this between the limits $x = 0$ and $x = l$, there results

$$f = \frac{w}{2H} l^2,$$

whence $$H = \frac{wl^2}{2f} \quad . \quad . \quad . \quad . \quad . \quad . \quad . \quad . \quad (121)$$

This is the same as the coefficient outside of the parenthesis in the expression for Q_2 in eq. (119).

But $$P = \sqrt{H^2 + w^2 x^2},$$

and substituting in which the value just found for H, we get

$$P = w\sqrt{\frac{l^4}{4f^2} + x^2}. \quad . \quad . \quad . \quad (122)$$

The value for H may be deduced directly by moments. For we have

$$H \times AX = wVX \times \tfrac{1}{2}VX,$$

or $$Hf = \frac{wl^2}{2},$$

whence $$H = \frac{wl^2}{2f}.$$

These expressions show that P is least at V and greatest at A, and that H is the same throughout. The value for H is independent of the form of the curve of mean fibre, whether parabolic, circular, or other shape.

Curved beams are frequently constructed so that the curve assumed by the mean fibre under the action of the load is that of a parabolic arc, the vertex being at the highest point.

In this case, the direction of P coincides with that of the tangent to the curve of mean fibre; the bending moment at each cross section is zero; and the strain is one of compression produced by P.

If the two halves abut against each other at V, or are hinge-jointed at this point, the assumption that the bending moment at this section is zero, is a correct one.

CURVED BEAMS WITH THE ENDS FIRMLY FIXED.

229. The curved beam in the foregoing discussion has had the analogous position of a straight beam resting on two supports. In each of these cases the beam has been regarded as continuous between the points of support, and the horizontal distance between these points as constant.

If, in addition, the condition be imposed that *the cross-sections at the points of support be fixed* so that they shall not move under the action of the external forces, the case becomes analogous to that of a straight beam whose ends are firmly imbedded in a wall. And there will be, as in that case, an unknown moment at the points of support, whose value must be found before the strains on the beams can be determined. Having found this, the processes of obtaining the strains and calculating the dimensions of the beam are analogous to those already used.

PART III.

FRAMING.

CHAPTER VIII.

230. The **art of construction** consists mainly in giving to a structure the proper degree of strength with the least amount of material necessary for the purpose. If any piece be made stronger than is necessary, the superfluous weight of this piece will in general be transmitted to some other part, and the latter, in consequence, will be required to sustain a greater load than it should. Hence, the distribution and sizes of the different parts of a structure should be determined before combining the parts together.

A **frame** is an arrangement of beams, bars, rods, etc., made for sustaining strains. The art of arranging and fitting the different pieces is called **framing**, and forms one of the subdivisions of the **art of construction**. It follows, then, from the previous remark, that the object to be attained in framing is to arrange the pieces, with due regard to *lightness* and *economy of material*, so that they shall best resist, *without change of form in the frame*, the strains to which the latter may be subjected.

231. The principal frames employed by engineers are those used in bridges, centres for arches, coffer-dams, caissons, floors, partitions, roofs, and staircases.

The materials used in their construction are generally timber and iron. The latter, in addition to superior strength, possesses an advantage over wood in being susceptible of receiving the most suitable form to resist the strains to which it may be subjected.

When the principal pieces of a frame are of timber, the construction belongs to that branch of framing known as **carpentry.**

The combination of the pieces, and the shape of a frame

will depend upon the purposes for which the frame is to be adapted and upon the directions of the straining forces.

One of the main objects in the arrangement of a frame is to give the latter such a shape that it will not admit of change in its figure when strained by the forces which it is intended to resist. This is usually effected by combining its parts so as to form a series of triangular figures, each side of the latter being a single beam. If the frame has a quadrilateral shape, secondary pieces are introduced either having the positions of the diagonals of the quadrilateral, or forming angles with the upper and lower sides of the frame. These secondary pieces are called **braces.** When they sustain a strain of compression they are termed **struts**; of extension, **ties.**

The strength, and hence the dimensions, of the pieces will be regulated by the strains upon the frame. Knowing the strains and the form of the frame, the amount of stress on each piece can be deduced, and from this the proper form and particular dimensions of each piece.

The arrangement of the frame should be such that, after being put together, any one piece can be displaced without disconnecting the others.

When practicable, the axes of the pieces should be kept in the plane of the forces which act to strain the frame, and the secondary pieces of the frame should be arranged to transmit the strains in the direction of their lengths. The pieces are then in the best position to resist the strains they have to support, and all unnecessary cross-strains are avoided.

The essential qualities of a frame are, therefore, *strength, stiffness, lightness*, and *economy of material.*

JOINTS.

232. The **joints** are the surfaces at which the pieces of a frame touch each other; they are of various kinds, according to the relative positions of the pieces and to the forces which the pieces exert on each other.

Joints should be made so as to give the largest bearing surfaces consistent with the best form for resisting the particular strains which they have to support, and particular attention should be paid to the effects of contraction and expansion in the material of which they are made.

In planning them the purpose they are to serve must be kept in mind, for the joint most suitable in one case would oftentimes be the least suitable in another.

JOINTS IN TIMBER WORK.

233. In frames made of timber, the pieces may be joined together in three ways; by connecting them,

1. End to end;
2. The end of one piece resting upon or notched into the face of another; and
3. The faces resting on or notched into each other.

I. **Joints used to unite beams end to end,** the axes of the beams being in the same straight line.

The joint used to lengthen a beam is either a **plain** or a **scarfed** joint. There are two cases: one in which the beam is subjected to compressive or tensile strains, and the other in which it is subjected to a cross strain.

234. *First.* Suppose the pieces are required to resist strains in the direction of their length.

Plain or Butt Joints.

A **plain** joint is one in which the two pieces abut end to end, as shown at *cd* (Fig. 42), the surface of the joint being perpendicular to the axis of the beam. The ends of the pieces being brought together are fastened to prevent displacement by any lateral movement. This fastening is usually effected by bolting to the beam, on each side of the joint, pieces of wood or iron. A joint fastened in this way

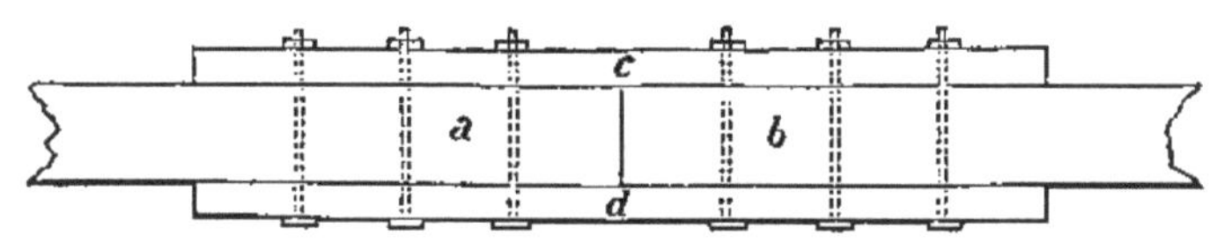

FIG. 42—Represents the manner in which two beams *a* and *b* are fished by side pieces *c* and *d* bolted to them.

is said to be **fished**, and is sometimes called a **fish joint.**

A plain joint is a good one when the only strain is that of compression. It is recommended, in this case, to place fish-pieces on all four of the sides of the beam, to prevent any lateral displacement of the ends that might be caused by shocks.

If the strain be one of tension, it is evident that the strength of this joint (Fig. 42) depends entirely upon the strength of the bolts, assisted by the friction of the fish-pieces against the sides of the timber. Such a joint would seldom be used for tension.

A better fastening for the joint would be that in which the fish-pieces were let into the upper side of the beam, as

FIG. 43—Represents a joint to resist extension, iron rods or bars being used to connect the beams instead of wooden fish-pieces.

shown in Fig. 44.

Sometimes the beam and the fish-pieces have shallow notches made in them, into which keys or folding wedges of

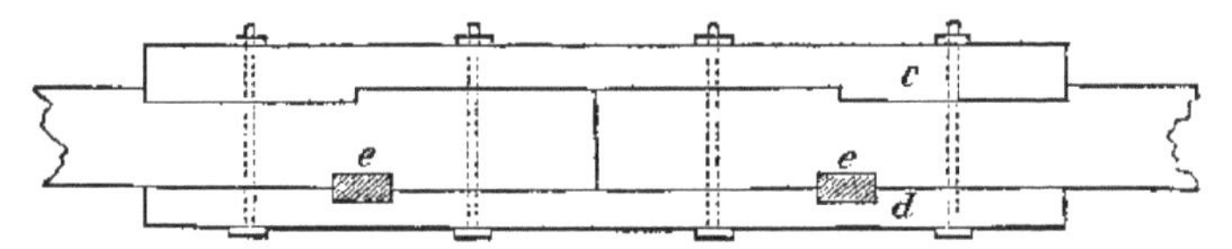

FIG. 44—Represents a fished joint in which the side pieces *c* and *d* are either let into the beams or secured by keys *e*, *e*.

hard wood as *e*, *e* (Fig. 44) are inserted.

Scarf Joints.

When the ends of the pieces overlap, the joint is called a **scarf joint**. The ends of the pieces are fastened together by bolts, to keep them in place. An example of a simple scarf joint is shown in Fig. 45, that is sometimes used when the beam is to be subjected only to a slight strain of exten-

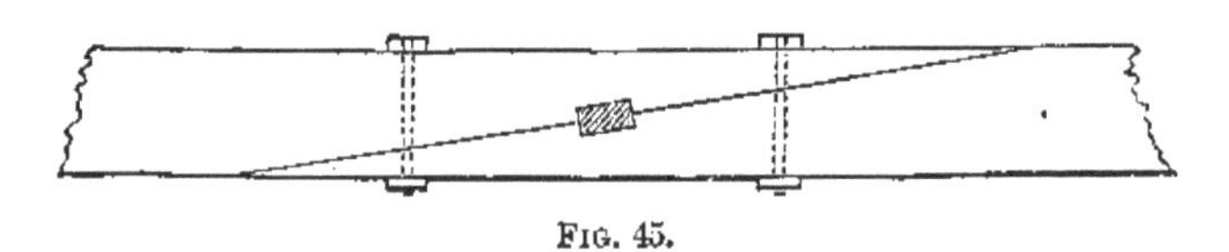

FIG. 45.

sion. A key or folding wedge is frequently added, notched equally in both beams at the middle of the joint; it serves to bring the surfaces of the joint tightly together.

A better scarf joint is made by cutting the ends in such a manner as to form projections on one which fit into corresponding indentations in the other, as shown in Fig. 46

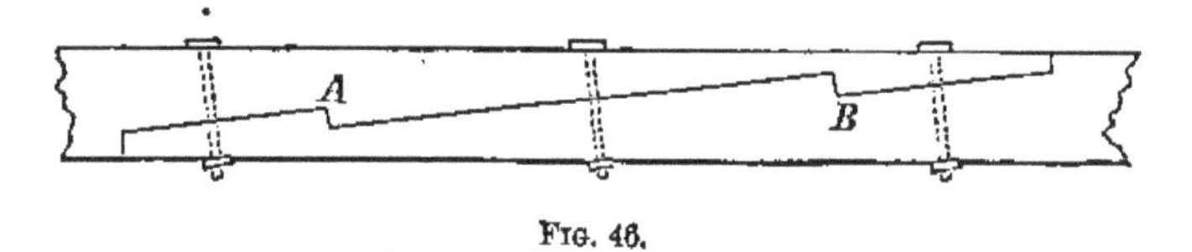

FIG. 46.

The total lap shown in this figure is ten times the thickness of the timber, and the depth of the notches at A and B are each equal to one-fourth that of the beam. The bolts are placed at right angles to the principal lines of the joint.

This is a good joint to resist a strain of tension, since the notches at A and B allow one-half of the cross-section of the beam to be utilized in resisting the tensile strain.

Another form of scarf joint is shown in Fig. 47. A joint made in this shape is serviceable to resist either a compressive, or a tensile strain on the beam.

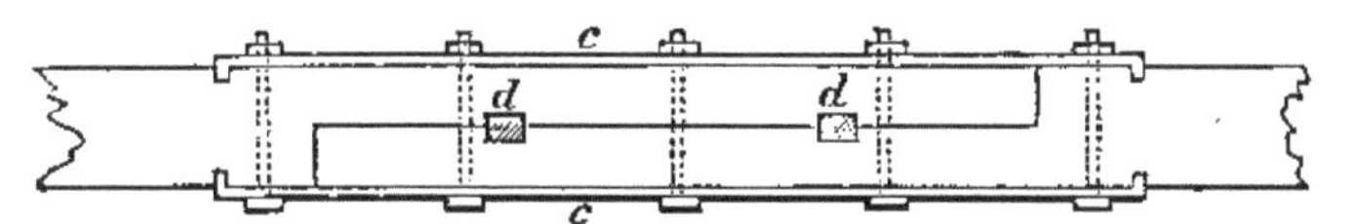

FIG. 47—Represents a scarf-joint secured by iron fish-plates *c*, *c*, keys *d*, *d*, and bolts.

235. *Second.* Suppose the pieces are required to resist a cross strain.

In this case the scarf joint is the one generally used. The joint may be formed by simply halving the beams near their ends, as shown in Fig. 47, and fastening the ends of the beams by fish-pieces bolted upon the upper and lower sides of the ends. Keys of hard wood are used to resist the longitudinal shear along the lap of the joint.

A more usual and the better form of joint for this case is shown in Fig. 48.

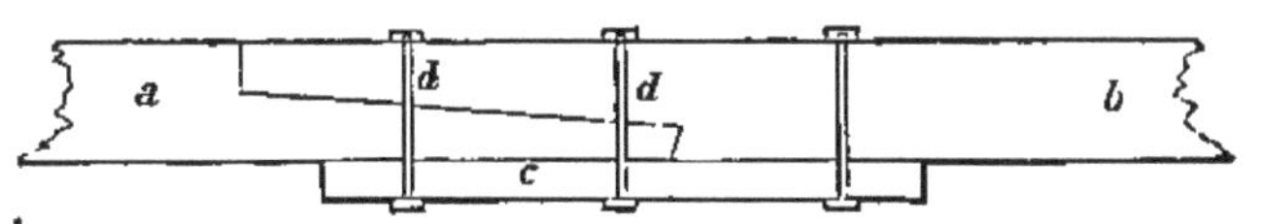

FIG. 48—Represents a scarf-joint for a cross-strain, fished at bottom by a piece of wood *c*.

In the upper portion of this joint the abutting surfaces are perpendicular to the length of the beam and extend to a depth of at least one-third and not exceeding one-half that of the beam. In the bottom portion they extend one-third of the depth and are perpendicular to the oblique portion joining the upper and lower ones.

The lower side of the beam is fished by a piece of wood or iron plate, secured by bolts or iron hoops, so as to better resist the tensile strain to which this portion of the beam is subjected.

Third. Suppose the beam required to resist a cross-strain combined with a tensile strain.

The joint, frequently used in this case, is shown in Fig. 49.

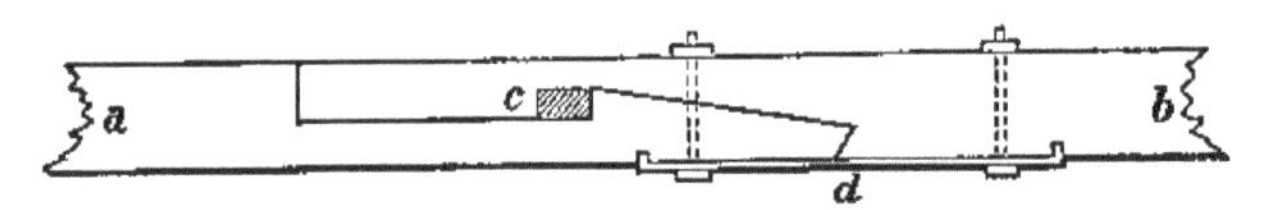

FIG. 49—Represents a scarf-joint arranged to resist a cross-strain and one of extension. The bottom of the joint is fished by an iron plate; and a folding wedge inserted at *c* serves to bring all the surfaces of the joint to their bearings.

II. **Joints of beams,** the axes of the beams making an angle with each other.

236. In the previous cases the axes were regarded as being in the same straight line. If it be required to unite the ends and have the axes make an angle with each other, this may be done by halving the beams at the ends, or by cutting a mortise in the centre of one, shaping the end of the other to fit, and fastening the ends together by pins, bolts, straps, or other devices. The joints used in the latter case are termed **mortise** and **tenon joints.** Their form will depend upon the angle between the axes of the beams.

Mortise and Tenon Joints.

237. When the axes are perpendicular to each other, the mortise is cut in the face of one of the beams, and the end of the other beam is shaped into a tenon to fit the mortise, as shown in Fig. 50.

When the axes are oblique to each other, one of the most common joints consists of a triangular notch cut in the face of one of the beams, with a shallow mortise cut in the bottom of the notch, the end of the other beam being cut to fit the notch and mortise, as shown in Fig. 51.

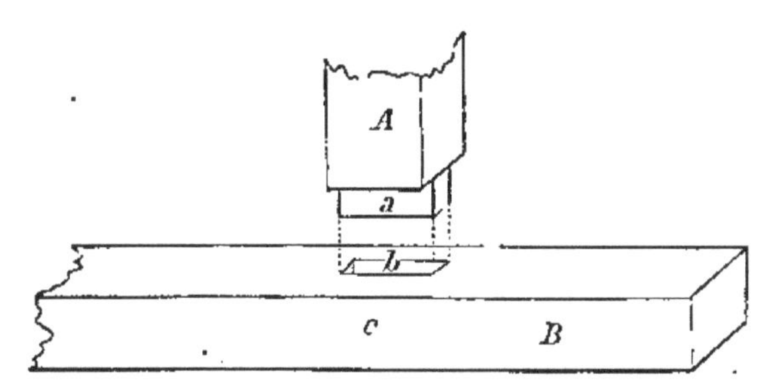

FIG. 50—Represents a mortise and tenon joint when the axes of the beams are perpendicular to each other.
a, tenon on the beam *A*.
b, mortise in the beam *B*.
c, pin to hold the parts together.

In a joint like this the distance *ab* should not be less than one-half the depth of the beam A; the sides *ab* and *bc* should be perpendicular to each other when practicable; and the

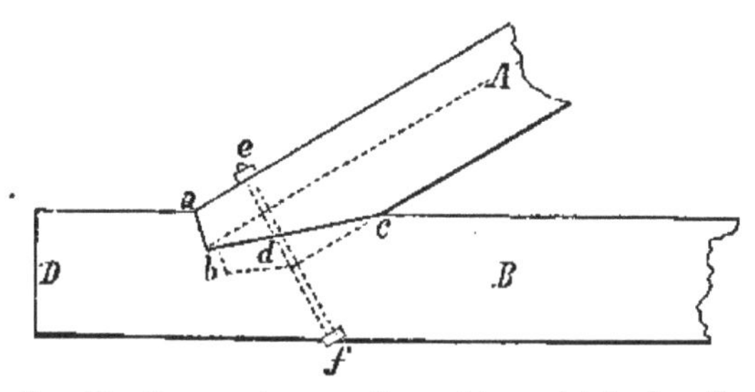

FIG. 51—Represents a mortise and tenon joint when the axes of the beams are oblique to each other.

thickness of the tenon *d* should be about one-fifth of that of the beam A. The joint should be left a little open at *c* to allow for settling of the frame. The distance from *b* to the end D of the beam should be sufficiently great to resist safely the longitudinal shearing strain caused by the thrust of the beam A against the surface *ab*.

Denote by

H the component of the thrust, parallel to the axis of the beam B D;

b the breadth in inches of the beam B D;

l the distance in inches from b to the end of the beam at D; and

S the resistance per square inch in the beam B to longitudinal shearing.

The total resistance to shearing will be $S \times bl$, hence

$$S \times bl = H, \text{ from which we have}$$

$$l = \frac{H}{S \times b}.$$

The value of S for the given material, Art. 166, being substituted in this expression, will give the value for l, when the strain just overcomes the resistance of the fibres. In this case the factor of safety is ordinarily assumed to be at least four. Therefore the value of l, when the adhesion of the fibres is depended upon to resist this strain, will be

$$l = \frac{4H}{S \times b},$$

S being taken from the tables.

A bolt, ef, or strap, is generally used to fasten the ends more securely.

In both of these cases the beam A is subjected to a strain of compression, and is supported by B. If we suppose the beams reversed, A to support B, the general principles for forming the joints would remain the same.

Suppose the axes of the beams to be horizontal, and the beam A to be subjected to a cross-strain, the circumstances being such that the end of the beam A is to be connected with the face of the other beam B.

In this case a mortise and tenon joint is used, but modified in form from those just shown.

To weaken the main or supporting beam as little as possible, the mortise should be cut near the middle of its depth; that is, the centre of the mortise should be at or near the neutral axis. In order that the tenon should have the greatest strength, it should be at or near the under side of the joint.

Since both of these conditions cannot be combined in the same joint, a modification of both is used, as shown in Fig. 52.

The tenon has a depth of one-sixth that of the cross-beam A, and a length of twice this, or of one-third the depth of the beam. The lower side of the cross-beam is made into a shoulder, which is let into the main beam, one half the length of the tenon.

Double tenons have been considerably used in carpentry

As a rule they should *never* be used, as both are seldom in bearing at the same time.

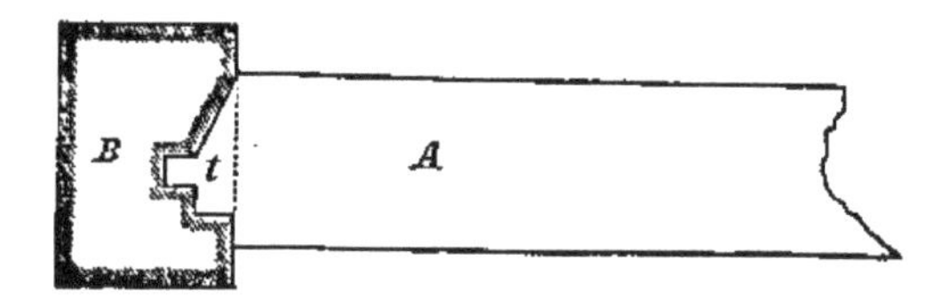

FIG. 52.—A, the cross-beam.
B, cross-section of main beam.
t, the tenon.

III. Joints used to connect beams, the faces resting on or notched into each other.

238. The simplest and strongest joint in this case is made by cutting a notch in one or both beams and fastening the fitted beams together.

If the beams do not cross, but have the end of one to rest upon the other, a **dove-tail** joint is sometimes used. In this joint, a notch trapezoidal in form, is cut in the supporting beam, and the end of the other beam is fitted into this notch.

On account of the shrinkage of timber, the dove-tail joint should never be used except in cases where the shrinkage in the different parts counteract each other.

It is a joint much used in joiner's work.

239. The joints used in timber-work are generally composed of plane surfaces. Curved ones have been recommended for struts, but the experiments of Hodgkinson would hardly justify their use. The simplest forms are as a rule the best, as they afford the easiest means of fitting the parts together.

FASTENINGS.

The fastenings used to hold the pieces of a frame together at the joints may be classed as follows:

1. **Pins**, including nails, spikes, screws, bolts, and wedges;
2. **Straps and tiebars**, including stirrups, suspending-rods, etc.; and
3. **Sockets.**

These are so well known that a description of them is unnecessary.

General Rules to be observed in the Construction of Joints.

241. The following general rules should be observed in the construction of joints and fastenings for frames of timber:

I. To arrange the joints and fastenings so as to weaken as little as possible the pieces which are to be connected.

II. In a joint subjected to compression, to place the abutting surfaces as nearly as possible perpendicular to the direction of the strain.

III. To give to such joints as great a surface as practicable.

IV. To proportion the fastenings so that they will be equal in strength to the pieces they connect.

V. To place the fastenings so that there shall be no danger of the joint giving way by the fastenings shearing or crushing the timber.

JOINTS FOR IRON-WORK.

242. The pieces of an iron frame are ordinarily joined by means of rivets, pins, or nuts and screws.

Riveted Joints.

243. A **rivet** is a short, headed bolt or pin, of iron or other malleable material, made so that it can be inserted into holes in the pieces to be fastened together, and that the point of the bolt can be spread out or beaten down closely upon the piece by pressure or hammering. This operation is termed **riveting**, and is performed by hand or by machinery. By hand, it is done with a hammer by a succession of blows. By machinery, as ordinarily used, the heated bolt is both pressed into the hole and riveted by a single stroke. If a machine uses a succession of blows, the operation is then known as **snap-riveting**. By many it is claimed that machine riveting possesses great superiority over that by hand, for the reason that the rivets more completely fill the holes, and in this way become an integral part of the structure. It is doubtful if it possesses the advantage of superior strength to any marked degree. It does certainly possess, however, the advantage of being more quickly executed without damage to the heads of the rivets.

The holes are generally made by punching, are about one-twentieth of an inch larger than the diameter of the rivet, and are slightly conical. The diameter of the rivet is generally greater than the thickness of the plate through which the hole is to be punched, because of the difficulty of punching holes of a smaller size. Punching injures the piece when the latter is of a hard variety of iron, and for this reason engineers often require that the holes be drilled. Drilling seems to be the better method, especially when several thicknesses of plates are to be connected, as it insures the precise matching of the rivet holes. The appearance of the iron around a hole made by punching gives a very fair test of the quality of the iron.

When two or more plates are to be riveted, they are placed together in the proper position, with the rivet-holes exactly over one another, and screwed together by temporary screw-bolts inserted through some of the holes. The rivets, heated red-hot, are then inserted into the holes up to the head, and by pressure or hammering, the small end is beaten down fast to the plate. In a good joint, especially when newly riveted, the friction of the pieces is very great, being sufficient to sustain the working-load without calling into play the shearing resistance of the rivets. In calculating the strength of the frame, this amount of strength due to friction is not considered, as it cannot be relied on after a short time in those cases where the frame is subjected to shocks, vibrations, or great changes of temperature.

Number and Arrangement of Rivets.

244. The general rule determining the number is *that the sum of the areas of the cross-sections of the rivets shall be equal to the effective sectional area of the plate after the holes have been punched.* This rule is based on the theory that the resistance to shearing strain in the rivet is equal to the tenacity of the plate.

To determine the proper distance between the rivets in the direction of any row, so that the strength of the rivets in any single row shall be equal to the strength of the section of the plate along this row after the holes have been punched, let

d, be the diameter of the rivet;
c, the distance between the centres of consecutive rivets;
a, the area of cross-section of the rivet;
A', the effective area, between two consecutive rivets, of the cross-section of the plate along the row of rivets; and
t, the thickness of the iron plate.

It has been assumed that

$$T = S,$$

and the rule requires that

$$TA' = S \times a, \text{ or } \frac{T}{S} = \frac{a}{A'} = 1.$$

We have

$$\frac{a}{A'} = \frac{\frac{1}{4}\pi d^2}{t(c-d)} = 1,$$

whence

$$c = \frac{1}{4}\frac{\pi d^2}{t} + d, \quad . \quad . \quad . \quad . \quad . \quad (123)$$

for the distance from centre to centre of the consecutive rivets in any one row.

English engineers, in practice, use rivets whose diameters are $\frac{5}{8}$, $\frac{6}{8}$, $\frac{7}{8}$, 1, $1\frac{1}{4}$, and $1\frac{1}{2}$ inches, for iron plates $\frac{1}{4}$, $\frac{5}{16}$, $\frac{3}{8}$, $\frac{1}{2}$, $\frac{5}{8}$, and $\frac{3}{4}$ inches thick, respectively, and take the distance from centre to centre at 2 diameters for a strain of compression, and $2\frac{1}{2}$ diameters for extension. The distance of the centre of the extreme rivet from the edge of the plate is taken between $1\frac{1}{2}$ and 2 diameters.

Instead of assuming the resistance to shearing in the rivet equal to the tenacity of the iron plate, a better rule would be *to make the product arising from multiplying the sum of the areas of the cross-sections of the rivets, by the amount of shearing strain allowed on each unit, equal to the maximum strain transmitted through the joint.*

If the strain was one of compression in the plates and the ends exactly fitted, the only riveting required would be that necessary to keep the plates in position. As the workmanship rarely, if ever, admits of so exact fitting, the rivets should be proportioned by the rules just given.

245. **The head** of a rivet is usually circular in form, with a diameter not less than twice the diameter of the rivet.

The thickness of the head at its centre should be not less than half the thickness of the rivet.

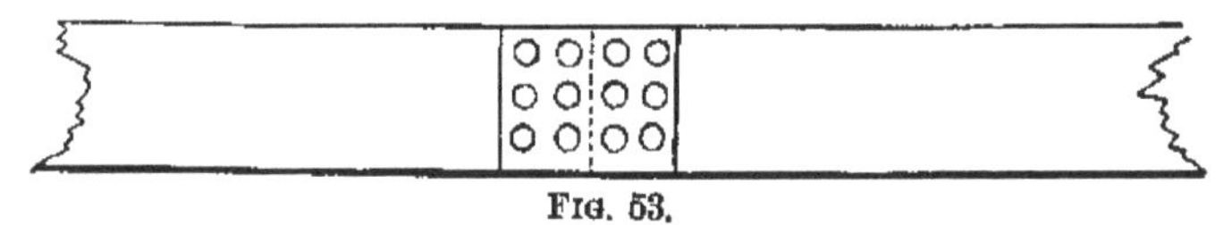

FIG. 53.

246. Various methods are used in the arrangement of the rivets. The arrangement often used for lengthening a plate is shown in Fig. 53. This method is known as "**chain riveting.**"

Fig. 54 shows another method used for the same purpose in which the number of rivets is the same as in the previous example, but there is a better disposition of them.

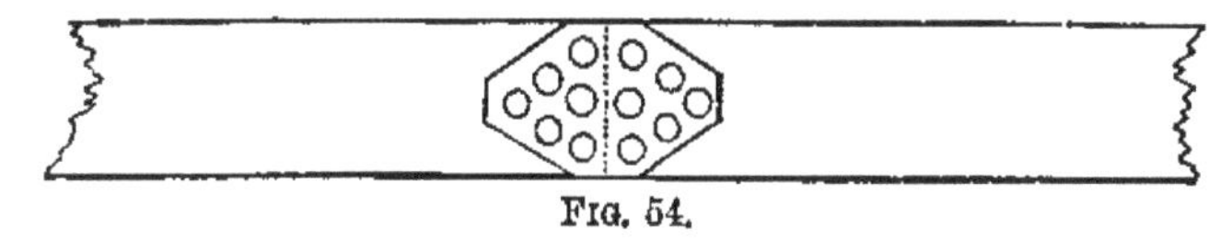

FIG. 54.

Figs. 55 and 56 show the arrangement of the rivets often used to fasten ties to a plate.

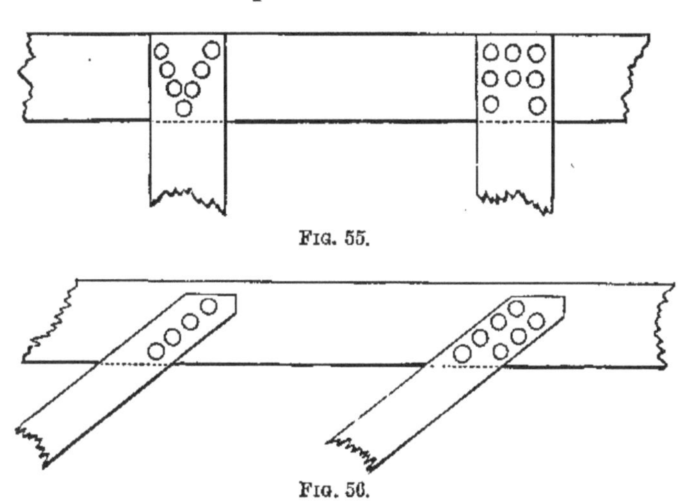

FIG. 55.

FIG. 56.

Figs. 57, 58, and 59 show in plan the forms of several kinds of riveted joints.

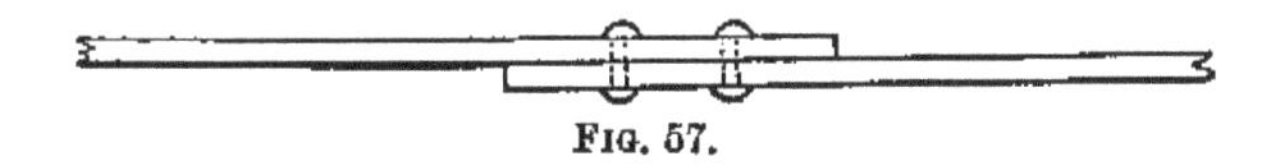

FIG. 57.

Fig. 57 shows the single shear-joint or single lap-joint.

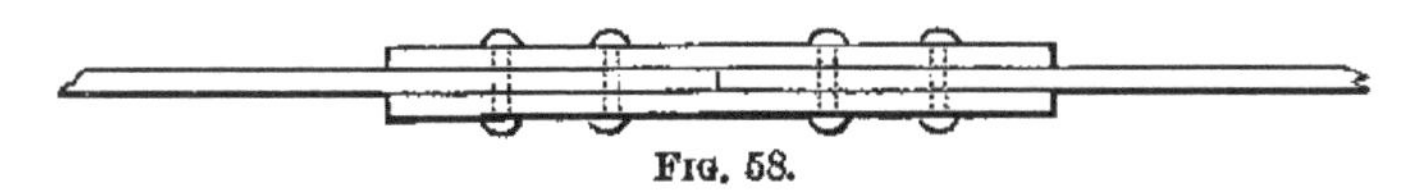

FIG. 58.

Fig. 58 is a plain joint fished. In this example the fish or cover plates are placed on each side, and have a thickness of half that of the plates to be connected; sometimes only one cover plate is used.

FIG. 59.

When several plates are to be fastened together, the method shown in Fig. 59 is the one ordinarily used.

Eye-bar and Pin Joints.

247. A simple and economical method of joining flat bars end to end when subjected to a strain of extension, is to connect them by pins passing through holes or eyes made in the ends of the bars.

When several are connected end to end, they form a flexible arrangement, and the bars are often termed links.

This method of connecting is called the **eye-bar and pin,** or **link and pin** joint, and is shown in plan in Fig. 60.

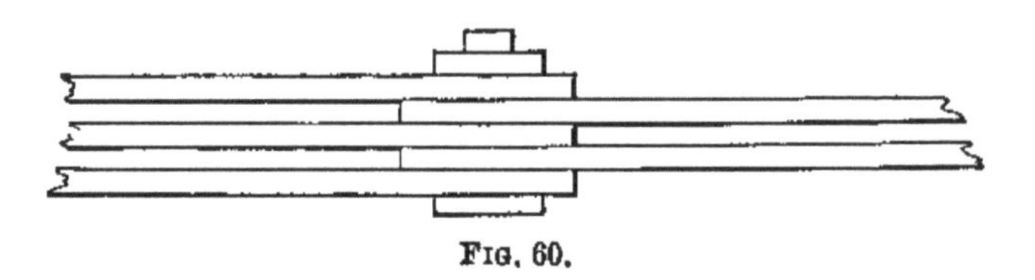

FIG. 60.

The bar should be so formed at the end that it would be no more liable to break there than at any other point. The following are the dimensions in the case where the head has the same thickness as the bar.

If the width of the bar be taken as equal to........ 1.
The diameter of the eye should equal............. .75.
Depth of head beyond the eye should equal........ 1.
Sum of the sides of the head through eye should equal 1.25.
Radius of curve of neck should equal............. 1.5.

Hence, for a bar eight inches wide, the dimensions would be as shown in Fig. 61.

By this rule the pin has a diameter which gives a sufficient bearing surface, the important point to be considered.

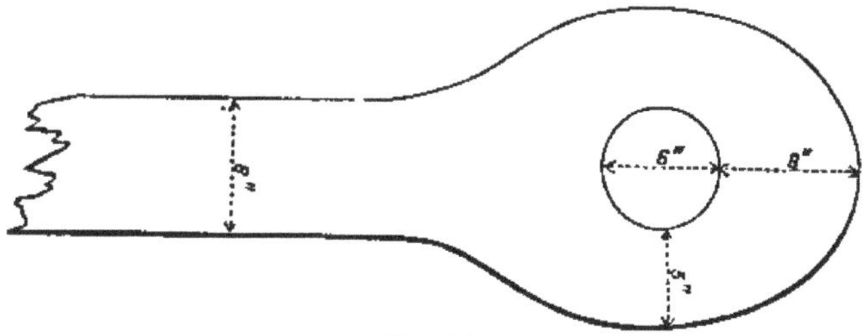

FIG. 61.

There should be a good fit between the pin and eye, especially in structures subjected to shocks, hence the conditions of manufacture and the quality of material and workmanship should be of the best kind.

Screw-bolt Joints.

248. The connection by **nut and screw** is simple and economical.

The strength of a bolt or rod on which a screw is made, when subjected to a shearing strain, is determined as in the case of rivets or pins. In case of a tensile strain the strength is measured by the area of cross-section of the spindle inside the thread.

The resistance offered to stripping by the nut depends upon the form of the thread and the depth of the nut. In order that this resistance should be equal to that offered by the bolt to being pulled apart, the length of the nut should be *at least equal to one-half* the diameter of the screw.

The following proportions have been recommended by the Franklin Institute:

Diameter of bolt in inches.	No. of threads per inch.	Six-sided nut.—Length of		Depth of head.	Depth of nut.
		Long diameter.	Short diameter.		
$\frac{1}{2}$	13	1	$\frac{7}{8}$	$\frac{7}{16}$	$\frac{1}{2}$
$\frac{3}{4}$	10	$1\frac{7}{16}$	$1\frac{1}{4}$	$\frac{5}{8}$	$\frac{3}{4}$
1	8	$1\frac{7}{8}$	$1\frac{5}{8}$	$\frac{13}{16}$	1
$1\frac{1}{2}$	6	$2\frac{3}{4}$	$2\frac{3}{8}$	$1\frac{3}{16}$	$1\frac{1}{2}$
2	$4\frac{1}{2}$	$3\frac{5}{8}$	$3\frac{1}{8}$	$1\frac{9}{16}$	2
$2\frac{1}{2}$	4	$4\frac{1}{2}$	$3\frac{7}{8}$	$1\frac{15}{16}$	$2\frac{1}{2}$
3	$3\frac{1}{2}$	$5\frac{3}{8}$	$4\frac{5}{8}$	$2\frac{5}{16}$	3

SIMPLE BEAMS.

249. One of the most common and simple use of frames is that in which the frame is supported at its extremities and subjected only to a transverse strain.

When the distance between the points of support, or the *bearing*, is not very great, frames are not necessary, as beams of ordinary dimensions are strong and stiff enough to resist the cross-strains arising from the load they support, without bending beyond the allowed limits. The load placed upon them may be uniformly distributed, or may act at a point; in either case the strains produced, and the dimensions of the beam to resist them, can be easily determined. (Arts. 177 and 179.)

The usual method is to place the beams in parallel rows, the distances apart depending on the load they have to support. The *joists* of a floor, the *rafters* of a roof, are examples of such cases.

The depth of a beam used for this purpose is always made much greater than its breadth, and arrangements should be made to prevent the beam twisting or bending laterally. It is usual to place short struts or battens in a diagonal direction between the joists of a floor, fastening the top of one joist with the bottom of the next by the battens to prevent them from twisting or yielding laterally.

SOLID BUILT BEAMS.

250. A **solid beam** is oftentimes required to be of a greater size than that possessed by any single piece of timber. To provide such a beam it is necessary to use a combination of pieces, consisting of several layers of timber laid in juxtaposition and firmly fastened together by bolts, straps, or other means, so that the whole shall act as a single piece. This is termed a **solid built beam.**

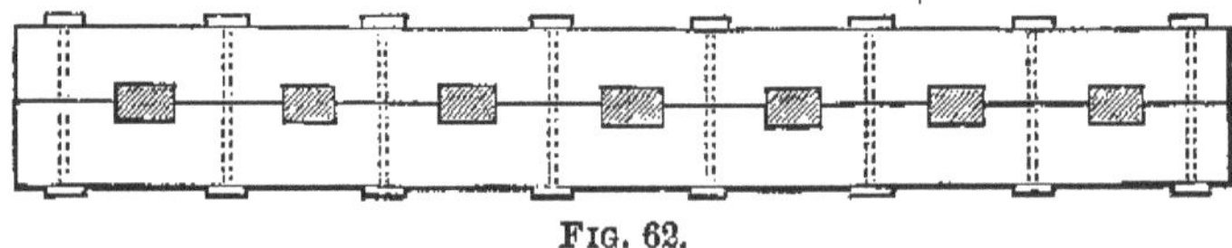

FIG. 62.

When two pieces of timber are built into one beam having twice the depth of either, *keys* of hard wood are used to resist the shearing strain along the joint, as shown in Fig. 62.

Tredgold gives the rule that the breadth of the key should be twice its depth, and the sum of the depths should be equal to *once and a third* the total depth of the beam.

It has been recommended to have the bolts and the keys on the right of the centre make an angle of 45° with the axis of the beam, and those on the left to make the supplement of this angle.

The keys are sometimes made of two wedge-shaped pieces (Fig. 63), for the purpose of making them fit the notches

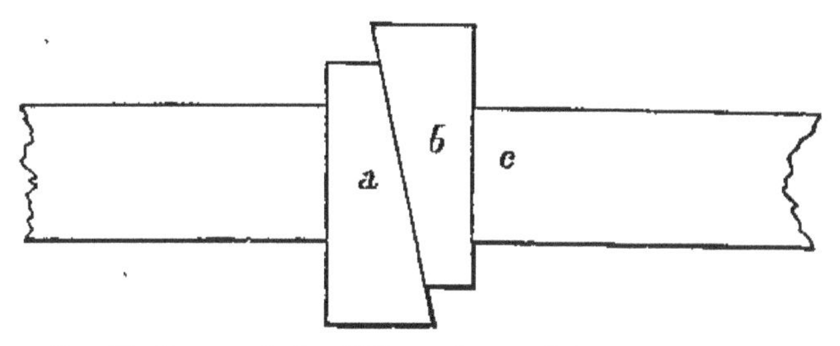

FIG. 63—Represents the *folding wedges*, *a*, *b*, let into a notch in the beam *c*.

more snugly, and, in case of shrinkage in the timber, to allow of easy readjustment.

When the depth of the beam is required to be less than the sum of the depths of the two pieces, they are often built into one by indenting them, the projections of the one fitting accurately into the notches made in the other, the two being firmly fastened together by bolts or straps. The built beam shown in Fig. 64 illustrates this method. In this particular example the beam tapers slightly from the middle to the ends, so that the iron bands may be slipped on over the ends and driven tight with mallets.

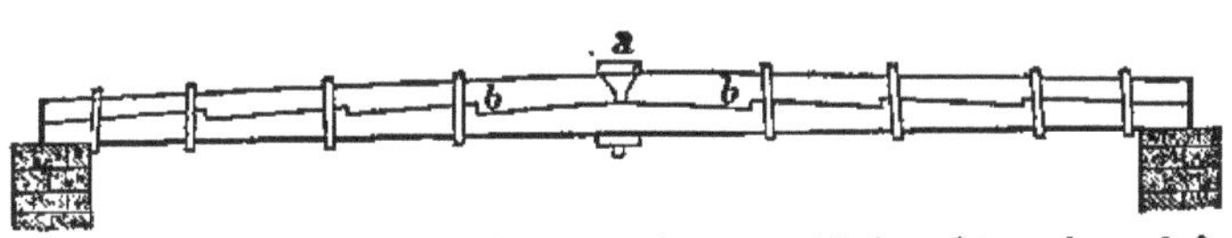

FIG. 64—Represents a solid built beam, the top part being of two pieces, *b*, *b*, which abut against a broad flat iron bolt, *a*, termed a *king-bolt*.

When a beam is built of several pieces in lengths as well as in depth, they should break joints with each other. The layers below the neutral axis should be lengthened by the scarf or fish joints used for resisting tension, and the upper ones should have the ends abut against each other, using plain *butt* joints.

Many builders prefer using a **built beam** of selected timber to a single solid one, on account of the great difficulty of getting the latter, when very large, free from defects; moreover, the strength of the former can be relied upon, although it cannot be stronger than the corresponding solid beam if perfectly sound.

FRAMING WITH INTERMEDIATE POINTS OF SUPPORT.

251. If the bearing be great, the beam will bend under the load it has to support, and to prevent this it will need intermediate points of support. These points of support may be below the beam, or they may be above it.

The simplest method, when practicable, is to place at suitable intervals under the beam upright pieces to act as **props** or **shores**.

When this cannot be done, but points of support can be obtained below those on which the beam rests, inclined struts may be used.

These may meet at the middle point of the beam, dividing it into two equal parts. The beam is then said to be **braced**, and is no longer supported at two points, but rests on three.

The struts may be placed so as to divide the beam (Fig. 65) into three parts, being connected with it by suitable joints.

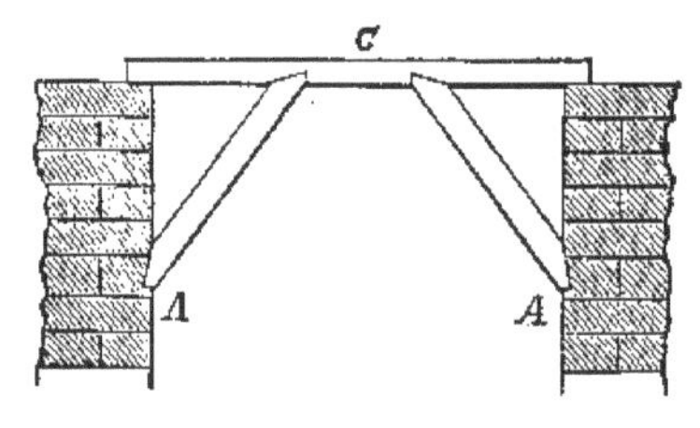

FIG. 65.

The bearing of the beam may be reduced by placing under it and on the points of support (Fig. 66) short pieces, termed **corbels**. These, when long, should be strengthened by struts, as shown in the figure.

In some cases the beam is strengthened by placing under

the middle portion a short piece, termed a **straining beam** (Fig. 67), which is supported by struts.

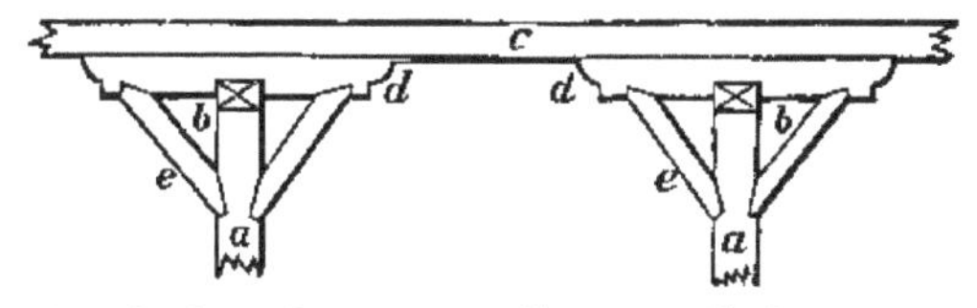

FIG. 66—A horizontal beam, *c*, resting on vertical posts, *a a*, with corbels, *d d*, and struts, *e e*.

These methods may be combined when circumstances require it, and the strains on the different parts can be determined. It is well to remember that placing equal beams over

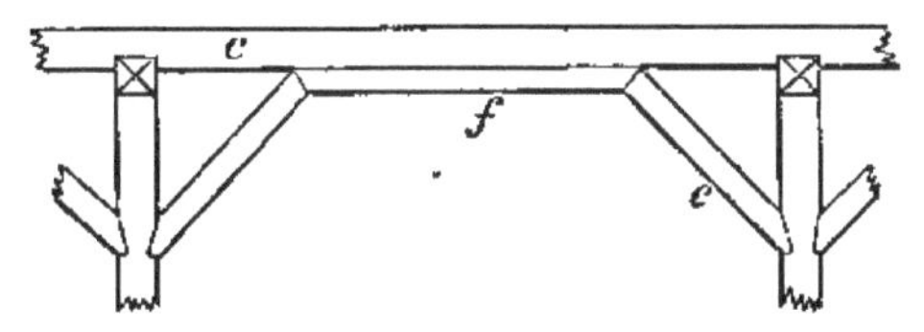

FIG. 67—A horizontal beam, *c*, strengthened by a straining beam, *f*.

each other only doubles the strength, unless they are firmly connected so as to act as one beam, in which case the combination follows the law already deduced, that is, the strength will be *four* times as great.

OPEN-BUILT BEAMS.

252. An **open-built beam**, or **truss**, is a frame in which two beams, either single or solid built, with openings between them, are connected by cross and diagonal pieces, so that the whole arrangement acts like a single beam in receiving and transmitting strains.

These frames are largely used in bridge building, and their details will be considered under that head.

The **king-post truss** is one of the simplest forms of frames belonging to this class.

This truss is employed when there are no points of support beneath the beam which can be used, but when the middle of the beam can be sustained by suspension from a point above.

The arrangement consists of two inclined pieces framed

into the extremities of the beam, and meeting at an angle above, from which the middle of the beam is supported by a third piece. This combination is shown in Fig. 68.

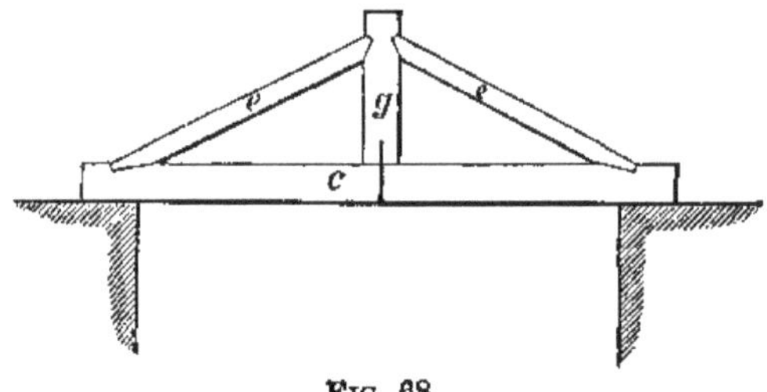

Fig. 68.

The construction is simple and the frame is rigid. It is frequently employed in roofs and in bridges of short span.

In the earlier constructions the third piece, g, was made of wood, and resembled a post, hence the name of **king-post.** The strain it sustains is one of tension, and in modern constructions an iron-rod is generally used. It would be better if a more appropriate name were given, since the term post conveys to the mind an impression that the strain is one of compression.

When the suspension piece is made of timber, it may be a single piece framed into the struts, and the foot connected with the beam by a bolt, an iron stirrup, or by a mortise and tenon joint; or it may be composed of two pieces bolted together, embracing the heads of the struts and the supported beam. In the latter case, these pieces are called **bridle-pieces,** two of which are shown in Fig. 69.

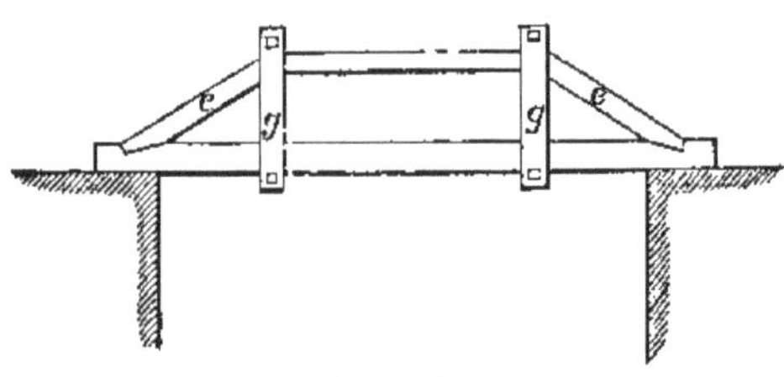

Fig. 69.

When two points of support are necessary, the arrangement known as the **queen-post truss** may be used. It consists of two struts framed into the extremities of the beam, and abutting against a short straining beam (Fig. 69). The suspension pieces are either of iron or wood, single or double, as in the king-post truss.

The remarks just made about the name "post" apply also to this combination.

Both of these trusses may be inverted, thus placing the points of support beneath the beam. This change of position changes the character of strains on the different parts, but does not affect their amount, which is determined in the same way in both cases.

Points of support above and beneath may be obtained by the use of curved beams.

METHODS OF CALCULATING STRAINS ON FRAMES.

253. It has been previously stated that to prevent a change of form in a quadrilateral frame, secondary pieces are introduced for the purpose of dividing the frame into two or more triangular figures.

In all frames where rigidity is essential to stability, this introduction of braces is necessary, as the triangle is the only geometrical figure which, subjected to a straining force, possesses the property of preserving its form unaltered as long as the lengths of its sides remain constant.

The triangular is the simplest form of frame, and will be first used in this discussion.

254. As a preliminary step, let **the strains in an inclined beam**, arising from a force acting in the plane of its axis, be determined.

For example, take

An inclined beam with the lower end resting against an abutment and the upper end against a vertical wall, and supporting a weight, W, applied at any point.

Fig. 70 represents the case.

Denote by

l, the length of the axis, A B, of the beam;

$n \times l$, the distance from A to the point C, where W is applied;

a, the angle between A B, and vertical line through C.

Disregarding the weight of the beam, the external forces acting on it are the weight, W, and the reactions at A and B.

Suppose the reaction at B to be horizontal and represent it by H. Represent the horizontal and vertical components of the reaction at A, respectively by H′ and W′.

These forces are all in the same plane, and the analytical conditions for equilibrium are

$$H - H' = 0, \text{ and } W - W' = 0.$$

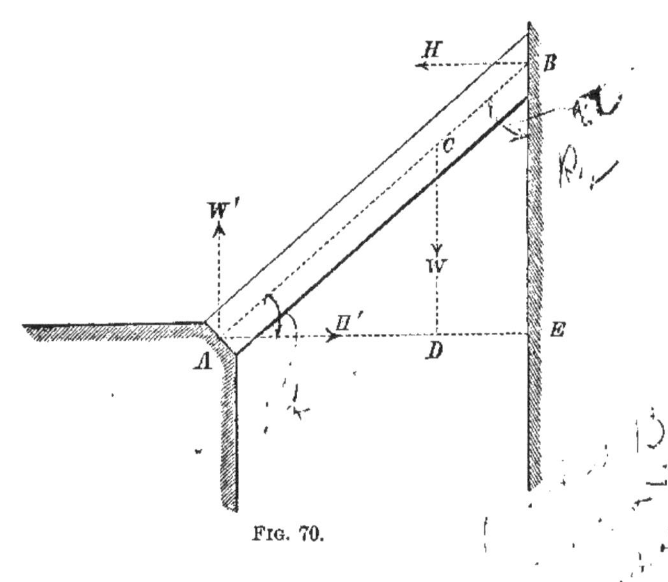

FIG. 70.

Taking the bending moment about A, we have

$$W \times AD - H \times BE = 0,$$

or, $$H \times l \cos a = W \times nl \sin a,$$

hence, $$H = nW \tan a. \quad \ldots \ldots \quad (124)$$

The forces H, H', W, and W' act in the plane of and obliquely to the axis, A B, and their effect is to produce deflection and compression of the fibres of the beam. The strain arising from deflection will be due to the algebraic sum of the perpendicular components, and that from compression will be due to the sum of the parallel ones. (Art. 217.)

Resolve W' and H' into components acting perpendicularly and parallel to the axis of the beam. Represent by P and P', and Q and Q', these components; see Fig. 71.

$$Ab = W' = W.$$
$$Ad = P, \ Ac = Q.$$
$$Ap = H' = nW \tan a$$
$$Am = P', \ An = Q'.$$

The perpendicular components Ad and Am act in opposite directions, hence the strain arising from deflection will be due to their difference, $P - P'$.

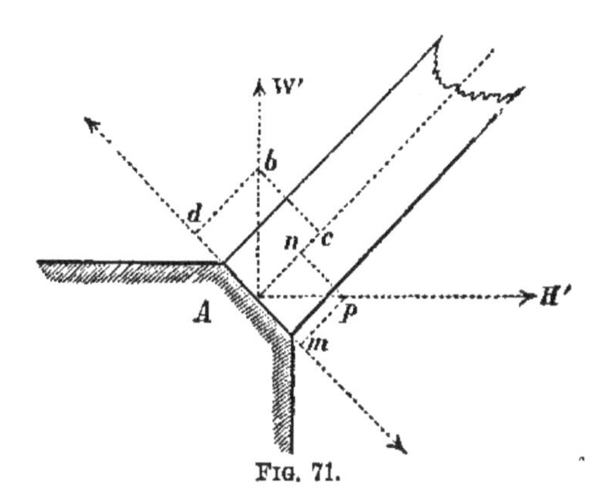

FIG. 71.

The parallel components Ac and An act in the same direction, hence the strain of compression will be due to their sum, $Q + Q'$.

Representing the force W, by the line Ab, we find the values of these components to be as follows:

$$P = W \sin a;\ P' = n\,W \tan a \cos a = n\,W \sin a;$$
$$Q = W \cos a;\ Q' = n\,W \tan a \sin a.$$

Suppose the cross-section of the beam to be a rectangle of uniform dimension, the sides of which are respectively b and d, the plane of the latter being taken parallel to the direction of the force, W, we have

$$Q + Q' = W \cos a + n\,W \tan a \sin a,$$

equal to the total compression on the segment from A to C; this sum divided by bd will be the amount of compression on the unit of area in any cross-section in this segment.

We also have

$$P - P' = (1 - n)\,W \sin a,$$

for the force perpendicular to the axis of the beam. Its moment for any section, at the distance, x, measured on the line A B, and lying between A and C, will be

$$(1 - n)\,W \sin a \times x.$$

Substituting in the expression for R′ (Art. 206), we have

$$R' = \frac{(1 - n)\, Wx \sin a}{\frac{1}{6}bd^2},$$

for the stress on the unit of area farthest from the neutral axis in any section produced by deflection, x being the lever arm.

For the segment of the beam, B C, it is seen that the strain of direct compression is due to the force

$$Q' = n\, W \tan a \sin a.$$

Giving values to n, from 0 to 1, we can place the force, W, at any point on the axis. And knowing b, d, and W, and substituting them in the foregoing expressions, we obtain the stresses in the beam.

Let us place it at the middle point, and suppose W and a to be given.

The value of n for the middle point is $\frac{1}{2}$; substituting which in the expressions for P, Q, etc., there obtains:

$$\frac{Q + Q'}{bd} = \frac{W \cos a + \frac{1}{2} W \tan a \sin a}{bd},$$

for the stress of compression on the unit of cross-section; and

$$\frac{(P - P')x}{\frac{1}{6}bd^2} = \frac{\frac{1}{2} W\, x \sin a}{\frac{1}{6}bd^2},$$

for the stress due to deflection on the unit of cross-section farthest from the neutral axis. Represent these by C′ and R′, respectively. To determine the greatest stress on the unit of area in any cross-section; first, determine R′ for the particular section and add to the value thus found that for C′, and the result will be the total stress on the unit, and hence the maximum stress in that section.

To determine the greatest stress produced by the force, W, upon the unit of surface of the beam: first, find the value of R′ for the **dangerous section** and then add to it the value of C′ for this section; the result will be the greatest stress.

Assuming limiting values for R′ and C′ and knowing b and d, the corresponding value for W can be deduced. Or, assuming R′ and C′ and having W given, we can deduce values for b and d.

Suppose the beam to be vertical, then $a = 0$, and we get

$$Q = W, \text{ and } Q' = 0,$$

or the compression in B C will be zero, and on A C equal to W. We also have $H' = 0$, or there is no horizontal thrust.

Suppose the beam horizontal, then $a = 90°$, and we get H' and Q', each equal to infinity.

From this it is seen that the compression on the beam and the horizontal thrust at the foot both decrease as a decreases, and the reverse.

255. **Uniformly loaded.**—Suppose the beam to be uniformly loaded, and let w be the load on a unit of length of the beam.

We have $$H = \tfrac{1}{2}wl \tan a.$$

The corresponding values for P, P′, Q, and Q′ are easily obtained.

256. Let it be required to determine **the strains on a triangular frame**, and take for example,

A frame made of three beams connected at the ends by proper joints and strained by a force acting in the plane of their axes and at one of the angular points.

Suppose the plane of the axes of the three beams to be vertical, and one of the sides, B C, to be horizontal, resting on fixed points of support at B and C.

Disregarding the weight of the frame itself, suppose the straining force to be a weight suspended from or resting on the point A. (Fig. 72.)

Represent by

W, the weight acting at A,
a, the angle B A D,
β, " " C A D.

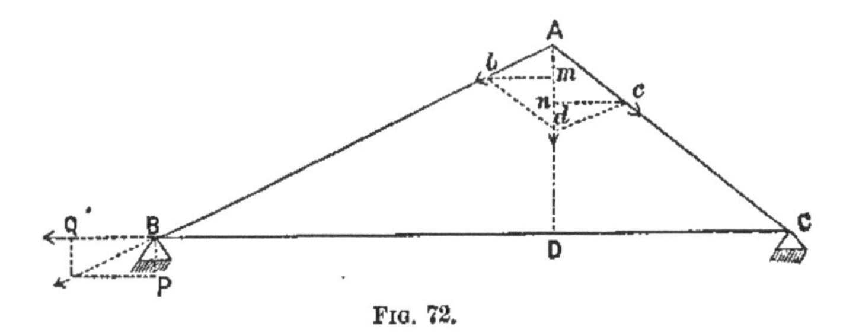

FIG. 72.

The weight, W, acts vertically downwards and is prevented from falling by the support at A. The pressure exerted by it at A is received by the inclined beams, A B, and A C, and is transmitted by them to the fixed points of support at B and C.

The weight, W, is therefore the resultant force acting on the frame, and the pressure on the inclined beams are its components in the directions of the axes of the beams.

Represent by A*d* the weight W, and construct the parallelogram A*bcd*. We have from the principle of the parallelogram of forces:

$$Ab = \frac{W \sin \beta}{\sin(a + \beta)} \text{ and } Ac = \frac{W \sin a}{\sin(a + \beta)}. \quad (125)$$

The strains produced by these components are compressive. Knowing the breadth and depth of the beams, the amount of stress on the unit of cross-section can be determined; or assuming a limit for this stress on the unit, the values for the breadth and depth of the beams may be deduced.

These components being transmitted along the axes of the beams to the points of support, B and C, may be resolved at these points into their horizontal and vertical components respectively.

Doing so, it is seen that the horizontal components are equal to *bm* and *cn*, and are equal to each other, but act in opposite directions. The value for these components is

$$bm = cn = W \frac{\sin a \sin \beta}{\sin(a+\beta)}. \quad . \quad . \quad (126)$$

Hence, they balance each other, producing a strain of extension on the beam, B C, the amount of which on the unit of cross-section, or dimensions of beam to resist which, may be determined. The vertical components are respectively equal to A*m* and A*n*, and act in the same direction. We have

$$Am = W \frac{\sin \beta \cos a}{\sin(a + \beta)}, \text{ and } An = W \frac{\sin a \cos \beta}{\sin(a + \beta)}. \quad (127)$$

They are resisted by the reactions at the points of support, which must be strong enough to sustain these vertical pressures. Adding A*m* to A*n* we find their sum is equal to W. It is well to observe that producing A*d* to D, we have the proportion, A*m* : A*n* :: C D : B D. That is, the vertical through A divides the side B C into two segments proportional to the vertical components acting at B and C.

257. The **common roof-truss**, in which A B is equal in length to A C, and the angle a equal to β, is the most usual form of the triangular frame.

For this case we would have

$$Ab = Ac = \frac{1}{2}\frac{W}{\cos a}, bm = \frac{1}{2}W \tan a, \text{and } Am = An = \frac{1}{2}W.$$

Represent by $2l$ the length of BC, d, the length of AD and h, the length of AB = AC, and substituting in the foregoing expression, we have

$$Ab = Ac = \frac{1}{2}W\frac{h}{d}, \text{and } bm = cn = \frac{1}{2}W\frac{l}{d},$$

which are fully given for any assumed value for W when either two of the quantities in the second members are known.

If, instead of a single weight, the frame had been strained by a uniform load distributed over the inclined pieces AB and AC, we may suppose the whole load to be divided into two equal parts, one acting at the middle point of AB and the other at the middle point of AC, the discussion of which would have been similar to that of the previous article.

If the frame be inverted (Fig. 73) the method of calculating the strains will be the same. Under this supposition the

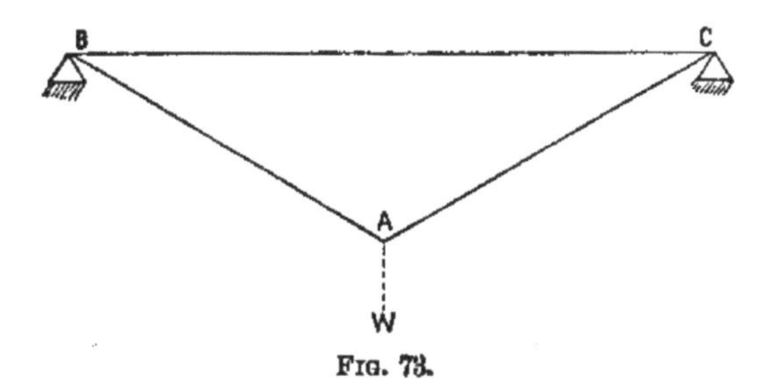

FIG. 73.

strains in the inclined pieces will be tensile instead of compressive, and in the horizontal piece BC will be compressive instead of tensile, the expression for the intensities remaining the same.

258. **The jib-crane.**—The machine known as the jib-crane, which is used for raising and lowering weights, is an example of a triangular frame. Its principal parts are a vertical post, BC; a strut, AC; and an arm or tie-bar, AB. (Fig. 74.)

Ordinarily, the whole frame allows a motion of rotation around the vertical axis, BC.

The weight, W, suspended from the frame at A is kept from falling by resistances acting in the directions A B and A C. There being an equilibrium of forces at A, the resultant, W, and the direction of the resistances being known, the intensities of these resistances are easily determined.

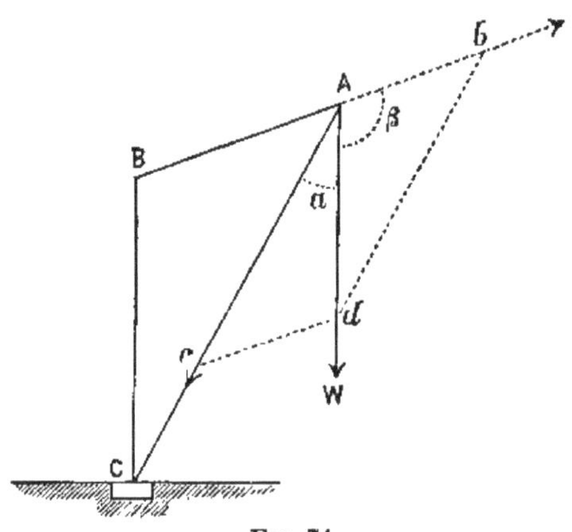

FIG. 74.

Represent W by A*d*, and construct the parallelogram A*bdc*. A*b* and A*c* will represent the intensities of the forces acting to keep W from falling.

From the parallelogram we have

$$Ac = W \frac{\sin \beta}{\sin (\alpha + \beta)}, \quad . \quad . \quad . \quad (128)$$

which, as it is seen, produces compression on the strut A C, and a transverse shearing strain at C on the post C B. The horizontal component of A C divided by the area of cross-section of the post B C, gives the shearing stress on the unit of cross-section at C.

We also have $$Ab = W \frac{\sin \alpha}{\sin (\alpha + \beta)},$$

for the stress acting in the direction of A B, tending to elongate it, and to produce a cross-strain on B C. The greatest bending moment is at C. Knowing the stresses, it is a simple problem to proportion the pieces so that the crane may be able to lift a given weight, or to determine the greatest weight which a given crane may lift with safety.

COMBINED TRIANGULAR FRAMES.

259. Open-built beams constructed by connecting the upper and lower pieces by diagonal braces are examples of combinations of triangular frames.

Triangular Bracing.

260. **Triangular bracing with load at free end.**—Take a beam of this kind and suppose it placed in a horizontal position, *one end firmly fixed, the other free to move, and strained by a force acting at the free end.* Suppose the triangles formed by the braces to be equilateral (Fig. 75) and disregard the weight of the beam.

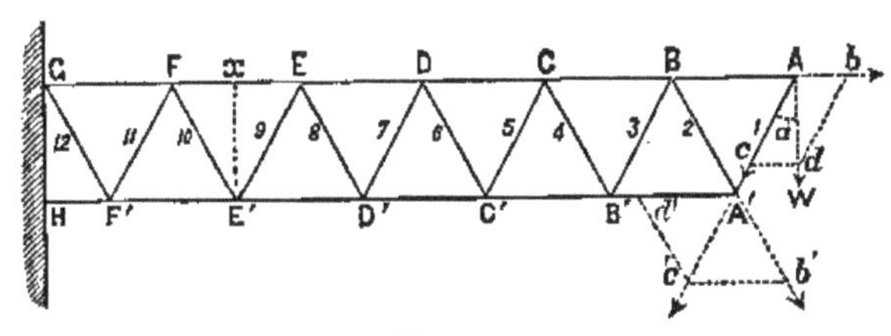

FIG. 75.

Represent by W the force acting at A, in the plane of the axes of the pieces of the frame and perpendicular to A G.

The force W acting at A is supported by the pieces A B and A A', and produces a stress of compression in A A' and tension in A B. Laying off on A W the distance Ad to represent W, and constructing the parallelogram Abcd, we have Ac and Ab representing the intensities of these stresses.

From the parallelogram there results

$$Ac = \frac{W}{\cos a}, \qquad \text{and} \qquad Ab = W \tan a.$$

The compressive force Ac is transmitted to A' and there supported by the pieces A'B and A'B'. Resolving this force at A' into its components acting in the directions of A'B and A'B', we have $A'd' = 2W \tan a$, which produces compression in A'B', and $A'b' = \frac{W}{\cos a}$, which produces tension in A'B.

This tension A' b' is transmitted by the brace to B. Resolving it into its components in the directions B B′ and B C, we have

$$\text{Compression on B B}' = \frac{W}{\cos a},$$

$$\text{Tension on B C} = 2W \tan a.$$

The tension at A is transmitted through the beam to B, hence the tension at B is equal to the sum of them, or

$$\text{Tension at B} = 2W \tan a + W \tan a = 3W \tan a.$$

Continuing this process, we find that the force W, strains all the diagonals equally, but by forces which are alternately compressive and tensile, and the expression for which is $\frac{W}{\cos a}$. In this case the braces numbered odd in the figure are compressed, and those even are extended.

The stresses in the upper and lower beams are cumulative, receiving equal increments, each equal to $2W \tan a$, at each point of junction of the brace with the beam. Hence, in this case, for the upper beam we have

$W \tan a$ for A B, $3W \tan a$ for B C, $5W \tan a$ for C D, etc.,

and for the lower,

$2W \tan a$ for A′B′, $4W \tan a$ for B′C′, $6W \tan a$ for C′D′, etc.

Having determined the stresses in the different parts of the frame produced by a force W, it is easy to find the greatest weight that such a frame will support, or to proportion its different parts to resist the strains produced by a given load.

The triangles taken were equilateral. If we denote by d the altitude E′x of one of these triangles, or depth of the beam; by l, the length of one of the sides F E, or distance between the vertices of two adjacent triangles, which we will call a **bay**; and express the values of $\cos a$ and $\tan a$ in terms of these; then we have $\cos a = \frac{d}{l}$, and $\tan a = \frac{l}{2d}$. Substituting which in the foregoing expressions, there obtains $\frac{l}{d}W$ for the stress in the diagonal, and, $\frac{l}{d}W$ for the increment to be added at each point of junction.

To find the stress in any segment; as, for example, E F.

The tension on A B is $W \tan a = \frac{l}{2d}W$, to which add four equal increments, there being four bays between A and the segment E F, and we have, for the tensile stress in E F,

$$9W \tan a, \text{ or its equal } \frac{9l}{2d}W.$$

261. Triangular Bracing Strained by a Uniform Load. —Suppose the strains on the same beam to be caused by a weight uniformly distributed over either the upper or lower beam of the frame.

Let A E F A′ (Fig 76) be an open-built beam supporting a load *uniformly distributed over the upper beam* A E.

Denote by w the weight distributed over any one segment.

We may, without material error, suppose the whole load divided into a number of equal parts, each equal to that resting on the adjacent half segments, acting at the points A, B, C, etc., where the braces are connected with the beam, A E.

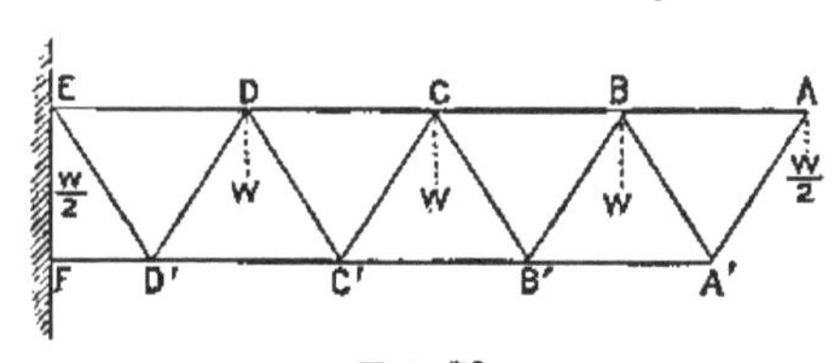

FIG. 76.

Since there are four of these bays, the total load is $4w$, the action of which may be considered to be the same as that produced by the weight w acting at each of the points B, C, and D, and $\frac{1}{2}w$ at A and E.

The strains on A B, A A′, A′B, and A′B′ are due to the weight $\frac{w}{2}$ acting at A, and are determined as in the preceding case.

The strains on B C, B B′, B′C, and B′C′ are due to the action of the weight w acting at B, increased by the strains due to $\frac{w}{2}$ acting at A.

The strains on the remaining parts are due to the weight acting at each vertex, increased by those transmitted from the points to the right of them.

Hence it is seen that the stresses in each of the pieces in any pair of diagonals are equal in amount, but different in kind, and increase as they go from the point of application to the points of support for each set; and that the stresses in the segments of the upper ond lower beams increase in the same direction. The rate of increase can be easily determined.

13

METHOD OF SECTIONS.

262. The stresses in the different pieces of a frame may be obtained by using the principle of moments, or, as it is usually called, the "method of sections." This method consists in supposing the frame to be divided by a section cutting not more than *three* pieces of the frame, and taking the intersection of two of these pieces as a centre of moments.

It is evident that the stresses in the two pieces passing through the centre of moments will have no moments to oppose those of the extraneous forces acting to turn the frame around the assumed centre, and that these external moments must be held in equilibrium by the moment of the stress in the third piece. If the moment of the stress in the third piece, with respect to the assumed centre, be placed equal to the bending moment of the extraneous forces with respect to the same point, an equation will be found that must be true for equilibrium, and which, when solved, will give the intensity of the stress in the third piece whenever the position of this piece and the bending moments are known.

Let it be required to find by this method the stress in the segment E F (Fig. 75).

Intersect the frame by a vertical plane perpendicular to the axis between x and E, and let T′ be the stress in the piece E F. This plane will cut the pieces E F, E E′, and E′ D′, and no others. Assume E′ to be the centre of moments.

The resultant of the stress T′ is supposed to act along the axis of the piece F E. Its moment with respect to E′ will be $T' \times E'x$.

Since there is an equilibrium,

$$T' \times E'x = W \times Ax, \text{ or, } T' \times d = W \times 4\tfrac{1}{2}l; \text{ hence}$$

$$T' = 4\tfrac{1}{2}\frac{l}{d}\,W, \text{ the same value before deduced.}$$

In a similar manner, assuming E as a centre, the intensity of the stress in E′ D′ may be obtained.

This method, in many cases, is a convenient one and its use is simply a matter of choice.

Vertical and Diagonal Bracing.

263. Suppose the triangles, instead of being equilateral, to be right-angled, as in Fig. 77, and the beam strained by a load, W, as in the preceding case.

The stresses in the upper and lower beams would be respectively tensile and compressive, and cumulative as in the preceding case.

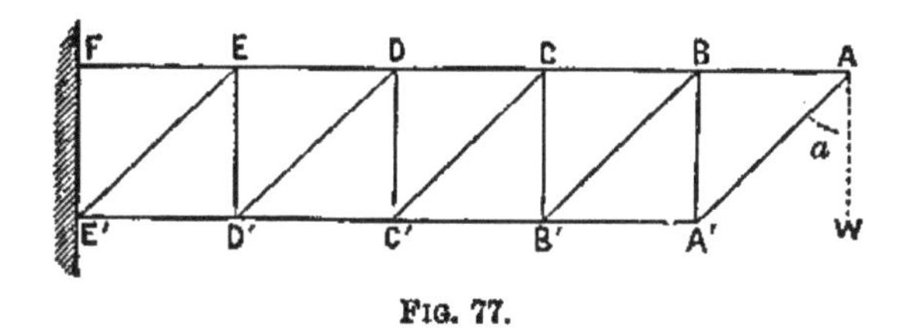

FIG. 77.

The expression for the equal increment would be

$$W \tan a.$$

The force acting on the diagonals woud be compressive and equal to

$$\frac{W}{\cos a}, \text{ same as in preceding case.}$$

The stress in the verticals would be tensile and equal to W for each.

Representing by

h, the length of a diagonal, A A',

l, the length of a segment, A B,

d, the length of a vertical, A'B, we can write

$$W \tan a = W \frac{l}{d}, \text{ and}$$

$$\frac{W}{\cos a} = W \frac{h}{d} = W \frac{\sqrt{l^2 + d^2}}{d}, \quad . \quad . \quad (129)$$

expressions more frequently used when calculating the stresses than the expressions involving the circular functions.

If, in the preceding cases, W had acted in the opposite direction, that is, pushed the point A upward instead of pulling it down, or the same thing, the frame had been turned over so that the upper beam became the lower, the stresses would have been determined in the same manner with similar results, excepting that the inclined pieces would have been extended instead of compressed, and the verticals compressed instead of extended.

ANGLE OF ECONOMY.

264. It has been shown that the stress on the unit of cross-section of a brace, strained by a force as W (Fig. 77) varies with the angle made by the brace with the straining force.

It is plain that of two braces of the same material, for the same stress on the unit and the same span, the more economical brace will be the one that contains the less amount of material; or, for the same stress and the same amount of material, the one that gives the wider span.

Suppose the stress on the unit of cross-section and the span to be fixed, it is required *to find the angle that a brace shall make with the straining force so that the amount of material in the brace shall be a minimum.*

Let B C be the fixed span (Fig. 78) and 2W the intensity of the straining force acting vertically to be transmitted by braces to the points B and C considered as fixed. Let $h =$ the length of A B = A C, $2l =$ the length of B C, and $d =$ the distance A D.

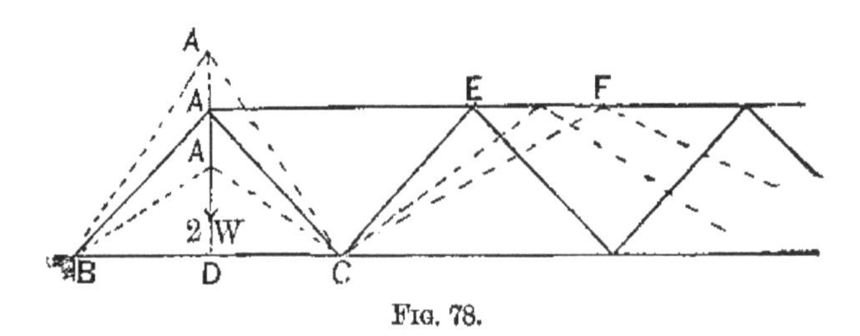

Fig. 78.

The straining force produces a compressive stress in each brace equal to $W\frac{h}{d}$.

Suppose the resistance offered by the brace to vary directly with the area of its cross-section (Art. 164) and let b^2 be the area of cross-section, and C′, the assumed compressive stress allowed on the unit. We can then form the following equation:

$$W\frac{h}{d} = b^2 \times C', \quad . \quad . \quad . \quad . \quad (130)$$

from which we obtain

$$b^2 = \frac{W}{C'} \times \frac{h}{d},$$

and

$$b^2h = \frac{W}{C'} \times \frac{h^2}{d}, \text{ for the volume of the brace.}$$

Substituting $d^2 + l^2$ in this expression for h^2, we have

$$\text{Volume of brace} = \frac{W}{C'} \times \frac{d^2 + l^2}{d}. \quad . \quad . \quad . \quad (131)$$

The value of $d = l$ makes this function a minimum. Hence, it is seen that the volume of the brace is a minimum when the angle which it makes with the straining force is equal to 45°. This angle is called "the angle of economy" of the brace.

In this discussion, the length of the bay or span has been fixed. A similar result would have been obtained if d, the depth of the truss, had been fixed and the length of the bay B C determined.

The resistance in a tie to tension varies directly with the area of cross-section, however long the piece may be, and therefore the angle above obtained is the true angle of economy for ties in all cases. This is not true for struts, as experiment has shown (Art. 202) that when the diameter is small in comparison to its length, the resistance to compression becomes also a function of its length, which latter dimension must be duly considered.

The angle of economy for a strut when its length exceeds its diameter more than fifteen or thirty times can be determined by taking the formulas deduced from Hodgkinson's experiments for finding the strength of pillars, and following the steps just described.

Merrill, in his "Iron Truss Bridges," gives the angle of economy for a cast-iron strut in a triangular frame at 27° 51′, or the depth of the frame to be a little greater than one-fourth of the span. In diagonal bracing with vertical ties (Art. 236) he gives the angle of economy for the struts to be 39° 49′ with the vertical.

PART IV.

MASONRY.

CHAPTER IX.

265. **Masonry** is the art of erecting structures in stone, brick, and mortar.

It is classified, from the nature of the material used, into **stone, brick,** and **mixed** masonry; from the manner in which the material is prepared, into **cutstone, ashlar, rubble,** and **hammered** masonry; and from the mode of laying the blocks, into **irregular** and **regular** masonry.

MASONRY STRUCTURES.

266. **Masonry structures** are divided into classes according to the kind of strains they are to sustain. Their forms and dimensions are determined by the amount and kind of strains they are required to resist. They may be classed as follows:

1st. Those which sustain only their own weight; as walls of enclosures.

2d. Those which, besides their own weight, are required to support a vertical pressure arising from a weight placed upon them; as the walls of a building, piers of arches, etc.

3d. Those which, besides their own weight, are required to resist a lateral thrust; as a wall supporting an embankment, reservoir walls, etc.

4th. Those which, sustaining a vertical pressure, are subjected to a transverse strain; as lintels, areas, etc.

5th. Those which are required to transmit the pressure they directly receive to lateral points of support; as arches.

WALLS.

267. **Definitions.**—In a wall of masonry the front is called the **face**; the inside or side opposite, the **back**; the layer of stones which forms the front is called the **facing**, and that of the back, the **backing**; the portion between these, forming the interior of the wall, the **filling**.

If a uniform slope is given to the face or back, this slope is termed the **batter**.

The section made by a vertical plane passed perpendicular to the face of the wall is called the **profile**.

Each horizontal layer of stone in the wall is called **a course**; the upper surface of the stone in each course, the **bed** or **build**; and the surfaces of contact of two adjacent stones, the **joints**.

When the stones of each layer are of equal thickness throughout, the term **regular coursing** is applied; if unequal, **irregular** or **random coursing**. The particular arrangement of the different stones of each course, or of contiguous courses, is called **the bond**.

Walls.—The simplest forms of walls are those generally used to form an inclosing fence around a given area, or to form the upright inclosing parts of a building or room.

RETAINING WALLS.

268. **A retaining wall** is the term used to designate a wall built to support a mass of earth in a vertical position, or one nearly so. The term **sustaining** is sometimes applied to the same case. In military engineering, the term **revetment** wall is frequently used to designate the same structure.

The earth sustained by a retaining wall is usually deposited behind and against the back after the wall is built. If the wall is built against the earth in its undisturbed position, as the side of an excavation or cutting, it is called a **face-wall**, and sometimes **breast-wall.**

Reservoir walls and dams are special cases of retaining walls, where the material to be supported is water instead of earth.

Counterforts are projections from the back of a retaining wall, and are added to increase its strength. The projections from the face or the side opposite to the thrust are called **buttresses.**

AREAS, LINTELS, AND PLATE-BANDS.

269. The term **area** is applied to a mass of masonry, usually of uniform thickness, laid over the ground enclosed by the foundations of walls.

The term **lintel** is applied to a single stone, spanning an interval in a wall; as over the opening for a window, door, etc.

The term **plate-band** is applied to the lintel when it is composed of several pieces. The pieces have the form of truncated wedges, and the whole combination possesses the outward appearance of an arch whose under surface is plane instead of being curved.

ARCHES.

270. **An arch** is a combination of wedge-shaped blocks, called **voussoirs** or arch-stones, supporting each other by their mutual pressures, the combination being supported at the two ends. (Fig. 79.)

These blocks are truncated towards the angle of the wedges by a **curved surface**, generally normal to the joints between the blocks.

The supports against which the extreme voussoirs rest are generally built of masonry.

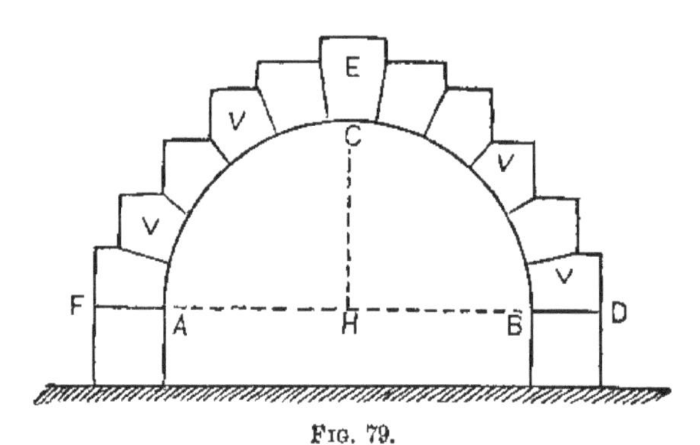

Fig. 79.

If this mass of masonry, or other material, supports two successive arches it is called a **pier**; if the pier be strong enough to withstand the thrust arising from either of the arches alone, it is called an **abutment pier**; the extreme piers which support an embankment, generally of earth, on one side, and an arch on the other, are called **abutments.**

The inner surface of the arch is called the **soffit**; its outer surface, the **back.** The sides of the arch are called **reins**; the end surface, the **face**, and sometimes the **head** of the arch. The connection of the arch with the pier is called the **impost**; if the top surface of a pier is sloped to receive the end of the arch, this surface is called a **skewback.**

The highest stones of a pier, or the stones on which an arch rests, are called **cushion stones**; the highest stone of the arch is called the **keystone.**

The line in which the soffit of the arch intersects the pier is called the **springing line.** The line of intersection of the face of the arch with the soffit is the **intrados**; with the back of the arch, the **extrados.** The chord, A B (Fig. 79) is termed the **span**, and the height, H C, of the keystone above this line, is termed the **rise.** The **length** of the arch is that of the springing line. The highest line of the soffit, that projected at C, is called the **crown.** The line in the plane of the springing lines projected at H, symmetrically disposed with respect to the plan of the soffit on that plane, is the **axis** of the arch. The courses of stones parallel to the head of the arch are called **ring-courses.** The courses which run lengthwise of the arch are termed **string-courses.** The joints between the different ring-courses are called **heading joints.** Those between the different string-courses are termed **coursing** or **bed-joints.**

A wall standing on an arch and parallel to the head is called a **spandrel-wall.**

271. **Classification.**—Arches may be classified according to the direction of the axis with respect to a vertical or horizontal plane, or according to the form of the soffit.

A **right** arch is one whose axis is perpendicular to the heads. The arch is called **oblique or askew,** when the axis is oblique to the heads; and **rampant,** when the axis is oblique to the horizontal plane.

Arches are termed **cylindrical, conical, warped,** etc., according as the soffit is cylindrical, conical, etc.

272. **The cylindrical arch.**—The cylindrical is the most usual and the simplest form of the arch. A section taken at right angles to the axis is called **a right section.**

These arches are classified according to the shape of the curve cut out of the soffit by the plane of right section.

If the curve be a semicircle, the arch is called a **full centre** arch; if a portion of a semicircle, a **segmental** arch.

When the section gives a semi-ellipse, the arch is called an **elliptical** arch; if the curve resembles a semi-ellipse, but is composed of arcs of circles tangent to each other, the term **oval** of three, five, etc., centres, according to the number of arcs used, is applied to designate it.

273. **Groined** and **Cloistered Arches.**—The intersection of cylindrical arches having their axes in the same plane, and having the same rise, form the arches known as groined and cloistered.

The **groined** arch (Fig. 80) is made by *removing* from each cylindrical arch those portions of itself which lie within the corresponding parts of the other arch; in this way, the two soffits are so connected that the two arches open freely into each other.

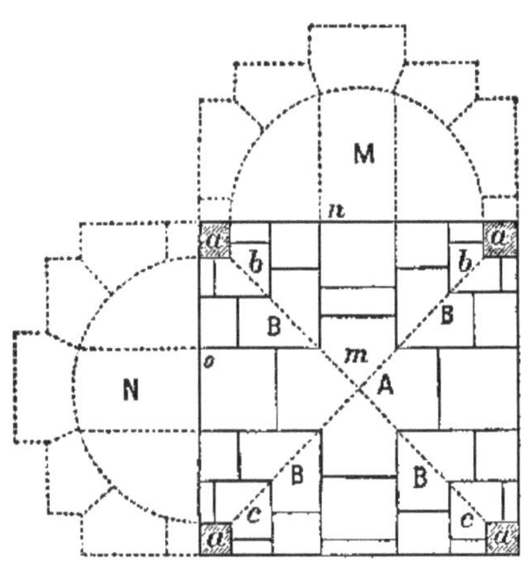

FIG. 80—Represents the plan of the soffit and the right sections M and N of the cylinders forming a groined arch.
aa, pillars supporting the arch.
bc, groins of the soffit.
om. *mn*, edges of coursing joint.
A, key-stone of the two arches formed of one block.
B, B, groin stones, each of one piece, situated below the key-stone, and forming a part of each arch.

The curves of intersection of the soffits form the edges of salient angles and are termed **groins**, hence the name of the arch.

The **cloistered** arch (Fig. 81) is made by *retaining* in each cylindrical arch only those portions of itself which lie within the corresponding portions of the other arch; thus, a portion of the soffit of each arch is enclosed within the other, these portions forming a four-sided vaulted ceiling.

Fig. 81—Represents a horizontal section through the walls supporting the arch and plan of the soffit of a cloistered arch.
B, B, the walls of the enclosure or abutments of the arches.
ab, curves of intersection of the soffits.
c, *c*, groin stones.

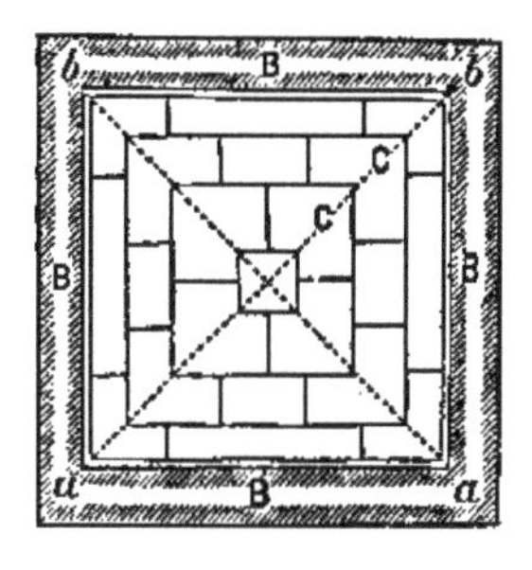

This arch was much used in forming the ceilings of the cells of monasteries; from their object and use is derived the term **cloistered.**

274. **Annular arches.**—An annular arch is one that may be generated by revolving the right section of an arch about a line lying in the plane of the section, but not intersect-

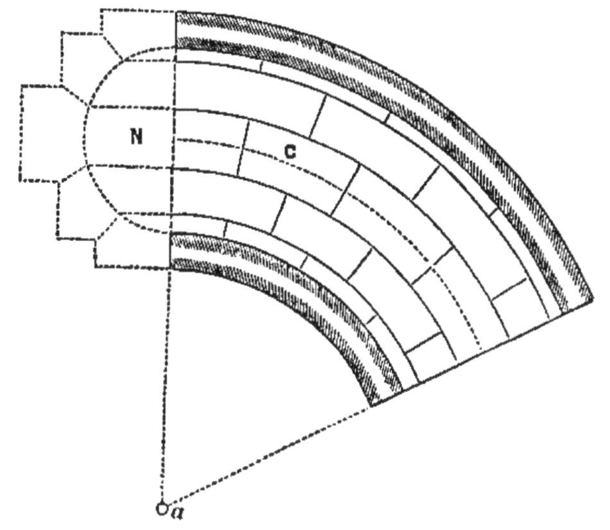

Fig. 82.—N, right section of an annular arch.
C, plan of soffit.

ing it. This line is usually vertical and also perpendicular to the span of the arch. (Fig. 82.) The axis is curved being described by the centre of the curve of right section. The coursing joints are conical, and the heading joints are plane surfaces.

275. **Domes.**—An arch whose soffit is the surface of a hemisphere, the half of a spheroid, or other similar surface, is called a **dome**. The soffit may be generated by revolving the curve of right section about the rise for 360°, or about the span for 180°. In the first case the horizontal section at the springing lines is a circle, in the other it is the generating curve.

The plan may be any regular figure. Fig. 83 represents a plan and vertical section of a circular dome.

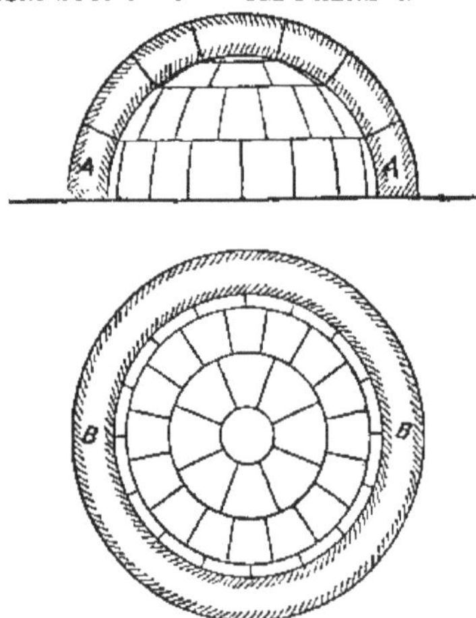

FIG. 83.—A, vertical section and elevation of a circular dome. B, B, horizontal section and plan of its soffit.

276. **Conical arches.**—Their name explains their construction. They are but rarely used, in consequence of the varying sizes of the voussoirs.

277. **Arches with warped soffits.**—Arches, whose soffits are warped surfaces, are frequently used. The particular kind of warped surface will depend upon circumstances.

A common example of this class is an arch which has the same rise at the heads but unequal spans. The soffit in this case may be generated by moving a straight line so as to continually touch the curves of section of the soffit at the heads, and at the same time to remain parallel to the plane of the springing lines. A surface generated in this manner belongs to the class of warped surfaces having a plane directer. In particular cases it is a conoid, hence the name of **conoidal arches** is frequently applied to this kind.

Arches whose soffits may be thus generated possess the advantage of having straight lines for the edges of the joints running lengthwise in the soffit.

278. **Oblique or askew arches.**—An arch whose axis makes an angle with the head is called **oblique** or **askew.** In arches of this kind the chord of the arc of the head is the span. The angle of **obliquity** is the angle which the axis makes with a normal to the head.

MECHANICS OF MASONRY.

DISTRIBUTION OF PRESSURE.

279. The surface on which a structure rests is required to support the weight of the structure, and also the load it carries, or the thrust it may have to resist. It is necessary for stability that the resultant line of these pressures should pierce this surface within the limits of the base of the structure, and that all the forces acting within this area be compressive. The point in which this resultant pierces the surface is known as the **centre of pressure.**

Structures generally rest upon plane surfaces and the portion pressed is usually a simple plane figure. Since the pressure on this surface may vary from point to point, it is necessary to determine what the pressure is at any point of the surface, and to find the limits within which the centre of pressure must be to have all the forces acting upon the surface compressive.

280. **Normal pressure.**—Suppose a series of blocks, of the form of rectangular parallelopipedons with equal bases, but (Fig. 84) whose altitudes increase in arithmetical progression, be placed side by side on a given plane area, A B C D. It is evident that the pressure on the area A B C D, is less on that part under block 1 than it is on the part under block 5, and that the pressure on any part, as B C 5, will be directly proportional to the altitude of the block resting upon it.

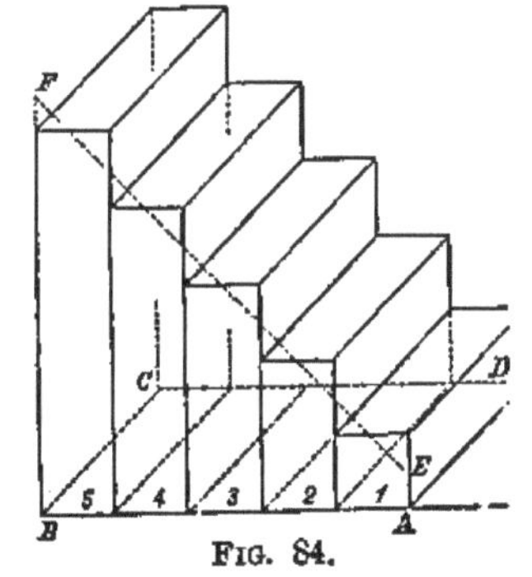

Fig. 84.

If these blocks be very thin, that is, the width of the bases measured in the direction of A B be infinitely small, and have altitudes that reach to the line E F drawn through the middle points of the upper sides of blocks 1, 2, 3, 4 and 5, the total pressure on the area A B C D will be the same as that produced by the five blocks. The pressure on the units of this area will not, however, be the same, being different for the two cases for most of them.

The pressure on each line of the surface parallel to B C, caused by the thin blocks, is directly proportional to the corresponding ordinate of the trapezoid B F E A, and the centre of pressure of each block will be found on the surface A B C D directly under the centre of gravity of the block. The centre of pressure of the entire mass will be found on the surface directly under the centre of gravity of the trapezoid forming the middle section of the thin blocks.

281. **Uniform pressure.**—If the blocks were all of the same size and of the same material, the pressure on a unit of

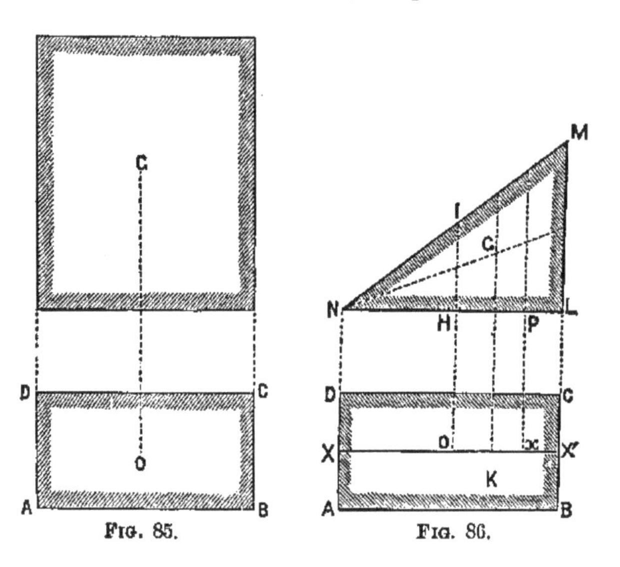

FIG. 85. FIG. 86.

area would be the same for every point pressed by it, and the centre of pressure would be directly under the centre of the base. Assuming the form of the base of a structure to be rectangular, the system of forces acting to produce a pressure that is uniformly distributed over the surface pressed may be represented by a rectangular parallelopipedon of homogeneous density, of which the rectangle is the base.

Suppose a rectangular surface, as A B C D (Fig. 85), to be pressed by such a system of forces, and P to be the resultant.

The centre of pressure would be at the centre, O, of the rectangle, and the pressure on each unit of area would be $\frac{P}{A}$.

282. **Uniformly varying pressure.**—Suppose the pressure to be zero along the line A D (Fig. 84), and to increase uniformly toward B C, along which the pressure is equal to B F. The system of forces producing this pressure may be represented by a wedge-shaped mass of homogeneous density, as shown in Fig. 86. The centre of pressure of any section parallel to A B, is below its centre of gravity and to the right of the middle point of its base at a distance equal to one-sixth of A B. The centre of pressure of the whole mass will therefore be on the line X X', and at a distance from O equal to one-sixth of A B.

The pressures on the different lines parallel to A D vary as the ordinates of the triangle, N L M. The pressure on the unit at O, the centre of the rectangle, is equal to $\frac{P}{A}$, the mean pressure on the surface of the rectangle, P being the resultant force.

To find the pressure P' on the unit, at the distance x from O measured on X X', we have, representing the sides of the rectangle by $2a$ and $2b$,

$$\frac{P}{A} : P' :: NH : NP, \text{ or } a : a + x,$$

whence,
$$P' = \frac{P}{A}\left(\frac{x}{a} + 1\right) \quad . \quad . \quad . \quad . \quad . \quad (132)$$

283. **Uniformly varying pressure combined with uniform pressure.**—If we suppose the wedge-shaped mass of the last case placed upon the rectangular parallelopipedon of the previous case, so that the base of the wedge shall exactly coincide with the upper base of the parallelopipedon, the corresponding pressure upon the base may be represented by Fig. 87. In this case, the centre of pressure will be, as before, below the centre of gravity of the mass representing the system of forces and to the right of the centre of base, O, a

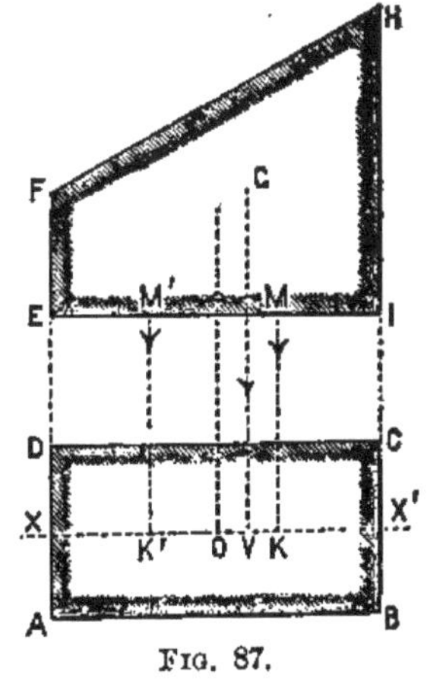

Fig. 87.

distance less than one-sixth of A B. Represent the resultant pressure by P, the distance O V by x', and divide the middle line X X′ into three equal parts, and let K and K′ be the points of division. Resolve the resultant P into two parallel components P_1 and P_2, acting at the points K and K′.

If P_1 acted alone, from what we have shown, we find the pressure upon any unit due to its action to be

$$P' = \frac{P_1}{A}\left(\frac{x}{a} + 1\right),$$

in which P' is the pressure due to P_1; in the same way the pressure P'' due to P_2 acting alone would be

$$P'' = \frac{P_2}{A}\left(-\frac{x}{a} + 1\right) = -\frac{P_2}{A}\left(\frac{x}{a} - 1\right).$$

The pressure P_x due to P will be equal to their sum, or

$$P_x = \frac{P_1}{A}\left(\frac{x}{a} + 1\right) - \frac{P_2}{A}\left(\frac{x}{a} - 1\right). \quad . \quad (133)$$

To find the value of P_1 and P_2 in terms of P, represent these parallel components as acting at M and M′. From the principle of parallel forces, we have

$$P_1 \times \frac{2a}{3} = P \times \left(\frac{a}{3} + x'\right) \text{ and}$$

$$P_2 \times \frac{2a}{3} = P \times \left(\frac{a}{3} - x'\right).$$

From which, finding the value of P_1 and P_2, and substituting in the expression for P_x, we have

$$P_x = \frac{P}{A}\left(1 + \frac{3x'x}{a^2}\right), \quad . \quad . \quad . \quad (134)$$

for the pressure on the unit of area at the distance x from the centre of the base measured on the line X X′.

284. Suppose the load, instead of being uniform along lines parallel to X X′, was uniform along lines parallel to some line making an angle with it. If we know the centre of pressure, the pressure on any unit of area of the base may be determined.

Let the centre of pressure be at any point, as V in the rectangle, and let the co-ordinates of this point be denoted by x' and y' (Fig. 88).

Through V draw a straight line V_1V_2, so that V shall be its middle point. The point V_1 would have for its abscissa $2x'$, and V_2 for its ordinate $2y'$.

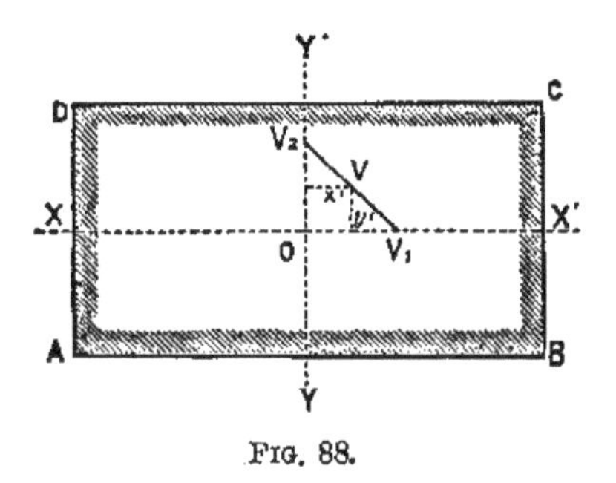

FIG. 88.

The resultant P being resolved into two parallel components acting at V_1 and V_2, these will be each equal to $\frac{P}{2}$.

From the preceding we have the pressure at any point produced by a force at V_1 to be

$$P_x = \frac{P}{2A}\left(1 + \frac{3 \times 2xx'}{a^2}\right),$$

and for that produced by the force at V_2 to be

$$P_y = \frac{P}{2A}\left(1 + \frac{3 \times 2y'y}{b^2}\right),$$

and hence the total pressure on the unit of area due to P acting at V, at the point whose co-ordinates are x and y, will be

$$P_{x,y} = \frac{P}{A}\left(1 + \frac{3x'x}{a^2} + \frac{3y'y}{b^2}\right). \quad . \quad (135)$$

The pressure at the different points of the base may be determined in a similar way when the base is a circle, ellipse, lozenge, etc.

285. **General solution.**—It is evident that there is a tendency to produce rotation about some right line in the base whenever the resultant pressure pierces the plane of the base in any point excepting its centre of figure. Regarding the base as a cross-section, this right line will be its neutral axis.

And since the condition is imposed that all the forces acting within the base shall be compressive, it is evident that this neutral axis must remain outside of, or at least tangent to, the base. If the neutral axis should intersect the base, it is plain that the portion of the base on the same side with the centre of pressure would be compressed, while the portion of the base on the other side would be subjected to a strain of extension, a condition which is not allowable.

The centre of pressure of any section is the centre of percussion of the plane area representing it. Hence, the general solution obtained from mechanics for obtaining the centres of percussion and axes of rotation for any plane figure may be applied to these cases.

The normal pressure upon the base is generally produced by a uniformly distributed load, by a uniformly varying one, or by a combination of the two, placed upon the structure. These are the cases which have been considered.

286. **Symmetrical base.**—In general the blocks used in building have a plane of symmetry, and these loads above named are symmetrically distributed with respect to this plane and to the base of the block. It follows, therefore, that the resultant pressure pierces the base in its axis or middle line.

For such cases the expression for the pressure on any point will be of the general form,

$$P_x = \frac{P}{A}\left(1 + \frac{Kx'x}{a^2}\right), \quad . \quad . \quad . \quad (136)$$

in which K is a positive coefficient depending upon the figure of the base. We have found it equal to 3 for the rectangle; we would find it equal to 4 for the ellipse or circle, and 6 for the lozenge, $2a$ being the longest diameter. Hence we conclude that the pressure is more equally distributed over a rectangular base than over a circular, elliptical, or lozenge-shaped one.

In the general expression for P_x it is seen that in the rectangle if x' is greater numerically than $\pm \frac{1}{3}a$, that the corresponding values of $x = \mp a$ give negative values for P_x. That is, there will be no pressure on the opposite edge; on the contrary, there will be tension, and the joint will open or tend to open, along this line. If $x' = \pm \frac{1}{3}a$ the values of P_x for $x = \pm a$ are 0; that is, there is no pressure on the edge. Hence, if the pressure is to be distributed over the entire base, the resultant must pierce it within the limits of $\pm \frac{1}{3}a$.

287. **Oblique pressure.**—In a large number of cases,

especially in structures of the third and fifth classes, the resultant pressure has its direction oblique to the plane of the base.

This resultant may be resolved at the centre of pressure into two components, one normal to the plane of the base and the other parallel to it. The former is the amount of force producing pressure on the base, and is to be considered as in the preceding cases. The latter does not produce pressure, but acts to slide the base along in a direction parallel to its plane. The effect of sliding will be alluded to in future articles.

MASONRY STRUCTURES OF THE FIRST AND SECOND CLASSES.

288. The strains which these structures sustain are produced by vertical forces.

For stability, the resultant pressure should pierce the plane of the base at a distance from its middle line not greater than one-sixth the thickness of the wall at its base.

The wall having to support a load, either its own weight alone, or its weight with a load placed upon it, the largest stones should be placed in the lower courses, and all the courses so arranged that they shall be perpendicular, or as nearly so as practicable, to the vertical forces acting on the wall. Great care should be taken to avoid the use of continuous vertical joints.

The thickness of the wall will depend upon the load it has to support and the manner of its construction.

STRUCTURES OF THE THIRD CLASS.

289. **Retaining walls**, besides supporting their own weight, are required to resist a lateral thrust which tends to turn them over.

Observation has shown that if we were to remove a wall or other obstacle supporting a mass of earth against any one of its faces, a portion of the embankment would tumble down, separating from the rest along a surface as B R (Fig. 89), which may be considered a plane; and that later more and more of the earth would fall, until finally a permanent slope as B S is reached.

The line B R is called **the line of rupture**, the line B S the **natural slope**, and the angle made by the natural slope with the horizontal is termed **the angle of repose.** The angle C B R is called the **angle of rupture.** If dry sand be poured out of a vessel with a spout upon a flat surface, the sand will form a conical heap, the sides of which will make

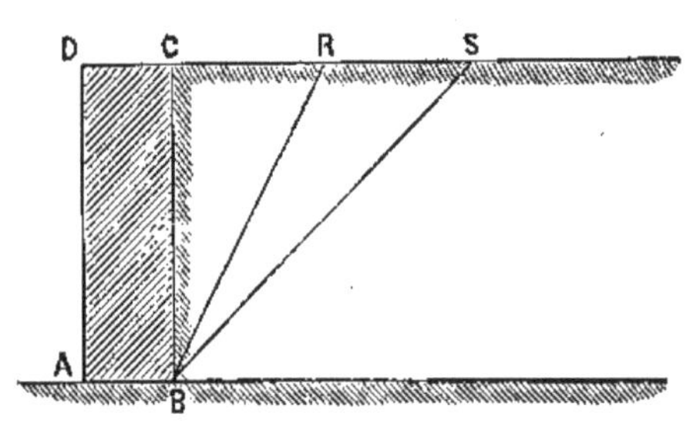

FIG. 89.

a particular angle with the horizontal, and it will be found that the steepness of this slope cannot be increased, however judiciously the sand may be poured, or however carefully it is heaped up. This slope or angle of repose varies for different earths, being as much as 55° for heavy, clayey earth, and as little as 20° for fine dry sand.

This prism of earth C B R, which would tumble down if not sustained, presses against the wall, producing a horizontal thrust, and the wall should be made strong enough to resist it.

290. Two distinct problems are presented: the *first* being to ascertain the intensity of the thrust exerted against the wall by the earth; and the *second*, to determine the dimensions of a wall of given form so as to successfully resist this thrust.

The intensity of the thrust depends upon the height of the prism, and upon the angle of rupture.

The angle of rupture, or the tendency in the earth to **slip**, is not only different for the various kinds of earth, but is different in the same earth, according as it is dry or saturated with water, being greater in the latter case.

The manner in which the earth is *filled in*, behind the wall, affects the intensity of the thrust, the latter being less when the earth is well rammed in layers inclining from the wall than when the layers slope towards it.

Therefore, in calculating the amount of resistance the wall should have, the effect produced by the maximum prism of pressure under the most unfavorable circumstances should be considered. The greatest pressure that earth can produce against the back of the wall is when the friction between its grains are destroyed, or when the earth assumes the form of mud. The pressure under these circumstances would be the same as that produced by a fluid whose specific gravity was the same as earth.

291. Retaining walls may yield by **sliding** along the base or one of the horizontal joints; by **bulging**; or by **rotation** around the exterior edge of one of the horizontal joints.

If the wall be well built and strong enough to prevent its being overturned, it will be strong enough to resist yielding by the other modes.

Hence, the formulas used in determining the thickness of a retaining wall are deduced under the supposition that the only danger to be feared is that of being overturned.

Having determined the horizontal thrust of the prism of pressure, its moment in reference to any assumed axis can be obtained.

A wall to be stable must have the moment of its weight about the axis of rotation greater than the moment of the overturning force about the same line.

The term stability in this subject differs slightly in its meaning from that previously given it. A mass is here said to be stable when it resists without sensible change of form the action of the external forces to which it is exposed—the variations produced by these forces being in the reactions of the points of support and the molecular forces of the body, and not changing in any way the form of the mass.

The excess of moment in the wall, or factor of safety, as we have heretofore designated it, will vary in almost every special case, being much greater for a wall exposed to shocks than when it has to sustain a quiescent mass; greater for a wall poorly built, or of indifferent materials, than one of better material and well constructed. The formulas which are used give results which make this factor of safety at least equal to 2, or twice as strong as strict equilibrium requires.

RETAINING WALLS, with back parallel to the face.

292. Let it be required, *to find the thickness of a retaining wall, the upper surface of the embankment being horizontal and on a level with the top of the wall.* The wall being of uniform thickness, with vertical face and back.

Denote by (Fig. 90),

H, the height BC of the wall,
b, " thickness AB of the wall,
w, " weight of a unit of volume of the earth,
w', " " " same unit of volume of masonry,
a, " angle CBS of the natural slope with the vertical BC,
β, " angle SBF of the natural slope with the horizontal.

Let it be assumed that the density and cohesion of the earth are uniform throughout the mass. The pressure exerted against the wall may then be represented by a single

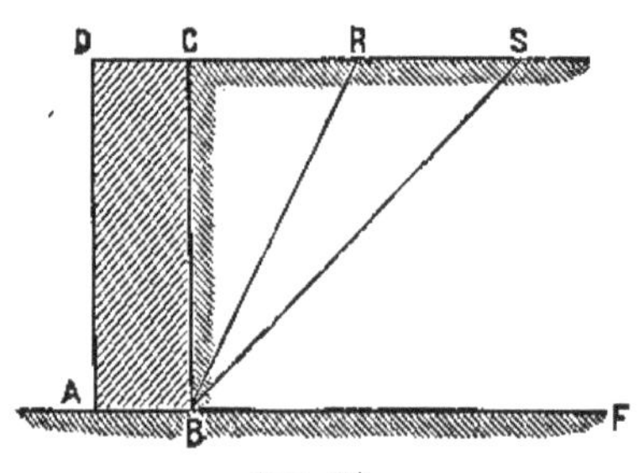

FIG. 90.

resultant force acting through the centre of pressure on the surface of the wall.

If we suppose the prism CBS to act as a solid piece, the friction along BS would be just sufficient to prevent sliding, and there would be no horizontal thrust. This is true for any prism making an angle less than β.

The horizontal thrust upon the back of the wall must therefore be due to a mass of earth, the lower surface of which makes a greater angle with the horizontal than β.

Let BR be a plane which makes an angle greater than β, and represent by ϕ the angle which it makes with the natural slope.

We may suppose two cases: one in which there is no friction existing between the prism and the plane which supports it; and the other, in which there is friction.

In the first case, the horizontal thrust would be equal to that of a fluid whose specific gravity is the same as that of the earth, or

$$\text{Hor. thrust} = \tfrac{1}{2}wH^2,$$

the centre of pressure being $\frac{2}{3}$H below C.

In the second case, the friction between the plane and prism is considered, and if we denote by P the horizontal component of the pressure acting to overthrow the wall, and neglect the adhesion and friction of the earth on the back of the wall, we have, supposing $\phi = \frac{1}{2}a$,

$$P = \tfrac{1}{2}wH^2 \tan^2 \phi \quad . \quad . \quad . \quad (137)$$

The moment of this force about the edge A will be

$$M = \tfrac{1}{2}wH^2 \tan^2 \phi \times \frac{1}{3}H.$$

The moment of the weight of the wall about the same line is

$$M = \frac{1}{2}w'Hb^2.$$

Equating these moments, we have

$$\frac{1}{2}w' Hb^2 = \frac{1}{2}w H^2 \tan^2 \phi \times \frac{1}{3}H;$$

whence,

$$b = H \tan \phi \sqrt{\frac{1}{3}\frac{w}{w'}}, \quad . \quad . \quad . \quad . \quad (138)$$

for the value of the thickness of base to give the wall to resist the pressure due to P.

It can be shown that the maximum prism of pressure will be obtained when the angle of rupture, C B R, is equal to $\frac{1}{2}(90° - \beta)$, or equal to $\frac{1}{2}a$. This has also been proved by experiment. Substituting for ϕ this value in the expression for b, and we get,

$$b = H \tan \frac{a}{2}\sqrt{\frac{1}{3}\frac{w}{w'}}. \quad . \quad . \quad . \quad . \quad (139)$$

The value for P may be put under the form,

$$P = \tfrac{1}{2}H^2 w \times \frac{1 - \sin \beta}{1 + \sin \beta}, \quad . \quad . \quad . \quad (140)$$

which is the form in which it frequently appears in other works when treating this subject.

Suppose B R to coincide with B S, then $\phi = 0$, and hence

$$P = 0,$$

a conclusion already reached.

293. General case.—The wall was assumed vertical in the preceding case. The general case would be where the back of the wall and the upper surface of the embankment were both inclined to the horizontal. Let B C (Fig. 91) be the back of the wall; C S, the upper surface of the embankment; B S, the line of natural slope; and ϕ and β represent the same angles as in preceding example. The pressure on the back of the wall is produced by some prism as C B R. The horizontal thrust produced by this prism is equal to its weight multiplied by the tan ϕ, or

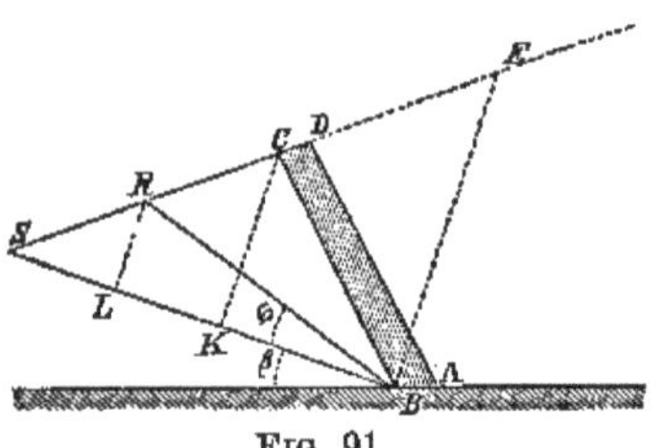

Fig. 91.

$$P = w \times \text{area C B R} \times \tan \phi.$$

Let it be required to find the maximum prism of pressure. This will be a maximum when the product of the area C B R and the tan ϕ is a maximum.

Draw through C and R perpendiculars to the line of natural slope B S. Represent the distance R L by x, the distance C K by a, and the distance B S by b.

The area C B R is equal to

$$\tfrac{1}{2}ab - \tfrac{1}{2}xb.$$

Substituting in the expression for P, we get

$$P = w \times \tfrac{1}{2}b(a - x)\tan \phi.$$

Represent the angle B S C by β', and we can write

$$P = w \times \tfrac{1}{2}b(a - x)\frac{x}{b - x\cot\beta'}.$$

This expression is in terms of a single variable x. Taking the factor $\dfrac{ax - x^2}{b - x\cot\beta'}$, and differentiating, and placing the differential coefficient equal to zero, we get

$$(b - x\cot\beta')(a - 2x) - (ax - x^2)(-\cot\beta') = 0,$$

whence

$$x^2\cot\beta' - 2bx = -ab. \quad . \quad . \quad (141)$$

This may be put under the form

$$ab - bx = bx - x^2\cot\beta' = x(b - x\cot\beta'),$$

or

$$ab - bx = x \times \text{B L}.$$

Whence,

$$\text{area C B S} - \text{area R B S} = \tfrac{1}{2}(x \times \text{B L}) = \text{area R B L},$$

and

$$\text{area R B L} = \text{area C B R},$$

or the thrust is a maximum when the area C B R is equal to the area B R L.

If C S is horizontal and B C is vertical, the triangle R B L is equal to R B C only when the line B R bisects the angle C B S. This result is the same as that of the previous case.

Substituting in the expression for P, the area R B L for the area C B R, we get

$$\text{P} = w \times \text{area R B L} \times \tan \phi.$$

Substituting for this area and for the tan ϕ, their values in terms of x, we get

$$\text{P} = \tfrac{1}{2}wx^2, \quad . \quad . \quad . \quad . \quad . \quad (142)$$

for the maximum thrust.

From equation (141) we find the value of x to be

$$x = b \tan \beta' - \sqrt{b \tan \beta' (b \tan \beta' - a)}.$$

We may write this value x under another form by drawing the line B E from B perpendicular to B S and representing it by c. We have $c = b \tan \beta'$, and substituting, we get

$$x = c - \sqrt{c(c-a)}.$$

Substituting this value of x in equation (142), we get

$$\text{P} = \tfrac{1}{2}w\left(c - \sqrt{c(c-a)}\right)^2, \quad . \quad . \quad (143)$$

for the horizontal thrust, produced by the maximum prism of pressure.

Knowing the horizontal thrust, its moment around the edge, A, can be obtained. The moment of the wall around the same line is easily found.

Equating these moments, the value of b can be deduced, giving the requisite thickness for an equilibrium.

294. These examples show the general method used to determine the thickness of retaining walls.

The specific gravity of the materials forming an embankment ranges between 1.4 and 1.9, and that of masonry between 1.7 and 2.5. The ratio of the weights $\frac{w}{w'}$, is therefore ordinarily between $\frac{2}{3}$ and 1. For common earth and ordinary masonry it is usual for discussion to assume $\frac{w}{w'} = \frac{2}{3}$, and $a =$ 45°. In practice it is recommended to measure the natural slope of the earth to be used, and to weigh carefully a given portion of the masonry and of earth, the latter being thoroughly moistened.

In military works, the upper surface of the embankment is generally above the top of the wall. The portion of the embankment above the level of the top is called the **surcharge**, and in fortifications rests partly on the top of the wall. When its height does not exceed that of the wall, the approximate thickness of the wall may be obtained by substituting, the sum of the heights of the wall and the surcharge, for H in the expression for the thickness already obtained.

The manner in which earth acts against a wall to overturn it cannot be exactly determined, hence, the thrust not being exactly known, the results obtained are only approximations. Nevertheless, a calculation right within certain limits is better than a guess, and its use will prevent serious mistakes being made.

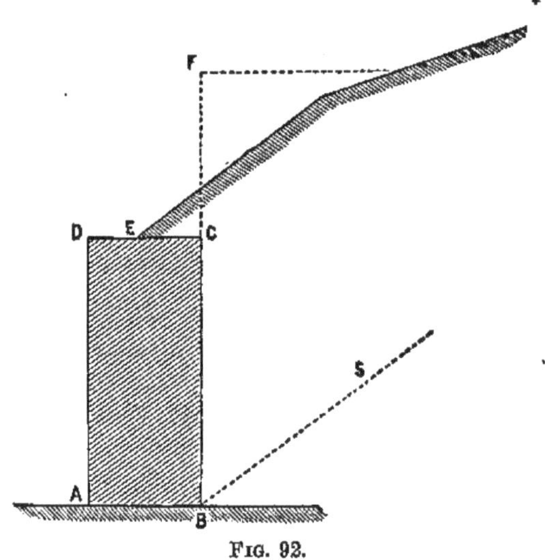

Fig. 92.

In our discussion the cohesion of the particles of earth to each other and their friction on the back of the wall have been disregarded. The results therefore give a greater thickness than is necessary for strict equilibrium, and hence errs on the side of stability.

295. Among the many solutions of this problem, those given

by M. Poncelet, and published in No. 13 "Du Mémorial de l'Officier du Génie," are the most complete and satisfactory.

In this memoir he gives a table from which the proper thickness of a retaining wall supporting a surcharge of earth may be obtained.

The principal parts of this table giving the thickness in terms of the height, for surcharges whose heights vary between 0 and twice the height of the wall, are as follows:

Represent by (Fig. 92).

H, the height B C of the wall;
h, the mean height of C F of surcharge;
a, the angle C B S made by the vertical with line of natural slope B S.
β, the angle of natural slope with the horizontal;
f, the coefficient of friction = cotan a;
u, the distance from foot of surcharge E to D outer edge of wall;
w, weight of unit of volume of earth;
w', weight of unit of volume of masonry.

TABLE.

Value of $\frac{h}{H}$	RATIO OF HEIGHT TO THICKNESS, OR $\frac{b}{H}$									
	When $w = w'$ and				$w = \frac{2}{3} w'$ $f = 1$ $\beta = 45°$		$w = \frac{2}{3} w'$			
	$f = 0.6$ $\beta = 31°$		$f = 1.4$ $\beta = 51° 25'$				$f = 0.6$ $\beta = 31°$		$f = 1.4$ $\beta = 41° 25'$	
	$u = 0$	$u = \frac{1}{3}H$	$u = 0$	$u = \frac{1}{3}H$	$u = 0$	$u = \frac{1}{3}H$	$u = 0$	$u = \frac{1}{3}H$	$u = 0$	$u = \frac{1}{3}H$
0	0.452	0.452	0.258	0.258	0.270	0.270	0.350	0.350	0.198	0.198
0.1	0.498	0.507	0.282	0.290	0.303	0.306	0.393	0.393	0.222	0.229
0.2	0.548	0.563	0.309	0.326	0.336	0.342	0.439	0.445	0.249	0.262
0.4	0.665	0.670	0.369	0.394	0.399	0.405	0.532	0.522	0.303	0.299
0.6	0.778	0.754	0.436	0.450	0.477	0.457	0.617	0.572	0.360	0.328
0.8	0.867	0.820	0.510	0.501	0.544	0.504	0.668	0.610	0.413	0.357
1	0.930	0.873	0.571	0.546	0.605	0.540	0.707	0.636	0.457	0.384
2	1.107	1.004	0.812	0.714	0.795	0.655	0.811	0.705	0.622	0.475

The thickness obtained by using this table are nearly double that of strict equilibrium. This *factor of safety* or excess of stability is that used by Vauban in his retaining walls which have stood the test of more than a century with safety.

The formula,

$$b = 0.845\,(\mathrm{H} + h)\sqrt{\frac{w}{w'}} \times \tan\left(45^\circ - \frac{\beta}{2}\right), \quad . \quad (144)$$

will give very nearly the same values as those given in the table.

RETAINING WALLS, **face and back not parallel.**

296. To transform a wall of rectangular cross-section into one of equal stability having a batter on its face and its back vertical, the usual form of cross-section of a retaining wall, we may use the following formula of M. Poncelet,

$$b' = b + \tfrac{1}{10}\,n\,\mathrm{H}. \quad . \quad . \quad . \quad (145)$$

in which (Fig. 93) b = the thickness, B d, of wall of rectangular cross-section,

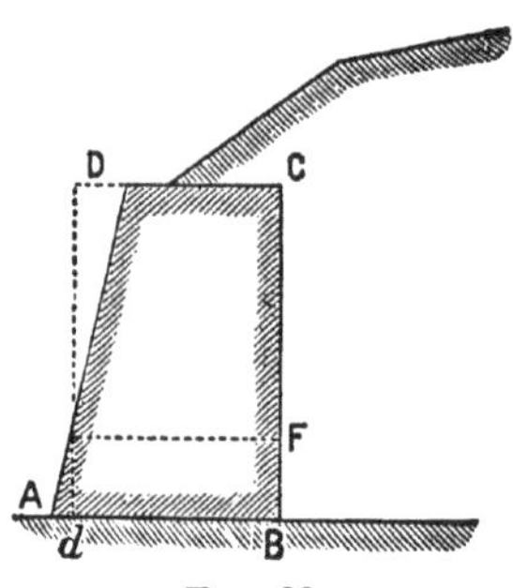

FIG. 93.

b' = the base, A B, of the equivalent wall with trapezoidal cross-section,

H = the height B C of the wall, and n = the quotient $\frac{\mathrm{A}d}{\mathrm{BF}}$.

The base of the rectangular wall for the height, H, is obtained from the previous formulas, then, knowing n, the value of b' is obtained from formula (145).

That is, the thickness of the equivalent trapezoidal wall at the base is equal to the thickness of the rectangular wall increased by one-tenth of the product obtained by multiplying the height of the wall by the quotient resulting from dividing the base of the slope by its perpendicular. This rule gives the thickness to within $\frac{1}{120}$ of the true distance for values of n less than $\frac{1}{4}$, and within $\frac{1}{80}$ for values less than $\frac{1}{6}$. Batters with a slope less than $\frac{1}{6}$ are seldom used.

297. **Counterforts.**—Counterforts are considered to give additional strength to a wall by dividing it into shorter lengths, these short lengths being less liable than longer ones to yield by bulging out or sliding along the horizontal courses; by the pressure being received on the back of the counterfort instead of on the corresponding portion of the wall, thus increasing the stability of the wall against overturning at those points; and by the filling being confined between the sides of the counterforts, the particles of the filling, especially in case of sandy material when confined laterally, becoming packed and thus relieving the back of the wall.

Counterforts are, however, of doubtful efficiency, as they increase the stability of the wall but slightly against rotation, and not at all against sliding. They certainly should not be used in treacherous foundations on account of the danger of unequal settling.

The moment of stability of a wall with counterforts may be found with sufficient accuracy for all practical purposes by adding together the moments of stability of one of the parts between two counterforts, and one of the parts augmented by a counterfort, and dividing this sum by the total length of the two parts.

Their horizontal section may be either rectangular or trapezoidal. The rectangular form gives greater stability against rotation, and costs less in construction; the trapezoidal form gives a connection between the wall and counterfort broader and therefore firmer than the rectangular, a point of some consideration where, from the character of the materials, the strength of this connection must mainly depend upon the strength of the mortar used for the masonry.

298. Counterforts have been used by military engineers chiefly for the retaining walls of fortifications. In regulating their form and dimensions, the practice of Vauban has been generally followed; this is to make the horizontal section of the counterfort trapezoidal, to make the length, *ef*, of the counterfort (Fig. 94) equal to *two-tenths of the height*

of the wall added to two feet, the front, *ab*, *one-tenth of the height added to two feet*, and the back, *cd*, equal to two-thirds of the front, *ab*.

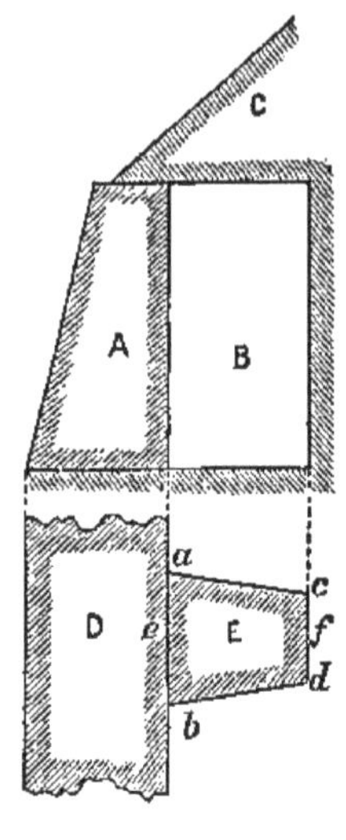

FIG. 94—Represents a section A and plan D of a wall, and an elevation B and plan E of a trapezoidal counterfort.

RESERVOIR WALLS AND DAMS.

299. These are retaining walls which are used to resist the pressure of a volume of water instead of earth, and they do not differ mathematically from the walls already discussed. Their dimensions are therefore obtained in the same way.

Their cross-section is generally trapezoidal.

Let A B C D (Fig 95) represent the cross-section of a reservoir wall, with a vertical water face B C, and let the upper surface of the water be at E F.

Represent by
h, the depth E B of the water;
h', the height B C of the wall;
b, b', the upper and lower bases A B and D C;
w, the weight of unit of volume of water;
w', the weight of unit of volume of masonry.

Lay off B H equal to one-third of B E, and draw the horizontal H. This gives the direction and point of application of the thrust on the wall produced by the pressure of the water. Its intensity is equal to $\frac{1}{2}wh^2$. The weight of the wall acts through the centre of gravity G, and is equal to $\frac{1}{2}w'h'(b + b')$. The moments around the edge at A can be determined and the values for b and b' found.

The resultant R of these pressures intersects the base A B between A and B. Stability requires that this should be so.

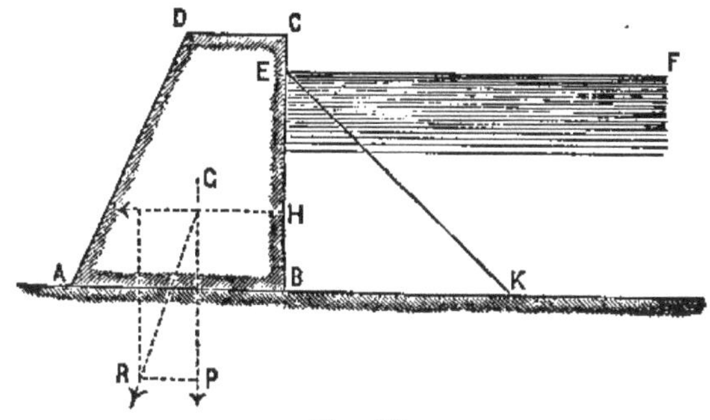

FIG. 95.

If the resistance to a crushing force were very great in the surface, A B, supporting the wall, it would make no difference how near the resultant came to the edge A. But as such is not the case, it should not come so near the edge as to produce a pressure along the latter sufficiently great to injure the resistance of the material.

The nearer the intersection is to the middle point of the base, the more nearly will the pressure on the foundation of the wall be uniformly distributed over it.

It is evident, from the figure, that the batter given to the face A D contributes greatly to the uniform distribution of the pressure. And it is easily seen that if the outer face had been made vertical, the resultant would have intersected the base much nearer to the edge A, producing a far greater pressure in that vicinity than in the former case.

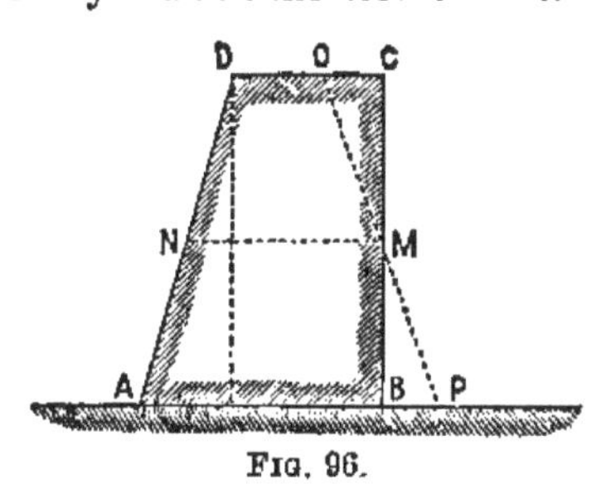

FIG. 96.

300. Reservoir walls are usually constructed with both their faces sloped Having found the thickness of the wall, as

above, the profile is easily transformed. For example, let A B C D (Fig. 96) be a cross-section of a wall in which b and b' have been determined by previous rule. Let M N be the thickness at the middle point of the inner vertical face. It is evident that if the thickness at top be diminished by O C, and that at the base be increased by the equal quantity B P, that the weight of the wall will remain the same, with an increase of stability.

STRUCTURES OF THE FOURTH CLASS.

301. Structures belonging to this class sustain a transverse strain. Since stone resists poorly a cross-strain, great caution must be used in proportioning the different parts of these structures. The rules for determining the strength of beams subjected to transverse strains can be applied.

STRUCTURES OF THE FIFTH CLASS.

302. **Arches** are the principal structures belonging to this class. They are used to transmit the pressure they directly receive to lateral points of support.

Arches are generally made symmetrical, hence the conditions of stability deduced for either half are equally applicable to the other.

303. **Modes of yielding.**—Arches may yield either by sliding along one of their joints, or by turning around an edge of a joint.

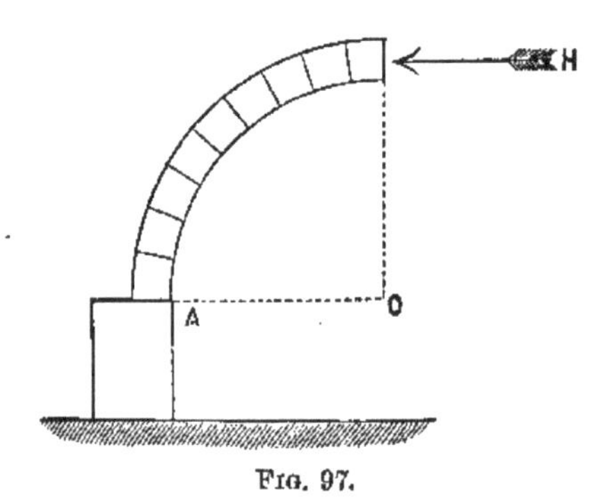

FIG. 97.

Suppose the arch to be divided into equal halves by its plane of symmetry, and let the right portion be removed

(Fig. 97). We may suppose the equilibrium preserved by substituting a horizontal force H for the half arch removed.

If the semi-arch were one single piece, the intensity of this force, H, could be easily determined, for the conditions of equilibrium would require the moment of the weight of the semi-arch around the springing line at A to be just equal to the moment of H about the same line.

The semi-arch not being a single piece, but composed of several, may separate at any of the joints, and therefore the difficulty of determining the values of H is increased.

CONDITIONS OF STABILITY to prevent sliding at the joints.

304. The resistance to sliding arises from the friction of the joints and from their adherence to the mortar.

Arches laid in hydraulic mortar, or thin arches in common mortar, may derive an increase of stability from the adhesion of the mortar to the joints, but in our calculations we should disregard this increase, and depend for stability upon the resistance due to friction alone.

It is found that friction, when the pressure is constant, is

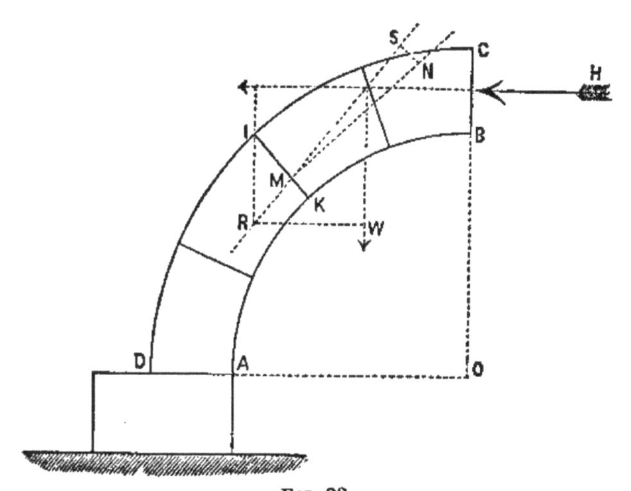

FIG. 98.

independent of the area of the surfaces in contact, and depends solely upon the nature and condition of the surfaces.

Let F be the resistance to sliding, produced by friction at any joint I K (Fig. 98). The external forces acting on this joint are the horizontal force H, and the weight of the mass K B C I. Denote by R the resultant of these forces, and construct it. This resultant pierces the plane of the joint I K at some point as M, and M N will be the normal component. Represent by P this normal component, and by S the component parallel to the joint. We have

$$F = f. P,$$

in which f is the coefficient of friction determined by experiment.

In order that sliding along this joint shall not take place, we must have

$$S < F, \text{ or } S < f. P, \quad \text{whence}$$

$$\frac{S}{P} < f.$$

But $\frac{S}{P}$ is equal to the tangent of the angle which the resultant R makes with the normal to the joint. Hence we conclude that when the angle made by the resultant of the pressures with the normal to the surface of the joint is less than the **angle of friction** of the blocks on each other, that there will be no sliding.

CONDITIONS OF STABILITY to prevent rupture by rotation.

305. Take any joint, as I K (Fig. 98). The arch may give way by opening at the back and turning around the lower edge at K, or by opening on the soffit and turning around the edge at I.

Let us suppose the first case, or that the arch opens at the back. Denote by x the lever arm of the weight W of the mass K B C I, and by y the lever arm of the force H, both x and y being taken with respect to the edge K.

For stability we must have

$$H \times y - Wx > 0.$$

Suppose the second case, or that the arch opens at K, and denote by u and v the lever arms of W and H with respect to I. We must have for stability

$$W \times u - H \times v > 0.$$

If we find the joints at which $W\frac{x}{y}$ is a maximum and

$W\frac{u}{v}$ is a minimum, then for stability the value of H must lie between these two values.

That is, the condition for stability against rupture by rotation around the edge of a voussoir requires *the thrust,* H *of the arch to be greater than the maximum value of* $W\frac{x}{y}$, *and less than the minimum value of* $W\frac{u}{v}$.

Joints of Rupture.

306. From observations made on the manner in which large arches have settled, and from experiments made in rupturing small ones, it appears that the ordinary mode of fracture is for the arch to separate into four pieces, presenting five joints of rupture.

Cylindrical arches in which the rise is less than half the span, and the full centre arch, yield by the crown settling and the sides spreading out. The vertical joint at the crown

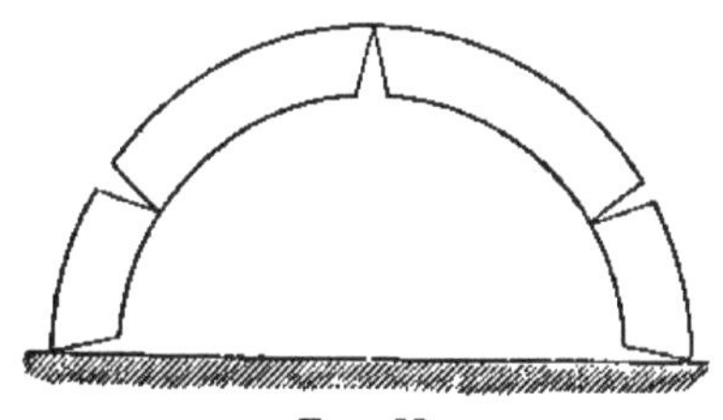

FIG. 99.

opens on the soffit, the reins open on the back, and if there be no pier, the joints at the springing line open on the soffit (Fig. 99).

The two lower segments revolve outwardly on the exterior edge of the joints, leaving room for the upper segments to revolve towards each other on the interior edges of the joints at the reins.

This is almost the only mode of yielding for the common cylindrical arch. If the thickness be very great compared with the span, the rupture will take place by sliding. As a rule, this mode of rupture never does take place for the reason that the arch will rupture by rotation around a joint before it will yield by sliding.

Very light segmental arches, full-centre arches which are slightly loaded at the crown and overloaded at the reins, and pointed arches, are liable to rupture, as shown in Fig. 100.

In this case the crown rises and the sides fall in; the open-

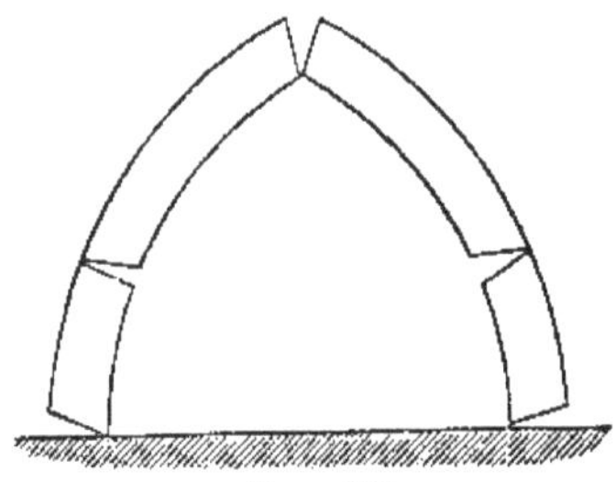

FIG. 100.

ing of the joints and the rupture occur in a manner exactly the reverse of that just described. This mode of rupture is still more uncommon than that by sliding; for all these reasons, the condition

$$H \times y - Wx > 0$$

is in general the one applied to test the stability of the arch.

Cylindrical Arch.

307. Let it be required to **find the conditions of equilibrium for a full centre arch.**

The strains in the arch are produced by the weight of the arch stones, the load placed upon the arch and the reactions at the springing lines.

The object of this discussion is to show how these external forces may be determined and how to arrange the joints and fix the dimensions of the voussoirs so as to resist successfully the action of these forces.

The joints are the weak places, since the separation of the parts at these points is not resisted by the material of which the arch is made.

As before stated, the arch may yield by sliding along one of the joints or by turning around an edge. The first mode of yielding may be prevented by giving the plane of the joint such a position, that its normal shall make with the resultant pressure an angle less than the angle of friction of the material of which the voussoirs are made.

This is usually effected by making the coursing joints normal to the ring courses and to the soffit of the arch.

Since there is little danger of the arch rupturing by the crown rising and the sides falling in, we make use of the formula

$$H \times y - Wx > 0.$$

The additional condition is imposed that the whole area of the joint must be subjected to compression. It therefore follows that the resultant of the external forces must pierce the joint within its middle third.

Since the form of the arch is known, the direction of the coursing joints chosen, and the limits of the resultant determined, it will only be necessary to find where the resultant pierces each joint and see if the angle it makes with the normal is less than the angle of friction, and that the resultant pierces the plane of the joint within the required limits.

Cylindrical Arch, *Unloaded.*

308. For simplicity, let us consider the arch to be a full centre, the extrados and intrados being parallel and the arch not loaded.

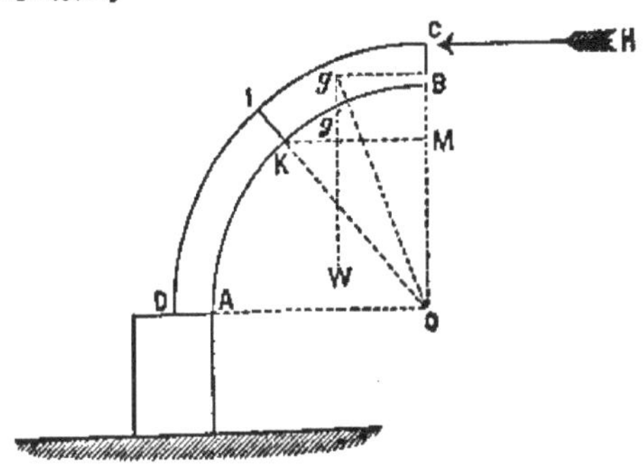

FIG. 101.

Let I K (Fig. 101) be a joint of the arch whose thickness in the direction of the length of the arch is unity.

Represent by

R, the radius of the extrados ;

r, the radius of the intrados ;

ϕ, the angle made by the joint I K with the vertical ;

W and H, same as in previous case ;

g, the centre of gravity of the ring K B C I ;

w, the weight of a unit of volume of masonry

The point of application of the thrust H, at the joint B C is somewhere above the middle of the joint, and when the arch begins to rupture it is at C (Fig. 101). The condition of stability for this case at the joint I K is

$$W\frac{x}{y} = H.$$

If the values of x and y be found in known terms, and substituted in this expression, the horizontal thrust can be determined.

To find these values of x and y, denote by u the distance of the centre of gravity g from O, and by u_1 and u_2 the distances of the centres of gravity of the sectors I O C and K O B from the same point. We have

$$u_1 \times \text{sector I O C} = u_2 \times \text{sector K O B} + u \times \text{ring K B C I}.$$

The areas of the sectors are $\frac{1}{2}R^2\phi$ and $\frac{1}{2}r^2\phi$, hence the area of K B C I is equal to $\frac{1}{2}\phi\ (R^2-r^2)$.

We find (Anal. Mech., par. 121, p. 96) the values of u_1 and u_2 to be $\frac{4}{3}R\frac{\sin\frac{1}{2}\phi}{\text{arc}\ \phi}$ and $\frac{4}{3}r\frac{\sin\frac{1}{2}\phi}{\text{arc}\ \phi}$.

Substituting for the areas, and for u_1 and u_2 their values as above, and solving with respect to u, we have

$$u = \frac{4}{3}\frac{R^3-r^3}{R^2-r^2}\cdot\frac{\sin\frac{1}{2}\phi}{\text{arc}\ \phi}.$$

Now x is equal to $KM - Mg' = r\sin\phi - Og\sin\frac{1}{2}\phi$, whence

$$x = r\sin\phi - \frac{4}{3}\frac{R^3-r^3}{R^2-r^2}\cdot\frac{\sin^2\frac{1}{2}\phi}{\text{arc}\ \phi}$$

$$= r\sin\phi - \frac{2}{3}\frac{R^3-r^3}{R^2-r^2}\cdot\frac{1-\cos\phi}{\text{arc}\ \phi}$$

and

$$y = R - r\cos\phi.$$

Hence, by writing k for $\frac{R}{r}$, we have

$$H = W\frac{x}{y} = r^2 w\ \frac{\frac{1}{2}\sin\phi\ (k^2-1)\ \text{arc}\ \phi - \frac{1}{3}(k^3-1)\ (1-\cos\phi)}{k-\cos\phi}, \quad . \quad . \quad (146)$$

an expression for the horizontal thrust, in terms of R, r, w, and ϕ, which force applied to the arch at C will prevent the rotation of the volume K C B I around the edge K.

This expression might be differentiated with respect to ϕ, and that value for ϕ obtained, which would make H a maximum. This maximum value thus found, if applied to the arch at C, would prevent its rotation around any edge on the soffit.

309. Instead of differentiating as suggested, it is usual in practice to take the above expression for H, calculate the values for every ten degrees, and select for use the greatest of these values. This greatest value thus obtained will differ but slightly from the true maximum.

If we assume $k = 1.2$, $r = 10$ feet, $R = 12$ feet, and $w = 150$ pounds, and find the values of H for the different values of ϕ for every ten degress from 10° to 90°; we may tabulate them as follows:

Values of ϕ.	Values of H in pounds.
10°	208
20°	670
30°	1,127
40°	1,450
50°	1,625
60°	1,675
70°	1,662
80°	1,490
90°	1,285

A calculation for $\phi = 57°$ gives $H = 1{,}672$, 63° gives 1,670, and 65° gives 1,661 pounds.

The angle requiring the maximum thrust is very nearly 60°.

310. The foregoing applies only to an unloaded full centre arch, its extrados and intrados being parallel. All arches carry loads which frequently rise above the arch to a surface either horizontal or nearly so. It is evident that if verticals be erected at the joints, and be produced until they meet the upper surface of the load, that they will define and limit the load resting on each voussoir. An analogous process to that just given will enable the student to determine the horizontal thrust in the arch thus loaded.

Prof. Rankine gives the following rule to find the approximate horizontal thrust in a full centre arch loaded as shown in the figure. (Fig. 102.)

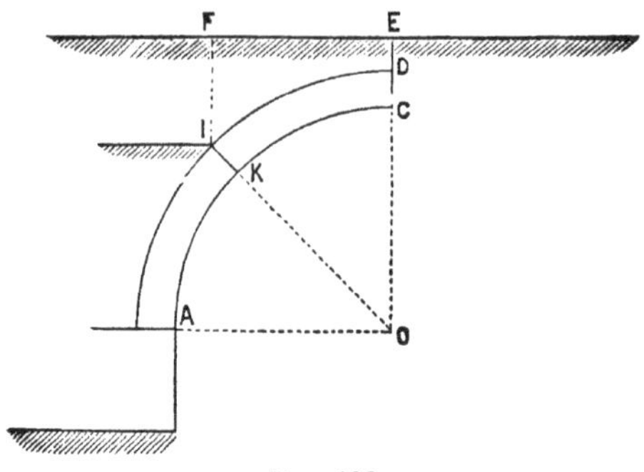

FIG. 102.

The horizontal thrust is nearly equal to the weight supported between the crown and that part of the soffit whose inclination is 45°.

The approximate thrust obtained by this rule seldom differs from the true horizontal thrust by so much as one-twentieth part.

Represent by (Fig. 102),

R, the radius O D of the extrados;

r, the radius O C of the intrados;

c, the distance D E, F E being horizontal;

w, the weight of a cubic foot of masonry;

w', the weight of a cubic foot of the load resting on the arch;

H, the horizontal thrust required.

Draw O K making an angle of 45° with the vertical; then, the horizontal thrust of the arch on the pier at A is stated to be nearly equal to the weight of the mass C K I F E, which lies between the joint I K and the vertical plane through C; hence,

$$H = w' R\,(.0644\,R + .7071\,c) + .3927\,w\,(R^2 - r^2). \quad (147)$$

for the value of the horizontal thrust.

The edge I is at the level to which it is advisable to build the backing solid, or at least to give the blocks a bond which will render the mass effective in transmitting the horizontal thrust.

In the case of a segmental arch, Rankine takes the weight of half the arch with its load, and multiplies it by the co-tangent of the inclination of the intrados, at the springing line, to the horizon; the result is the approximate value of H.

311. Having determined the value for H for the given arch, combine it with the external forces acting on the first voussoir at the crown and construct their resultant. The point in which this resultant pierces the joint will be the centre of pressure for that joint. Do the same for the other joints and the intensity of the resultant and the centre of pressure for each joint are known.

If these resultants be produced, a polygon will be formed, each angle of which will be on the resultant of the external forces, acting on the voussoir between the two joints to which the sides of the polygon correspond. A curve inscribed in this polygon tangent to its sides is called the **curve of pressure** of the arch, since a right line drawn through a centre of pressure, tangent to this curve, will give the direction of the resultant pressure for this point.

If normals be drawn through the centres of pressure a polygon will be formed whose sides give the direction of the components producing pressure on the joints. A curve tangent to its sides is called the **curve of resistance**, and is the locus of the centres of pressure of the joints.

For stability, the curve of resistance should pierce each joint in its middle third, and the curve of pressure should be so situated that right lines tangent to it drawn through the centres of pressure should make angles with the normals less than the angle of friction.

312. **Equation of the curve of resistance.**—Suppose the loads on an arch to be symmetrically disposed so that the resultant forces will lie in a vertical plane.

Equations (688) of Anal. Mechanics for this case will be

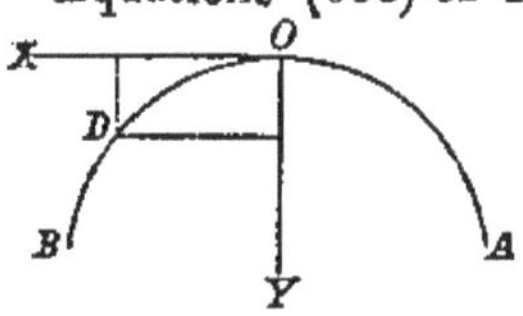

$$\left.\begin{aligned} H - C\frac{dx}{ds} &= 0, \\ W - C\frac{dy}{ds} &= 0, \end{aligned}\right\} \quad . \quad . \quad (148)$$

FIG. 103.

in which H is the horizontal thrust at O (Fig. 103); C the compressive stress on any section, as D; s, the length of any portion of the curve, as O D; and W the sum of the vertical forces acting on the portion considered.

The first of equations (148) shows that the horizontal component of the force of compression at any joint is equal to the horizontal thrust at the crown, or is the same at every section of the arch.

The second of these equations shows that the vertical component of the force acting at any joint is equal to the load between the vertical plane through the crown and the section considered.

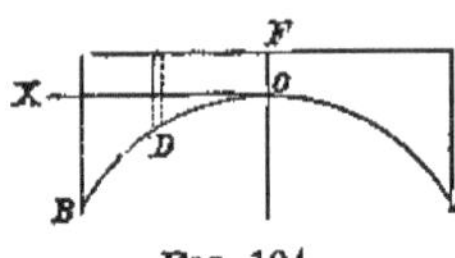

FIG. 104.

313. Suppose an arch loaded as shown in figure (104); the material being homogeneous and the weight of a unit of volume being represented by w. Represent O F by a.

The weight of the volume resting on the arch between the vertical section at D and the consecutive section is

$$(adx + ydx)\,w.$$

Taking this between the limits, 0 and x, we get

$$\left(ax + \int_0^x ydx\right)w,$$

for the load resting on O D. Substituting this in the second of equations (148) for W, we get

$$w\left(ax + \int_0^x ydx\right) - \mathrm{C}\frac{dy}{ds} = 0. \quad . \quad (149)$$

Combining this equation with the first of equations (148), we have

$$\frac{dy}{dx} = \frac{w}{\mathrm{H}}\left(ax + \int_0^x ydx\right),$$

whence, by differentiating, we get

$$\frac{d^2y}{dx^2} = \frac{w}{\mathrm{H}}(a + y). \quad . \quad . \quad (150)$$

Integrating this differential equation twice, we get the equation of the curve, and find it to be a transcendental line.

314. If the load had been placed on the arch so as to be a function of the first power of the abscissa, that is, if the load between the origin and any section whose abscissa is x, was wx, then equations (148) would have taken the form

$$\left.\begin{aligned} \mathrm{H} - \mathrm{C}\frac{dx}{ds} &= 0, \\ wx - \mathrm{C}\frac{dy}{ds} &= 0. \end{aligned}\right\} \quad . \quad . \quad (151)$$

Whence, by combination,

$$dy = \frac{w}{H} x dx,$$

and by integration,

$$y = \frac{w}{2H} x^2. \quad . \quad . \quad . \quad . \qquad (152)$$

which is the equation of a parabola.

315. **Polar equation of the curve of resistance.**— This equation is deduced by General Woodbury as follows:

Represent by (Fig. 105)

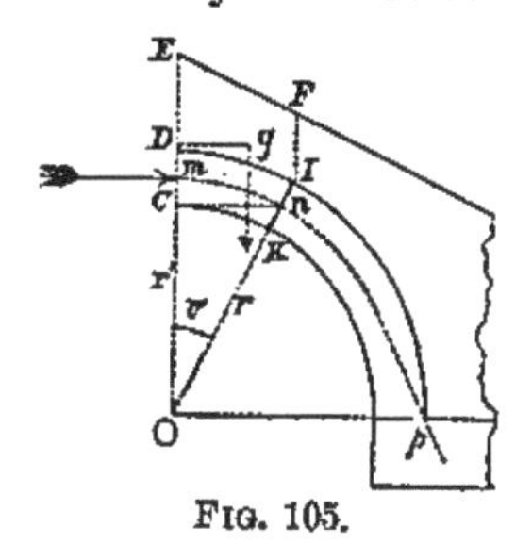

FIG. 105.

H, the horizontal thrust at m; mnp, the curve of resistance; r', the distance, Om, from pole to the point of application, m of the horizontal thrust; b, the horizontal distance between the centre of gravity of the segment E F I K C and the vertical through C; A, the area of the segment; v, the variable angle nOm, and r, the variable distance On.

For equilibrium, considering w equal to unity, we have

$$H(r' - r\cos v) = A(r\sin v - b),$$

whence

$$r = \frac{Hr' + Ab}{A\sin v + H\cos v}. \quad . \quad . \quad . \qquad (153)$$

Assuming any joint, the corresponding values of A and b for this joint are easily calculated. These being substituted, and H and v being known, the corresponding value of r is deduced. The curve may then be constructed by points.

A simple inspection of the curve of resistance will show where the weak points of the arch are, where the heaviest strains are exerted, and where the joints tend to open, whether on the soffit or on the back.

316. The deviation of the curve of pressure from the curve of resistance is not great, and no material error is ordinarily made when the points of the curve of pressure cut by the joints are taken as the centres of pressure for the joints.

In arches with the ordinary form of voussoirs, the curve of pressure lies below the curve of resistance, and the condition that it shall lie within the middle third of the joints is favorable to the stability of the arch.

When the weight of the voussoirs and the load on the arch are determined, as in Art. 313, by considering them composed of vertical laminæ, the curves of pressure and of resistance will coincide with each other.

Economy of material would indicate that the intrados and extrados should be similar curves.

317. **Depth of keystone.**—The form of the arch being assumed, the next step is to fix its thickness or **depth**. The power of the arch to resist the horizontal thrust at the crown will depend upon the strength of the material of which it is made and upon the vertical thickness (depth) of the key.

The pressure at the extrados of the key, which in general is the most exposed part of the joint, should not exceed $\frac{1}{10}$ the ultimate strength of the material. Admitting that the centre of pressure on this joint may be at one-third of the length of the joint from the extrados, we see that in order to keep within this limit of $\frac{1}{10}$, the mean pressure should not exceed $\frac{1}{20}$.

The celebrated Perronnet gave a rule for determining the thickness or depth of the key, which is very nearly expressed by the following formula:

$$d = \frac{r}{15} + 0.33. \quad . \quad . \quad . \quad . \quad (154)$$

d, the depth in metres; and

r, the radius of the semicircle, or intrados, in same unit.

Gen. Woodbury expressed this rule as follows:

$$d = 13 \text{ inches} + \tfrac{1}{23} \text{ the span.}$$

For arches with radius exceeding 15 metres, this rule gives too great a thickness.

Prof. Rankine gives

$$d = \sqrt{.12r},$$

in which r is the radius of curvature at the crown in feet. His rule is, "For the depth of the keystone, take a mean proportional between the radius of curvature of the intrados at the crown and a constant whose value for a single arch is .12 feet."

He recommends, however, in actual practice, to take a depth founded on dimensions of good examples already built.

318. **Thickness of piers and abutments.**—The stability of these may be considered by regarding them either as continuations of the arch itself clear to the foundation, or as walls whose moment about the axis of rotation is greater than the moment of the thrust of the arch.

In either case, the student will be able, by applying the principles already discussed, to determine the dimensions necessary to give the pier, in order that its moment around any edge shall exceed the moment of the thrust around the same axis.

The factor of safety is taken at about 2. In piers of great height this factor should be increased, while for small heights it may be reduced.

319. Thickness of abutment and depth of keystone for small arches.

The following empirical table is deduced from actual examples, and may be used for small arches if made of first-class masonry:

TABLE.

Span in feet.	Thickness of Abutment—for heights of				Depth of keystone in inches.
	10 feet.	15 feet.	20 feet.	25 feet.	
10	5	6	7	8	14
20	6	7	8	9	19
25	6½	7½	8½	9½	20
30	7	8	9	10	21
35	7½	8½	9½	10½	22
40	8	9	10	11	23
45	8½	9½	10½	11½	24
50	9	10	11	12	25

If the masonry be second-class, or be roughly dressed, the depth of the keystone should be increased about one-fourth.

Form of Cylindrical Arches.

320. As stated before, these arches may be **full centre, segmental, elliptical**, or **oval**.

Full centre arches offer the advantages of simplicity of form, great strength, and small lateral thrust. But where the span is considerable, they require a correspondingly great rise, which is often objectionable.

The segmental arch enables us to reduce the rise, but causes a greater lateral thrust on the abutments.

The oval affords a means of avoiding both the great rise and the great lateral thrust, and gives a curve of pleasing appearance.

Rampant and Inverted Arches.

321. The arch in the preceding cases has been supposed to have been upright, and either right or oblique. Rampant arches are frequently used; sometimes the axis is even vertical. A retaining wall with a semi-circular horizontal section would be an example. Arches are often constructed with their soffits forming the upper side. These are frequently used under openings, their object being to distribute the weight equally over the substructure or along the foundations. They are known as inverted arches, or **inverts**. The principles already laid down for the upright arch apply equally to them.

Wooden Arches.

322. This term, **wooden arch**, is quite often applied to a beam bent to a curved shape, its ends being confined so that the beam cannot resume its original form. In this shape the beam possesses under a load greater stiffness than when it is straight.

A single beam may be used for narrow spans, but built beams, either solid or open, must be used for wide ones.

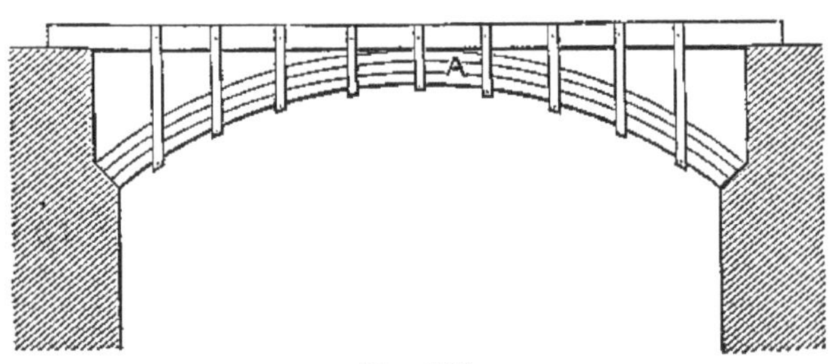

FIG. 106.

The load they support rests upon the top of the beams, as shown in Fig. 106, or is suspended from them, as shown in Fig. 107.

Although called arches, they are so only in form, as they are not composed of separate pieces held in place by mutual pressure. They are now more generally called by their proper name, **curved beams.**

If we assume that the beam resists by compression alone, the dimensions of the beam can be easily determined, in terms of the load, of the rise, and the span.

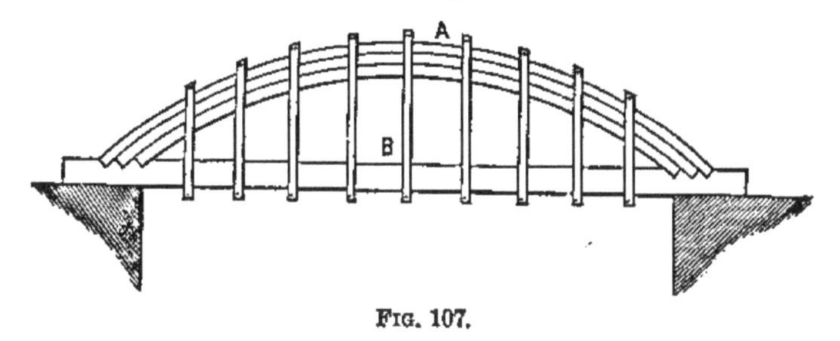

FIG. 107.

GRAPHICAL METHOD OF INVESTIGATION.

323. The graphical method by means of the curve of equilibrium is a method much used at the present time for obtaining the strains on the different parts of the arch.

This method of investigation will be alluded to in a future article.

CHAPTER X.

CONSTRUCTION OF MASONRY.

WALLS OF STRUCTURES.

Stone-masons class the methods of building walls of stone into **rubble work** and **ashlar work.**

I. Rubble Work.

324. The stones used are of different sizes and shapes, prepared by knocking off all sharp, weak angles of the blocks with a hammer. They are laid in the wall either dry or in mortar. If laid without reference to their heights, the masonry is known as **uncoursed rubble,** or **common rubble masonry.**

In building **a wall of rubble** (Fig. 108) the mason must be careful to place the stones so that they may fit one upon the other, filling the interstices between the larger stones by smaller ones. Care should be taken to make the vertical courses break joints.

If mortar is used, the bed is prepared by spreading mortar over the top of the lower course, and in this bed the stone is firmly imbedded. The interstices are filled with smaller

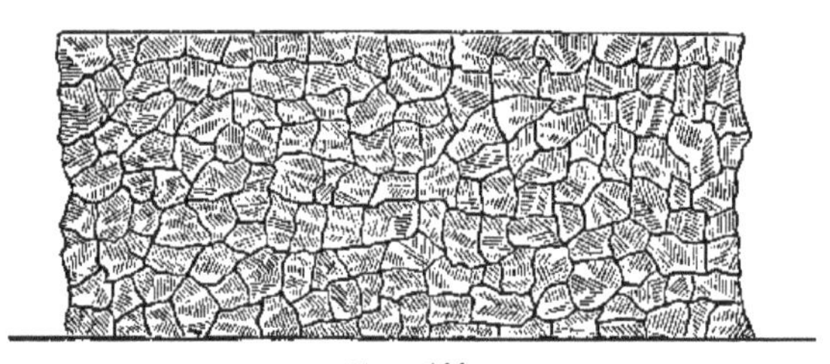

FIG. 108.

stones, or stone chippings, and mortar, and finally the whole course grouted.

The mean thickness of a rubble wall should not be less than one-sixth of the height; in the case of a dry stone wall, the thickness should never be less than two feet. It strengthens the wall very much to use frequently in every course, stones which pass entirely through the wall from the front to the back. These are called **throughs**. If they extend only part of this distance, they are called **binders**.

325. **Coursed rubble, or hammered** masonry.—When the stones are laid in horizontal courses, and each course levelled throughout before another is built upon it, the work is termed **coursed rubble**. As this requires the stones to be roughly dressed, or hammered into regular forms before they are laid, the work is frequently called **hammered**, or **dressed rubble**. The same care should be taken in building masonry of this kind as that required for common rubble. The mason must be particular in making the upper and lower surfaces of each stone parallel, and when laying the stones to keep a uniform height throughout each course. If a stone in the course is not high enough, other stones are laid on it till the required height is obtained.

The different courses are not all of the same height, but vary according to the size of the stone used. The only condition required is that each course shall be kept of the same height throughout.

At the corners, stones of large size, and more acurately dressed, are used. These are known as **quoins**, and are laid with care, serving as gauges by which the height of the course is regulated.

II. Ashlar Work.

326. The stones in this kind of masonry are prepared by having their beds and joints accurately squared and dressed. They are made of various sizes depending on the kind of wall to be built and the size of the blocks produced by the quarry. Ordinarily they are about one foot thick, two or three feet long, and have a width from once to twice the thickness. They are used generally for the facing of a wall, to give the front a regular and uniform appearance, and where, by the regularity of the masses, a certain architectural effect is to be produced.

Ashlar work receives different names, from the appearance of the face of the "ashlar," and from the kind of tool used in dressing it. If the block be smooth on its face, it is called **plane ashlar** (Fig. 109); if fluted vertically, **tooled ashlar**;

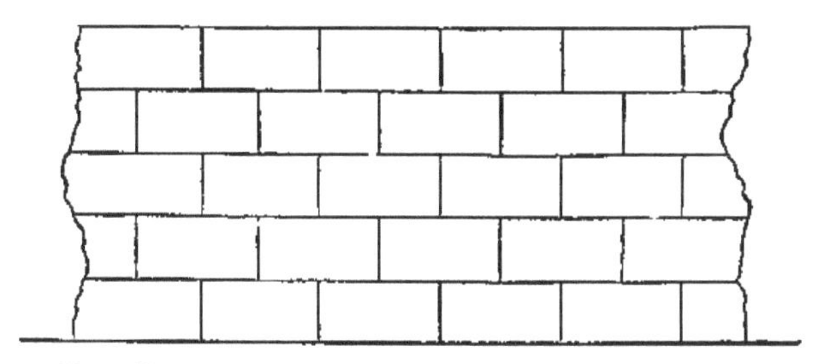

FIG. 109—Represents a wall with facing of plane ashlar.

if roughly trimmed, leaving portions to project beyond the edges, **rustic ashlar**, etc., etc. Rustic ashlar is known as rustic, rustic chamfered, rustic work frosted, rustic work vermiculated, etc.

Ashlars are laid in fine mortar or cement. Each one should be first fitted in its place dry, so that any inaccuracy in the dressing may be discovered and corrected before the stone is finally set in mortar.

To provide for a uniform bearing the stone should be accurately squared. Frequently the bed is made to slant downwards, from front to back, for the purpose of making close horizontal joints in front. This weakens the stone, as the weight is thrown forward on the edges of the stones, which chip and split off as the work settles.

327. Walls built with ashlar facing are backed with brick or rubble. Economy will decide which is to be used. In the construction, *throughs* of ashlars should be used to bind the backing to the facing. Their number will be proportioned to the length of the course. The vertical courses break joints, each vertical joint being as nearly as possible over the middle of the stone below.

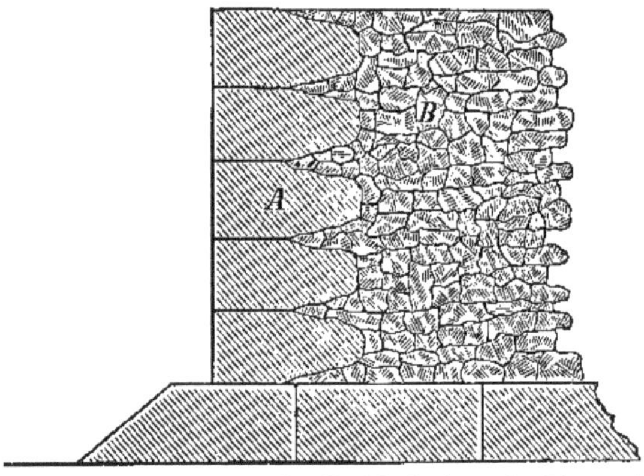

Fig. 110—Represents a section of wall with facing of ashlar and a backing of rubble.

When the backing is rubble, the method of slanting the bed may be allowed for the purpose of forming a better bond between the rubble and ashlar; but, even in this case, the block should be dressed true on each joint, to at least one foot back from the face. If there exists any cause which would give a tendency to an outward thrust from the back, then, instead of slanting off all the blocks towards the tail, it will be preferable to leave the tails of some thicker than the parts which are dressed.

Cut-stone Masonry.

328. Where great strength is required in the wall, each stone is prepared by cutting it to a particular shape, so that it can be exactly fitted in the wall; masonry of this kind is called **cut-stone**. In other words, every stone is an ashlar;

hence the terms cut-stone and ashlar masonry are often used one for the other.

Cut-stone masonry, when carefully constructed, is more solid and stronger than any other class. The labor required in preparing the blocks makes it the most expensive. It is, therefore, restricted in its use to those structures where great strength is indispensable.

Stone-cutting.

329. The usual method of dressing a surface is to cut draughts around and across the stone with a chisel, and then work down the intermediate portions between the draughts by the use of proper tools. The latter are usually the chisel, axe, and hammer.

No particular difficulty occurs in working a block of stone, the faces, beds, and joints of which are to be plane or even cylindrical surfaces; the only difference in method for the two being that a curved rule is used in one direction and a straight one in another for the cylindrical surface, while for the plane surface only one rule is used.

If the surfaces are to be conical, spherical, or warped, the operation is more difficult. It becomes necessary to bring the block to a series of plane or cylindrical surfaces, and then reduce them to the required form. To show how this can be done with the least waste of material is one of the objects of **"stereotomy."**

Strength of Masonry.

Strength.—The strength of masonry will depend on the *size* of the blocks, on the *accuracy* of the *dressing*, and on the *bond*.

330. **Size of stone.**—The size of the blocks varies with the kind of stone and the nature of the quarry.

Some stones are of a strength so great as to admit of their being used in blocks of any size, while others can only be used with safety when the length, breadth, and thickness of the block bear certain relations to each other.

The rule usually followed by builders, with ordinary stone, is to make the breadth at least equal to the thickness, and seldom greater than twice this dimension, and to limit the length to within three times the thickness. When the breadth or the length is considerable in comparison with the thick ness, there is danger that the block may break, if any unequal settling or unequal pressure should take place. As to the absolute dimensions, the thickness is generally not less than one foot, nor greater than two; stones of this thickness, with the relative dimensions just laid down, will weigh from 1,000 to 8,000 pounds, allowing, on an average, 160 pounds to the cubic foot. With these dimensions, therefore, the weight of each block will require a very considerable power, both of machinery and men, to set it on its bed.

From some quarries the formation of the stone will allow only blocks of medium or small size to be furnished, while from others stone of almost any dimensions can be obtained.

331. **Accuracy of dressing.**—The closeness with which the blocks fit is solely dependent on the accuracy with which the surfaces in contact are wrought or **dressed**; if this part of the work is done in a slovenly manner, the mass will not only open at the joints with an inequality in the settling, but, from the courses not fitting acurately on their beds, the blocks will be liable to crack from the unequal pressure on the different points of the block.

To comply with the first of the general principles to be observed in the construction of masonry, we should have, in a wall supporting a vertical pressure, the surfaces of one set of joints, the beds, horizontal. This arrangement will prevent any tendency of the stones to slip or slide under the action of the weight they support.

The surfaces of the other set should be perpendicular to the beds, and at the same time perpendicular to the face, or to the back of the wall, according to the position of the stones in the mass; two essential points will thus be attained; the angles of the blocks at the top and bottom of the course, and at the face or back, will be right angles, and the block will therefore be as strong as the nature of the stone will admit.

The greater the accuracy of the dressing, the more readily can these surfaces be made to fulfil these conditions.

When a block of cut stone is to be laid, the first point to be attended to is to examine the dressing, by placing the block on its bed, and seeing that the face is in its proper plane, and that the joints are satisfactory. If it be found that the fit is not accurate, the inaccuracies are marked, and the requisite changes made.

332. **Bond.**—Among the various methods used, the one known as **headers** and **stretchers** is the most simple, and offers, in most cases, all requisite solidity; in this method the vertical joints of the blocks of each course alternate with the

vertical joints of the courses above and below it, or **break joints** with them, and the blocks of each course are laid alternately with their greatest and least dimensions to the face of the wall; those which present the longest dimension along the face are termed stretchers, the others headers. (Fig. 111.)

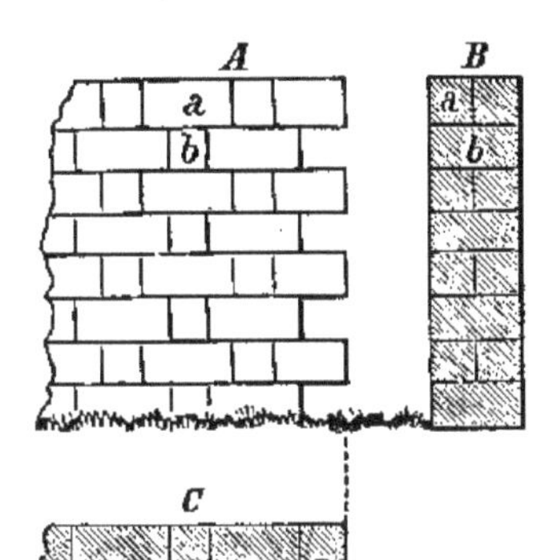

Fig. 111—Represents an elevation A, vertical section B, and horizontal section C, of a wall arranged as headers and stretchers.
a, stretchers.
b, headers.

By arranging the blocks in this manner the facing and backing of each course are well connected; and, if any unequal settling takes place, the vertical joints cannot open, as would be the case were they continuous from the top to the bottom of the mass, for each block of one course confines the ends of the two blocks on which it rests in the course beneath.

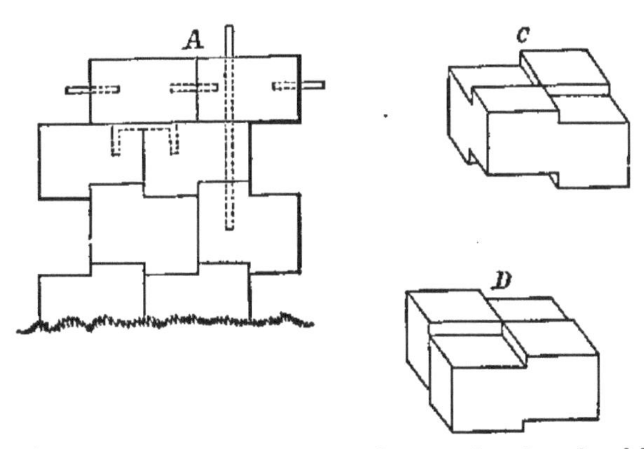

Fig. 112—Represents an elevation A, and perspective views C and D, of two of the blocks of a wall in which the blocks are fitted with indents, and connected with bolts and cramps of metal.

333. In masonry exposed to violent shocks, the blocks of each course require to be not only very firmly united with each other, but also with the courses above and below them. To effect this various means have been used. Sometimes the stones of different courses are connected by **tabling**, which consists in having the beds of one course arranged with projections (Fig. 112) which fit in corresponding indentations of the next course. Iron cramps in the form of the letter S, set with melted lead, are often used to confine two blocks together. Holes are, in some cases, drilled through several courses, and the blocks of these courses are connected by strong iron bolts fitted to the holes.

Light-houses, in exposed positions, are peculiarly liable to violent shocks from the waves. They are ordinarily, when thus exposed, built of masonry, are round in cross-section, and solid up to the level of the highest tide. The stones are oftentimes dove-tailed and dowelled into each other, as well as fastened together by metal bolts and cramps.

The manner of dove-tailing the stones is shown in plan in Fig. 113, which represents part of a course where this method is used.

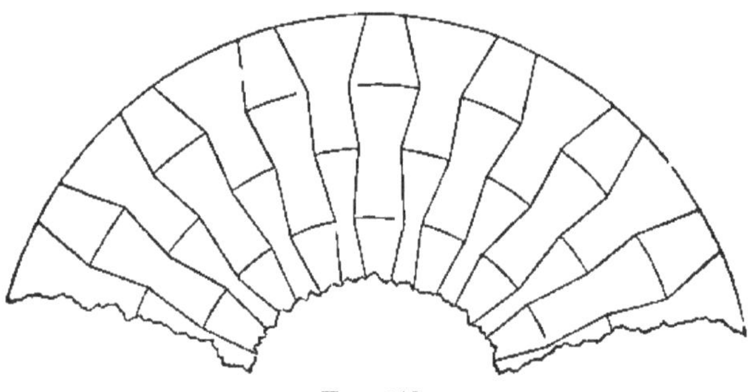

Fig. 113.

The chief use of the dove-tailing is to resist the tendency of the stones to *jump out* immediately after receiving the blow of the wave. This method was first used by Smeaton in building the Eddystone light-house. The light-house on Minot's Ledge, Massachusetts Bay, built under the superintendence of General B. S. Alexander, U. S. Corps of Engineers, by the Light-House Board, is a good example of the bond and metal fastenings used in such structures. (Figs. 114 and 115.)

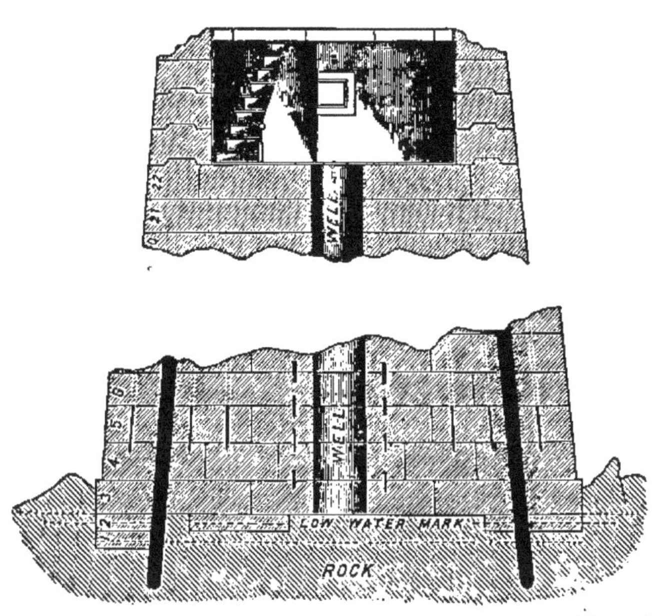

FIG. 114.—Vertical section showing foundation courses, metal fastenings, and the first story above the foundation courses.

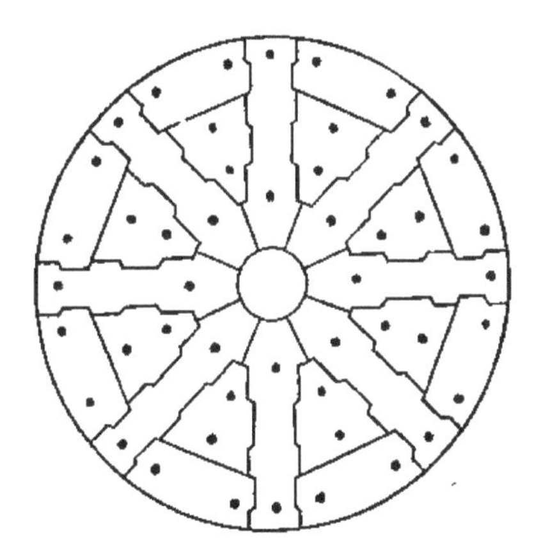

FIG 115.—Plan of twenty-second course, showing the method of dovetailing the stones.

Machinery used in Constructing Walls of Stone.

334. **Scaffolding.**—In building a wall, after having raised it as high as it can be conveniently done from the ground, arrangements must be made to raise the workmen higher, so that they can continue the work. This is effected by means of a temporary structure called **scaffolding.**

If the wall is not used to afford a support for the scaffolding, two rows of poles are planted firmly in the ground, parallel to the wall, and about four and a half feet apart. These uprights in each row are from twelve to fourteen feet apart, and from thirty to forty and even fifty feet in height, depending upon how high the wall is to be built.

Horizontal pieces are then firmly fastened to the uprights, having their upper surfaces nearly on the same level as the highest course of masonry laid. Cross pieces or joists are laid on these, and upon them a flooring of boards. Upon this platform the masons place their tools and materials and continue the work.

As the wall rises other horizontal pieces are used, and the joists and boards carried to the new level. Diagonal pieces are used between the rows to brace them together, and in each row to stiffen the supports.

The workmen ascend the scaffolding by means of ladders. The materials are hoisted by means of machinery placed on the scaffolding or detached from it.

335. **Crane.**—The movable or travelling **crane,** which is so arranged as to admit of being moved in the direction of the scaffolding and across it, is often used on the scaffolding for hoisting the stone.

Shears, which consist of two or more spars or stout pieces of timber, fastened together near the top, and furnished with blocks and tackles, are sometimes used.

The kind of machinery to be used in hoisting the stone will be determined by the size of the blocks to be lifted, the magnitude and character of the work, and the suitability of the site.

In the United States, the machine known as the "**boom derrick,**" or simply "**derrick,**" a modified form of crane, is much used in works of magnitude.

In the example shown in Fig. 116, the mast is held in a vertical position by four guys, generally wire ropes, fastened to a ring on the iron cap which is fitted to the top of the mast. Below this ring, and revolving freely on the cap, is a wrought-iron frame containing two sheaves or pulleys.

The "boom," or derrick, has its outer end supported by a topping-lift fastened to this wrought-iron frame. The other end fits into an iron socket with collar, or is fastened to a wooden frame which embraces the mast, and has a motion of rotation around it. The wooden frame bears two windlasses and a platform on which the men stand while working them. Two tackles are used, one suspended from the outer end of the boom, the other from the mast-head, the falls of both leading over the sheaves and thence to the windlasses.

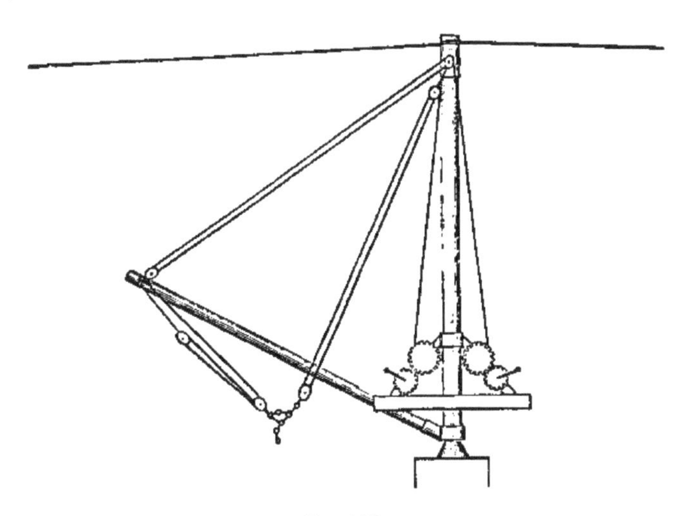

FIG. 116.

The lower blocks of the tackles are fastened to a triangular plate from which a hook is suspended. It is seen that by hauling upon or slacking the falls alternately, the stone suspended from the triangular plate can be placed at any point within the circle described by the outer end of the boom.

336. The blocks of stone are attached to the tackle in various ways. Some of the most usual methods are as follows:

I. By nippers or tongs, the claws of which enter a pair of holes in the sides of the stone.

II. By two iron pins let into holes, which they closely fit, sloping towards each other (Fig. 117). The force applied to the chain to lift the block, jams the pins in their holes.

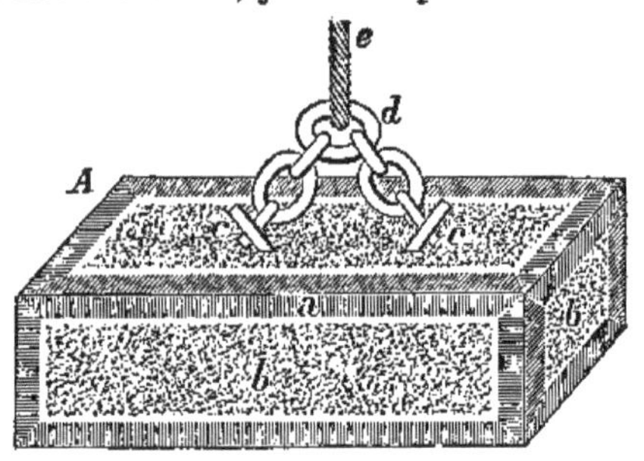

FIG. 117—Represents a perspective view of the tackling for hoisting a block of stone, A, with draughts around the edges of its faces, and the intermediate space axed or knotted.
a, draughts around edge of block.
b, knotted part between draughts.
c, iron bolts with eyes let into oblique holes cut in the block.
d and *e*, chain and rope tackling.

III. By a simple contrivance made of three pieces of iron, called a **lewis** (Fig. 118), which has a dove-tail shape, with the larger end downwards, fitting in a hole of similar shape. The depth of the hole depends upon the weight and the kind of stone to be raised. The tapering side-pieces, *n*, *n*, of the lewis are inserted and placed against the sides of the hole; the middle piece, *o*, is then inserted and secured in its place by a pin. The stone is then safely hoisted, as it is impossible for the lewis to draw out of the hole.

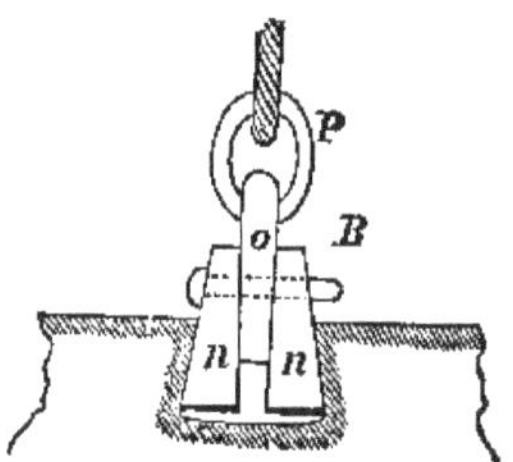

FIG. 118—Represents the common iron *lewis* B.
n, *n*, side pieces of the lewis.
o, centre piece of lewis, with eye fastened to *n*, *n* by a bolt.
P, iron ring for attaching tackling.

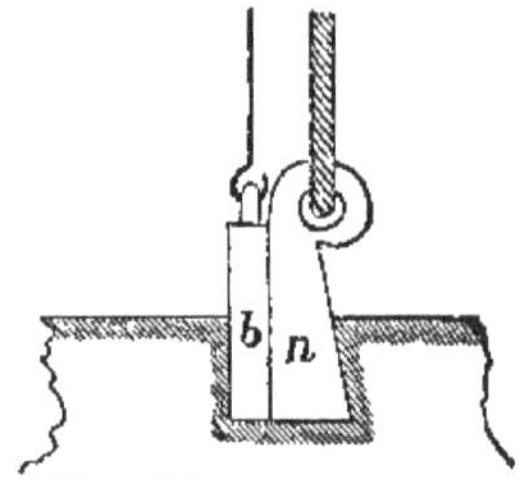

FIG. 119—A line attached to the straight piece, *b*, admits of the latter being drawn out, allowing the piece, *n*, to be removed.

Where it may not be convenient to reach the pin after the stone has been placed in position, a lewis of the form shown in (Fig. 119) may be used.

WALLS OF BRICK.

337. Bricks have been referred to in a previous chapter as artificial stones. It therefore follows that the general principles enunciated for the construction of stone masonry are the same for brick as far as they are applicable.

From the uniformity of size of brick, builders describe the thickness of a wall by the number of bricks extending across it. Thus, a wall formed of one thickness of brick lying on their broad side, with their length in the direction of the length of the wall, is said to be "half brick thick." If the thickness of the wall is equal to the length of one brick, the wall is called "one brick thick," etc.

The bond used depends upon the character of the structure. The most usual kinds are known as the **English** and **Flemish.**

338. **English bond.**—This consists in forming each course entirely of headers or of stretchers, as shown in Fig. 120.

Sometimes the courses of headers and stretchers occur alternately; sometimes only one course of headers for three or four courses of stretchers. The effect of the stretchers is to tie the wall together lengthwise, and the headers, cross-

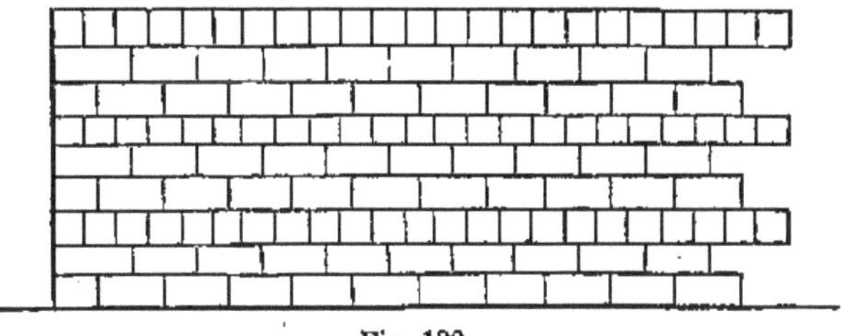

Fig. 120.

wise. The proportionate number of courses of headers to those of stretchers depend upon the relative importance of the transverse and longitudinal strength in the wall.

Since the breadth of a brick is nearly equal to half its length, it would be impossible, beginning at a vertical end or angle, to make this bond with whole bricks alone. This difficulty is removed by the use of a half-brick, made by cutting a brick in two longitudinally. A whole brick, used as a header, is placed at the corner; next to this is put a half-brick. This allows the next header to make the necessary overlap, which can be preserved throughout the course. These half-bricks are called **closers.**

339. **Flemish bond.**—This consists in laying headers and stretchers alternately in each course.

A wall built with this bond presents a neater appearance than one built in English bond, and is, therefore, generally preferred for the fronts of buildings. It is not considered as strong as the English, owing to there being, ordinarily, a less number of headers in it.

340. **Strengthening of bond.**—Pieces of hoop-iron or iron lath, so thin that they may be inserted in the joints without materially increasing their thickness, add to the strength of the bond, especially when hydraulic mortar is used. They are laid flat in the bed-joints, and should break joints. It is well to nick them at intervals and bend the ends at right angles for the length of two inches, inserting the bent extremities into the vertical joints.

This method was used by Brunel in forming the entrance to the Thames tunnel, and is sometimes designated as **hoop-iron bond.**

341. **Hollow masonry.**—Hollow brick walls are now extensively used in buildings.

The advantages of hollow walls are economy, lightness, and, particularly, freedom from dampness.

The bricks may be hollow, being laid in the usual way, but the usual method of forming the walls is to use ordinary brick, and so arrange them in the walls as to leave hollow spaces where required.

342. **Strength of brick masonry.**—The strength of brick masonry depends upon the same three conditions already given for stone. Hence, all misshapen and unsound bricks should be rejected.

With good bricks and good mortar a masonry of strength and durability nearly equal to stone is easily formed, and at less cost. Its strength is largely due to the strong adhesion of mortar to brick. The volume of mortar used is about *one-fifth* that of the brick.

343. **Laying the bricks.**—The strength of brick masonry is materially affected by the manner in which the bricks are laid. They should not only be placed in position, but pressed down firmly into their beds.

As bricks have great avidity for water, it would always be well not only to moisten them before laying, but to allow them to soak in water several hours before they are used.

By taking this precaution, the mortar between the joints will set more firmly.

To wet the bricks before they were carried on the scaffold would, by making them heavier, add materially to the labor of carrying. It is suggested to have arrangements on the scaffold where they can be dipped into water, and then handed to the mason as he requires them. The wetting is of great importance when hydraulic mortar or cement is used, for if the bricks are not wet when laid, the cement will not attach itself to them as it should.

Machinery of Construction.

344. **Scaffolding.**—In ordinary practice the scaffolds are carried up with the walls, and are made to rest upon them. The essential features are the same as those used for stone walls. It would be an improvement if an inner row of uprights were used instead of the wall to support the framework, for the cross-pieces, resting as they often do on a single brick in a green wall, must exert an injurious influence on the wall.

Machinery for hoisting the bricks, mortar, etc., are used in extensive works. For ordinary buildings the materials are carried up by workmen by means of ladders.

WALLS OF CONCRETE.

345. **Concrete masonry.**—Within recent years much attention has been paid to the construction of walls entirely of concrete.

Method of construction.—The concrete is moulded into blocks, as previously described, and then laid as in stone masonry; or it is moulded into the wall, the latter becoming a monolithic structure.

The walls in the latter case are constructed in sections about three feet high and ten or fifteen long. For this purpose a mould is used made of boards forming two sides of a box, the interior width of which is equal to the thickness of the wall. Its sides are kept in place by vertical posts, which are connected together and prevented from spreading apart by small iron rods, as shown in Fig. 121.

The concrete is shovelled into the mould in layers and rammed with a pestle. As soon as the mould is filled, the iron rods are withdrawn and the mould lifted up. A second section is formed in like manner on the top of the first, and the process goes on until the wall reaches the required height.

If scaffolding be required in their construction, one of the ordinary form may be used, or one like that shown in Fig. 121.

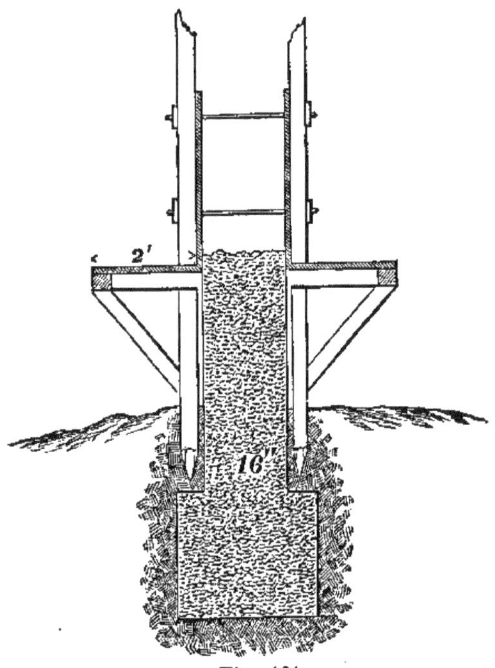

Fig. 121.

Tall's bracket scaffolding, in which the platforms are sustained by clamping them to the wall as it is built up, using the holes left when the iron rods are withdrawn, is an example of one of the devices used in the construction of concrete walls; so also **Clarke's adjustable frame**, in which the platform is supported by a frame from above, fastened to clamps embracing the wall. Hoisting apparatus suitable for the work is also employed.

Hollow walls.—In case the wall is required to be hollow, a piece of board of the thickness of the required space to be left open, and slightly wedge-shaped to admit of its being easily removed, is laid horizontally in the mould, and the concrete rammed in well around it. When the concrete is filled to the top of the board, it is drawn out, leaving the re quired air space. At regular intervals, ordinary bricks are laid as ties to connect together the outer and inner walls.

Flues, pipes, and other openings for heating, ventilating, conveying water, gas, smoke, etc., are constructed in a similar manner by using movable cores of the proper size and form.

Strength and advantages of concrete walls.—It is claimed that concrete walls are easier of construction, cheaper, and stronger than brick walls of the same thickness, and that they possess the great advantage in allowing air passages and flues to be easily constructed of uniform size and smooth interiors.

RETAINING AND RESERVOIR WALLS.

346. Especial care should be taken, in the construction of these walls, to secure a firm foundation, and to observe all the precautions mentioned in previous articles for laying masonry.

Thorough drainage must be provided for, and care be taken to keep water from getting in between the wall and the earth. If the water cannot be kept out, suitable openings through the masonry should be made to allow the water to escape.

When the material at the back of the wall is clay, or is retentive of water, a dry rubble wall, or a **vertical layer** of coarse gravel or broken stone, at least one foot thick horizontally, must be placed at the back of the retaining wall, between the earth and the masonry, to act as a drain.

In filling in the earth behind the wall, the earth should be well rammed in layers inclined downward from the wall.

Especial care should be taken to allow the mortar to harden before letting the wall receive the thrust of the earth.

Whenever it becomes necessary to form the embankment before the mortar has had time to set, some expedient should be employed to relieve the wall as far as possible from pressure. Instead of bringing the embankment directly against the back of the wall, dry stone or fascines may be interposed, or a stiff mortar of clay or sand with about $\frac{1}{20}$th in bulk of lime may be used in place of the dry stone.

347. **Form of cross-section of retaining walls.**—The rectangular and the trapezoidal forms are the most common. It is usual, in the latter case, to give the face a batter, varying between $\frac{6}{1}$ and $\frac{48}{1}$, and to build the back, or side in contact with the earth, vertical, or in steps. From experiments made with models of retaining walls, it was shown that as the wall

gave way, the prism of earth pressing against it did not revolve around any line, but settled suddenly and then rested until another shock. These experiments seem to confirm the practice of building the back in steps.

In some cases the wall is of uniform thickness with sloping or curved faces. (Figs. 122 and 123.)

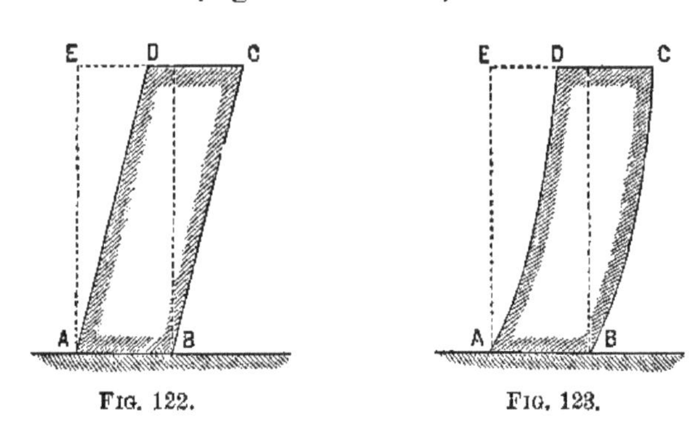

FIG. 122. FIG. 123.

It will be seen that, the weight remaining the same, the wall with sloped or curved faces has an increase of stability over the corresponding equivalent wall of rectangular cross-section.

The advantage of such forms, therefore, lies in the saving of material.

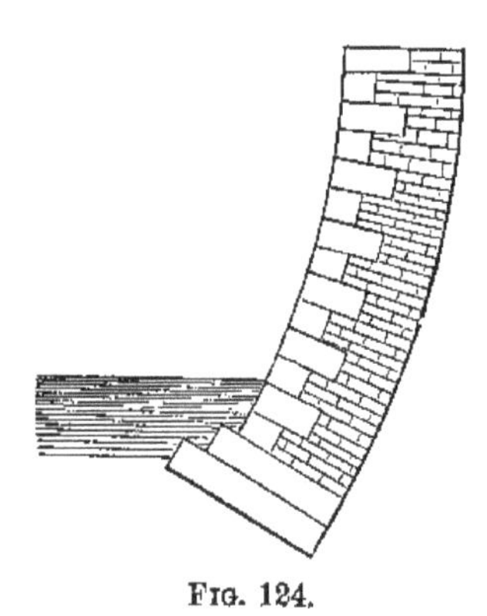

FIG. 124.

Walls with curved batter should have their bed-joints perpendicular to the face of the wall, so as to diminish the obliquity of pressure on the base. (Fig. 124.)

348. **Counterforts.**—Counterforts are generally placed along the back of the wall, 15 to 18 feet apart, from centre to centre; their construction is in every way similar to that recommended for retaining walls.

They should be built simultaneously with the wall, and be well bonded into it.

349. **Relieving arches.**—The name of **relieving arches** is given to a range of arches resting against the back of a retaining wall to relieve it from the pressure, or a part of the pressure, produced by the earth behind. (Fig. 125.)

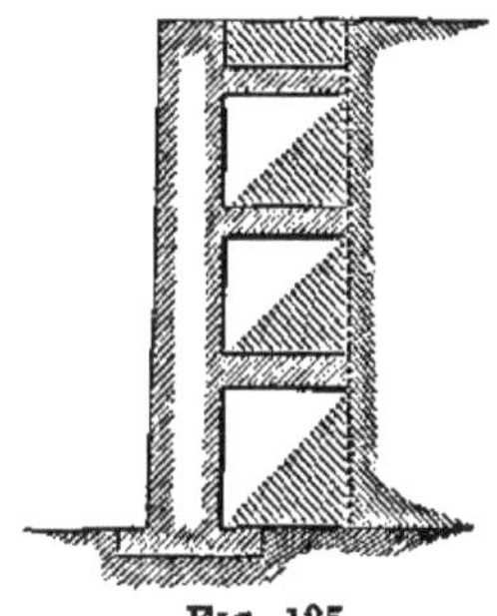

FIG. 125.

These arches have their axes placed at right angles to the back of the wall, and may have their fronts enclosed by the earth, as shown in the vertical section represented in Fig. 125. There may be one or several tiers of them.

Knowing the natural slope of earth to be retained, and assuming the length of the arch, its height can be deduced, or assuming the height, its length may be obtained, so that the pressure of the earth on the wall shall not exceed a given amount.

The distance between the centres of relieving arches is ordinarily about 18 or 20 feet. The thickness of the arch and piers will depend upon the weight they have to support.

AREAS, LINTELS, ETC.

350. These structures sustaining a vertical pressure either upwards or downwards, are subjected to a cross-strain.

Area.—It happens sometimes that an upward pressure is produced on an area by the presence of water; this pressure must be guarded against. The area of the new capitol at

Albany, N. Y., is several feet thick, and was made by first placing large flat stones over the surface, and then adding successive layers of broken stone and concrete.

Lintels.—The resistance to a transverse strain is very slight in stone; therefore the distance to be spanned by the lintel should be quite small, seldom exceeding *six* feet.

Plate-bands.—For a similar reason to that just given for lintels, the span of a plate-band should not exceed *ten* feet, and all pressure from above should be borne by some inter posing device.

ARCHES.

351. The form of the arch is generally assumed, and the number and thickness of the voussoirs are determined afterwards. The curves of right section of full centre, segmental, and elliptical arches require no further description, as the student has already learned the method of constructing these curves. The various ovals will be the only ones described.

Methods of Constructing Ovals.

352. The span and rise of an arch being given, together with the directions of the tangents to the curve at the springing lines and crown, an infinite number of curves, composed of arcs of circles, can be determined, which shall satisfy the conditions of forming a continuous curve, or one in which the arcs shall be consecutively tangent to each other, and the conditions that these arcs shall be tangent at the springing lines and the crown to the assumed directions of the tangents to the curve at those points. To give a determinate character to the problem, there must be imposed, in each particular case, certain other conditions upon which the solution will depend.

When the tangents to the curve at the springing lines and crown are respectively perpendicular to the span and rise, the curve satisfying the above general conditions will belong to the class of **oval** or **basket-handle** curves; when the tangents at the springing lines are perpendicular to the span, and those at the crown are oblique to the rise, the curves will belong to the class of **pointed** or **obtuse** curves.

The pointed curve gives rise to the **pointed** or **Gothic arch.**

If the intrados is to be an **oval** or **basket-handle**, and its rise is to be *not less than one-third of the span*, the *oval of three centres* will generally give a curve of a form more pleasing to the eye than will one of a greater number of centres; but if the *rise is to be less than a third of the span*, a curve of five, seven, or a greater odd number of centres will give the more satisfactory solution. In the pointed and obtuse curves, the number of centres is even, and is usually restricted to four.

353. **Three centre curves.**—To obtain a determinate solution in this case it will be necessary to impose one more condition which shall be compatible with the two general ones of having the directions of the tangents at the springing lines and crown fixed. One of the most simple conditions, and one admitting of a great variety of curves, is to assume the radius of the curve at the springing lines. In order that this condition shall be compatible with the other two, the length assumed for this radius must lie between zero and the rise of the arch; for were it equal to zero or to the rise there would be but one centre; and if taken less than zero or greater than the rise, then the curve would not be an oval.

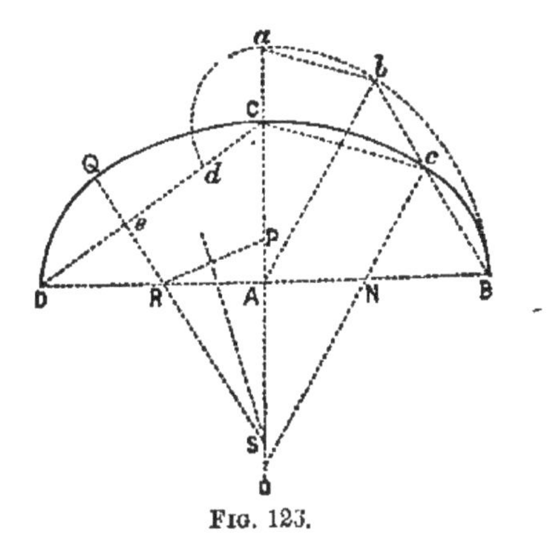

FIG. 123.

General Construction.—Let A D (Fig. 126) be the half span, and A C the rise. Take any distance less than A C, and set it off from D to R, along A D; and from C to P, along A C. Join R and P, bisect by a perpendicular. Prolong this perpendicular until it intersects C A produced. Then S, R, and a point on A B, distant from A equal to A R, will be the three centres of the required oval.

It is evident that there will be an infinite number of ovals for the same span and rise.

For, denote by R the radius S C of the arc at the crown, by r the radius R D at the springing line, by a the half span A D, and by b the rise A C.

There results from the right angled triangle S A R,

$$\overline{SR}^2 = \overline{AS}^2 + \overline{AR}^2,$$

or

$$(R - r)^2 = (R - b)^2 + (a - r)^2,$$

from which is obtained

$$R = \frac{a^2 + b^2 - 2ar}{2(b - r)},$$

which may be satisfied by an infinite number of sets of values of R and r.

354. **To construct an oval of three centres,** *with the condition that each of the three arcs shall be of* 60°.

Let B D be the span and A C the rise (Fig. 126). With the radius A B describe Bba of 90°; set off on it Bb = 60°; draw the lines ab, bB, and Ab; from C draw a parallel to ab, and mark its intersection c with bB; from c draw a parallel to Ab, and mark its intersections N and O with A B, and C A prolonged. From N, with the radius N B, describe the arc Bc; from O, with the radius Oc, describe the arc Cc. The curve BcC will be the half of the one satisfying the given conditions, and N and O two of the centres.

355. **To construct an oval of three centres** imposing the condition that the *ratio between the radii of the arcs at the crown* and *springing line* shall be a *minimum.*

Let A D be the half span, A C the rise (Fig. 126). Draw D C, and from C set off on it Cd = Ca, equal to the difference between the half span and rise. Bisect the distance Dd by a perpendicular, produced until it intersects C A prolonged. From the points of intersection, R and S, as centres, with the radii R D and S Q, describe the arcs D Q and Q C; and the curve D Q C will be the half of the one required.

For, from the triangle S A R, we get

$$\frac{R}{r} = \frac{a^2 + b^2 - 2ar}{(2b - 2r)r} \text{ for the ratio.}$$

Differentiating this expression, and placing its first differential coefficient equal to zero, $\dfrac{d\left(\dfrac{R}{r}\right)}{ar} = 0$, there results, after the terms are reduced,

$$r = \frac{a^2 + b^2 - (a-b)\sqrt{a^2+b^2}}{2a} = \frac{\sqrt{a^2+b^2}}{a}\left(\frac{\sqrt{a^2+b^2}-(a-b)}{2}\right),$$

but $\sqrt{a^2+b^2} = DC$, and $\sqrt{a^2+b^2} - (a-b) = Dd$, hence the given construction.

When the rise is less than one-third of the span, ovals of three centres are not of so pleasing a shape, and one of five or even a greater number of odd centres must be used.

356. **To construct an oval of five centres.**—This oval may be constructed as follows (Fig. 127):

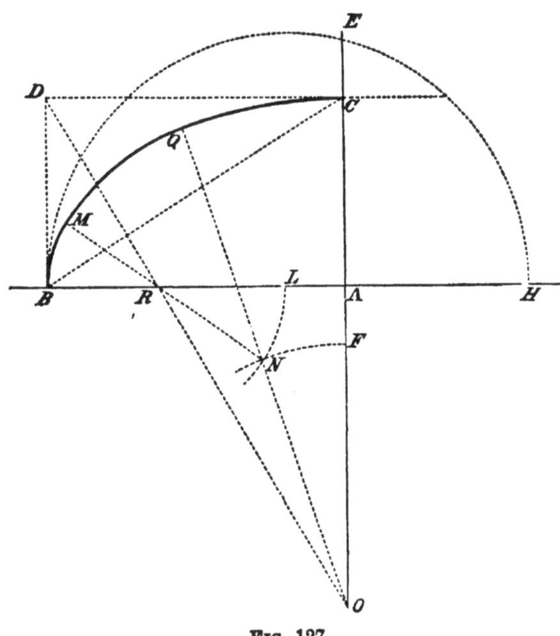

FIG. 127.

Let A B be the half span, and A C the rise of the arch. Erect at B a perpendicular to A B, and lay off B D equal to A C. Join B and C, and through D draw D O perpendicular to B C, and produce it until it intersects C A prolonged. Lay off A H to the right of A equal to A C, and on B H as a diameter describe the semicircle B E H. From A on A O lay off A F equal to C E, and with O as a centre and F O as a radius describe the arc F N. Lay off from B, on B A, a distance B L equal to AE, and with R as a centre and a radius equal to RL describe the arc LN.

The points O, N, and R are the centres, and OQ, NM, and RB = RM are the radii of the arcs forming the oval.

In other ways, by assuming conditions for the radii of the two consecutive arcs from the springing line, other ovals of five or a greater number may be constructed.

The curve of the intrados of Perronnet's fine bridge at Neuilly, over the Seine, is an oval of eleven centres, the radius at the springing line being 21 feet, and at the crown 159 feet, the span being 128 feet, and the rise 32 feet.

357. **Ovals of four centres,** or **obtuse and pointed curves.**—Their constructions are analogous to those already given for three centres. For example—

To construct an oval of four centres.—One method is as follows:

Let AB (Fig. 128) be the half span, AC the rise of the required curve and CD the direction of the tangent to it at

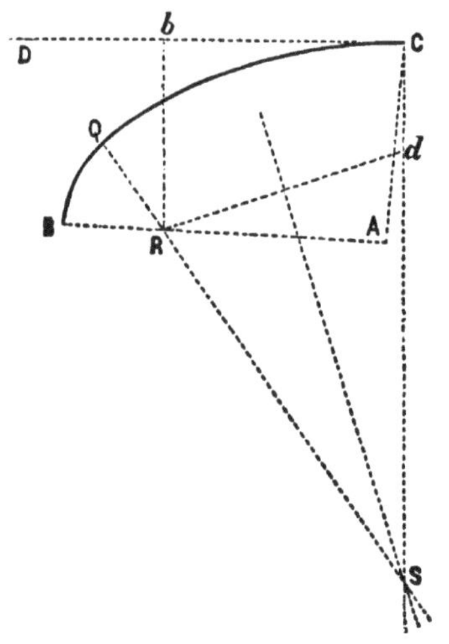

FIG. 128.

the crown. At C draw a perpendicular to CD. Take any point R on AB, such that RB shall be less than the perpendicular R*b* from R upon the tangent CD. From C, on the perpendicular to CD, set off C*d* equal to the assumed distance RB; draw R*d* and bisect it by a perpendicular, which prolong to intersect the one from C at the point S· through

S and R draw a line; from R, with the radius R B, describe an arc, which prolong to Q to intersect the line through S and R; from S, with the radius S Q, describe an arc which will be tangent to the first at Q and pass through C. The curve B Q C will be the half of the one required to satisfy the given conditions.

The four-centred **Tudor arch** is generally constructed as follows:

Let A B (Fig. 129) be the span, and divide it into four equal parts, the points of division being D, C, and D′.

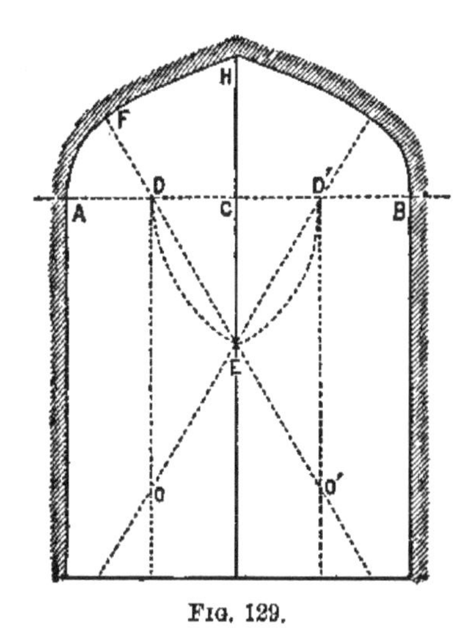

Fig. 129.

From D and D′, with a radius equal to D D′, describe arcs intersecting at E. Through E draw the lines D E and D′E, and produce them until they intersect the perpendiculars to the span through D and D′. With the radius D A describe the arc A F, and with the radius O′F the arc F H. The other half is drawn in a similar manner.

358. **Voussoirs.**—The form of intrados and depth of keystone being determined, the form of the extrados and the number of voussoirs are then fixed. The shape and dimensions of the voussoirs should be determined both by geometrical drawings and numerical calculation, whenever the arch is important, or presents any complication of form. The drawings should be made to a scale sufficiently large to determine the parts with accuracy, and from these, **pattern drawings**

may be constructed giving the parts in their true size. To make the pattern drawings, the side of a vertical wall or a firm horizontal area may be prepared with a thin coating of mortar, to receive a thin, smooth coat of plaster of Paris. The drawing is then made on this prepared surface by constructing the curve by points from its calculated abscissas and ordinates, or, where it is formed of circular arcs, the centres falling within the limits of the prepared surface, by using the ordinary instruments for describing such arcs. To construct the intermediate normals, whenever the centres of the arcs do not fall on the surface, an arc with a chord of about one foot may be set off each side of the point through which the normal is to be drawn, and the chord of the whole arc, thus set off, be bisected by a perpendicular. This construction will generally give a sufficiently accurate practical result for elliptical and other curves if of a large size.

From the pattern drawings thus constructed, **templets** and **bevels** are made which guide the stone-cutter in shaping the angles and surfaces of the voussoirs.

The methods of representing the voussoirs by projections, and from them deducing the true dimensions and forms of the joints, are discussed in "STONE CUTTING."

359. **Bond.**—The same general principles are followed in arranging the joints and bond of the masonry of arches, as in other masonry structures. The surfaces of the joints should be normal to the soffit, and the surfaces of any two systems of joints should be normal to each other at their lines of intersection. These conditions, with respect to the joints, will generally be satisfied by tracing upon the soffit its lines of least and greatest curvature and taking the edges of one series of joints to correspond with one of these systems of lines, and the edges of the other series with the other system, the surfaces of the joints being formed by the surfaces normal to the soffit along the respective lines in question. Whenever the surface of the soffit is a single curved surface, the joints will be thus either plane or developable surfaces.

Hence, in the right cylindrical arch the edges of one series of joints will correspond to the right line elements of the cylindrical surface, while those of the other will correspond to the curves of right section, the former answering to the line of least, and the latter of greatest curvature. The surfaces of the joints will all be plane surfaces, and, being normal to the soffit along the lines in question, will be normal also to each other.

In full centre and segmental arches, the voussoirs are

usually made of the same breadth, estimated along the curve of right section. In the right cylindrical arches of other forms of right section, it may not in many cases be practicable to give to all the voussoirs the same breadth, owing to the variable curvature of the right section; but the arrangement is the same throughout all the ring courses.

By this arrangement of the joints in the right arch, the joints are normal to each other and the coursing joints are very nearly perpendicular to the pressure they have to support.

360. **Oblique or askew arches.**—When the obliquity is considerable, this arrangement of the coursing joints cannot be used for the oblique arch, as the pressure would be very oblique to the coursing joints.

The best method for the coursing joints in this case, when the heading joints are taken parallel to the face of the arch, is to trace curves on the soffit at right angles to the edges of the heading joints, and take these curves as the edges of the coursing joints (Fig. 130). The projections of these edges on the plane of the springing lines are logarithmic curves, and give the name **logarithmic** to this method.

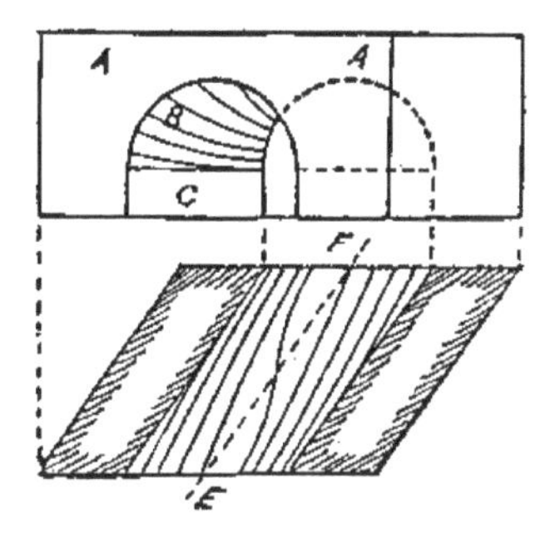

FIG. 130.—Elevation and plan of an oblique cylindrical arch, with the edges of the coursing joints constructed by the logarithmic method.

The logarithmic method makes the voussoirs in a course variable in width, and gives joints difficult to execute.

Another method is much used in preference, by which the coursing joints are kept parallel to each other. This method, known as the **helical method**, consists in tracing on the soffit for the edges of the heading joints, helices that are parallel to the helix which passes through the extremities of the span and rise of the face of the arch (Fig. 131).

Helices are then drawn on the soffit perpendicular to the first set and taken for the edges of the coursing joints.

The logarithmic and helical methods are used for arches

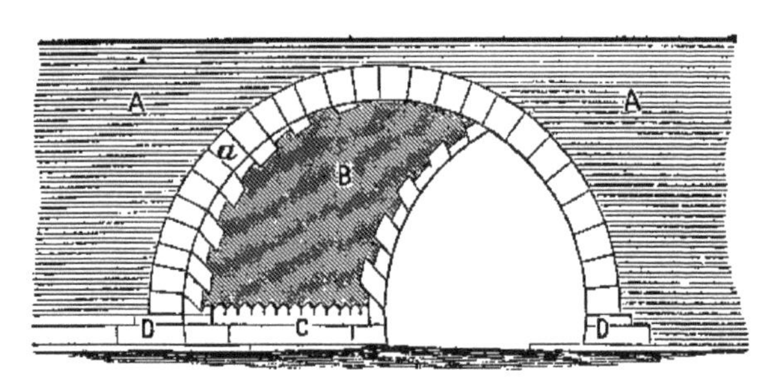

FIG. 131.—Elevation of an oblique cylindrical arch with helical joints.
a, voussoirs of cut stone.
c, *c*, bottom course of stone voussoirs cut to receive the brick courses.
C, face of the abutment.
D, ends of the abutments.

of considerable obliquity. When the obliquity is slight, the arch may be built of separate ribs, each rib slightly overlapping the one adjacent, or it may be a right arch supported on piers of trapezoidal cross-section.

CONSTRUCTION OF ARCHES.

Arches may be either of **stone, brick,** or **mixed** masonry.

361. **Arches of stone.**—In wide spans, and particularly in flat arches, cut stone alone should be used.

Rubble stone may be used for very small arches, which do not sustain much weight, or as a **filling** between a network of the ring and string courses of larger ones. In both cases the blocks should be roughly dressed with the hammer, and the best of mortar should be used.

362. **Arches of brick.**—Brick may be used alone or in combination with cut stone for arches of considerable size. The brick used may be wedge-shaped, or of the common form. There is no difficulty in wedge-shaped bricks accommodating themselves to the curved shape of the arch. In common brick this accommodation can be partially effected by making the joints thicker towards the extrados than towards the intrados.

Brick arches are often built in concentric rings, each half

a brick thick, the connection of the rings depending upon the tenacity of the mortar. Continuous joints are thus formed parallel to the soffit, and are liable to yield on the arch settling. The layers are called **shells**. This method should not be used in arches of more than thirty feet span. Another mode of construction is to lay the bricks in ordinary string courses. In this method continuous joints are formed, extending from the soffit outward; they are necessarily very open at the back, and must be filled with mortar, pieces of slate, or other material.

To obviate the defects of both methods as much as possible, the arch may be constructed by building partly in one way and partly in the other; or, as it is termed, in **shells and blocks** (Fig. 132). This method is to use blocks of brickwork built as solidly as possible, separated at short intervals by portions of concentric rings. The bricks in the blocks

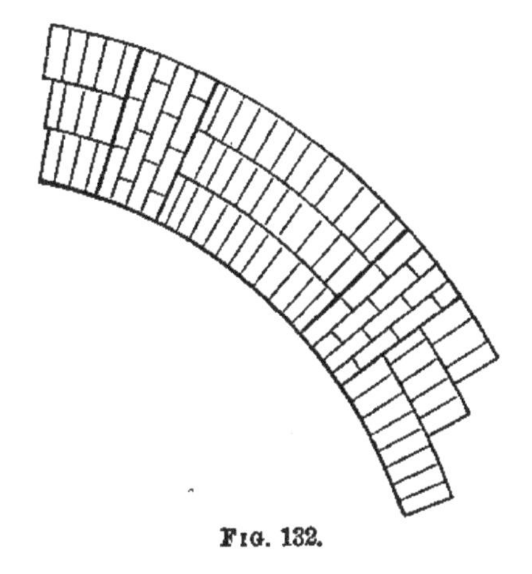

FIG. 132.

should be moulded or rubbed down to the proper form, especially in arches of importance. Pieces of hoop-iron laid in the joints would increase the strength of the bond.

363. **Arches of mixed masonry.**—When a combination of brick and cut stone is used, the ring courses of the heads, with some intermediate ring courses, the bottom string courses, the key-stone course, and a few intermediate string courses, are made of cut stone, the intermediate spaces being filled with brick (Fig. 133).

The voussoirs which form the ring course of the heads are

usually terminated by plane surfaces at the top and on the sides, for the purpose of connecting them with the horizontal

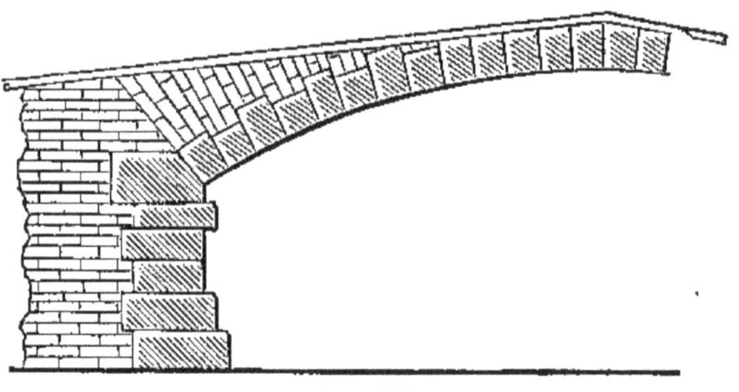

FIG. 133.

courses of the head which lie above and on each side of the arch (Figs. 134 and 135).

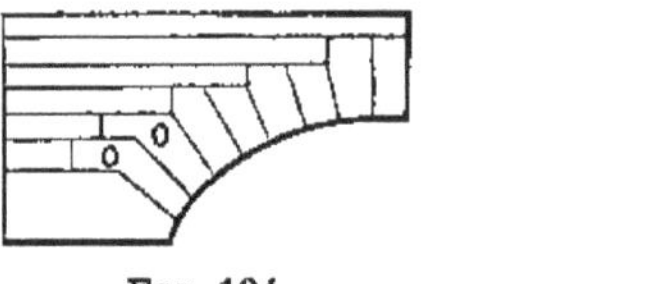

FIG. 134.

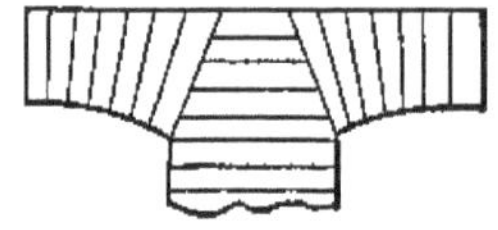

FIG. 135.

This connection may be made in various ways. The points to be observed are to form a good bond between the voussoirs and horizontal courses, and to give a pleasing architectural effect.

Sometimes the voussoir is so cut as to form an **elbow-joint,** as shown at O, O, in Fig. 134. This is objectionable both on account of waste of material in the cutting and from the liability of the stone to split when the arch settles.

364. **Cappings.**—When the heads of the arch form a part of the exterior of a structure, as when they are the faces of a wall or the outer portions of a bridge, then the top surface of the voussoirs of the ring courses, between the heads, is usually left in a roughly dressed state to receive the courses of masonry, termed the **capping,** which rest upon the arch between the walls of the head. Before laying the capping, the joints of the voussoirs on the back of the arch should be carefully examined, and, wherever they are found to be open from the settling of the arch, they should be filled.

The capping may be of brick, rubble, or concrete. When the arches are exposed to filtration of rain-water, as in bridges, casemates of fortifications, etc., the capping should be made water-tight.

The difficulty of forming water-tight cappings of masonry has led engineers to try a covering of asphalt laid upon concrete. This asphalt is put on as previously described, using sometimes several coats, care being taken to make the squares of each successive layer break joints with the preceding.

In a range of arches, like those of bridges or casemates, the top of the capping of each arch forms two inclined surfaces, like those of a common roof. The bottom of these surfaces, by their junction, form gutters where the water collects, and from which it is conveyed off in conduits, formed either of iron pipes or of openings made through the masonry of the piers.

When the space between the head walls above the capping is filled in with earth, a series of drains should be made running from the top or *ridge* of the capping, and leading into the main gutter drain. They are made of dry brick laid flat, with intervals, being covered by other courses of dry brick with open joints.

365. **Abutments and piers.**—The same care and precautions recommended in constructing retaining walls apply equally to the construction of abutments and piers.

When abutments, as in the case of buildings, require to be of considerable height, and would therefore demand extraordinary thickness if used alone to sustain the thrust of the arch, they may be strengthened by carrying them up above their connection with the arch, thus adding to their weight, as in the **battlements and pinnacles** of Gothic architecture; by adding to them ordinary, full, or arched buttresses, termed **flying buttresses**; or by using ties of iron below the key-stone to connect the voussoirs which are near the joints of rupture. The employment of these different expedients, their forms and dimensions, will depend on the character of the structure and the kind of arch. The iron tie, for example, cannot be hidden from view except in the plate-band, or in very flat segmental arches; and wherever its appearance would be unsightly some other expedient must be tried.

366. **Connection of the arch with its abutment.**—Care should be taken to make a firm connection between the lowest courses of the arch and the top of the abutment, particularly in the askew and segmental arches.

The top stone of the abutment, or cushion stone, should be well bonded with the stones of the backing; should be made thick enough to resist the pressure brought to bear on it; and made secure against any sliding.

Machinery Used in Construction.

367. Scaffolding and hoisting arrangements are necessary, and are in all things similar to those used for other stone masonry. In addition, strong frames called **centerings** are used. From the nature of an arch, formed as it is of separate pieces, it is evident that it could not be placed in position without some artificial support for the blocks to rest upon during construction. When the arch is completed the artificial support is removed, leaving clear the space arched over. This artificial support is called the **centre** or **centering** of the arch, and is made generally of wood.

A **centre** may be defined to be a wooden frame which supports the voussoirs of an arch while the latter is in progress of construction.

It consists of a number of vertical frames, termed **ribs**, upon which horizontal beams, called **bolsters**, are placed to receive the voussoirs of the arch. These ribs are placed from five to six feet apart, and have the upper or bearing surface curved to a figure parallel to that of the soffit of the arch. For an arch of considerable weight, the pieces forming the **back** of the centre on which the bolsters rest consist of beams of suitable lengths shaped to the proper curvature and abutting end to end, the joints between them being normal to the curved surface. The joints are usually secured by short pieces, or blocks, placed under the abutting ends and to which the pieces are bolted. The blocks are shaped so as to form abutting surfaces for struts which rest against them and against firm points of support beneath. To prevent the struts from bending, braces or bridle pieces are used, and the whole frame is firmly connected by iron bolts.

This is the general construction of a centre. The position of the points of support and the size of the arches will affect materially the combinations of the parts.

If for a light arch, as that thrown over a window or a door, planks instead of beams are used to form the back, and two ribs only are required. Their construction is shown in (Fig. 136).

In the figure, the centre is shown resting on the walls. If the intrados is to be tangent to the inner face of the walls,

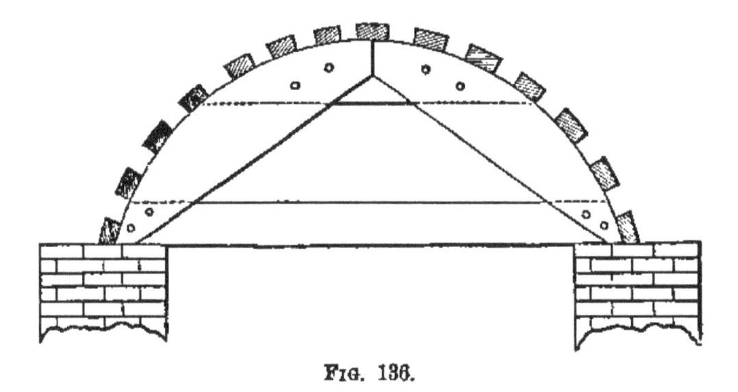

FIG. 136.

supports must be placed next to the wall, as shown in Fig. 137, to hold up the centre.

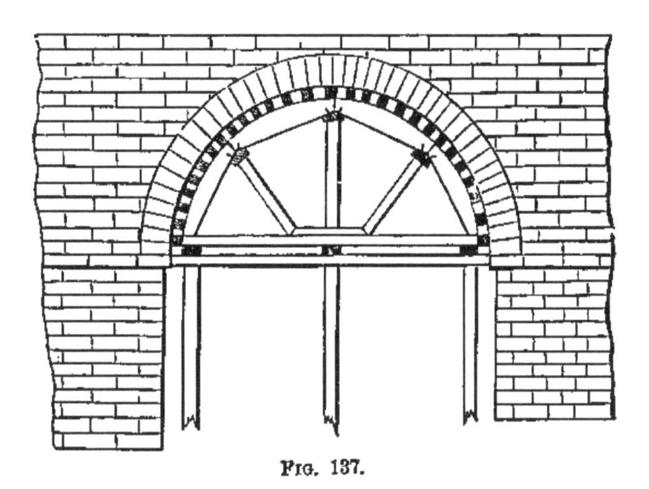

FIG. 137.

If the arch be heavier, an arrangement such as shown in Fig. 137 may be used, in which the back may consist of two or three thicknesses of plank nailed together, or of pieces of scantling of proper size.

The points to be considered in the construction of centres are, that the upper or bearing surface shall be correctly formed; that the centre shall be strong enough to bear the load which is to be placed upon it; that is, to support the weight of voussoirs, workmen, tools, etc., without sinking or changing its form during the construction of the arch; and that it may be easily and conveniently removed without injury when the arch is completed.

The most important centerings are those used in the construction of bridges of wide span, and of domes of important public buildings.

368. **General remarks.**—The rules given for laying ashlar or cut-stone masonry should especially be strictly observed in the construction of arches. The manner of laying the voussoirs which form the head of the arch demands peculiar care. The arch should be built up equally and simultaneously on the two sides of the centering, so that its construction should not be more rapid on one side than on the other. The load on the centering will in this way be kept symmetrical.

The centres, particularly of large arches, should not be removed until the mortar has set; it is recommended that, after removing the centre, the arch should be allowed to settle and assume its permanent state before any load is placed upon it.

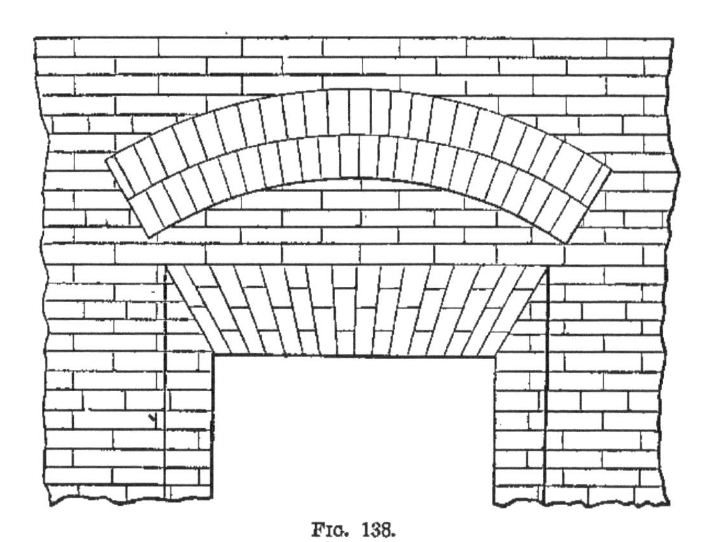

FIG. 138.

Very flat arches and plate-bands over doorways or wide openings in a wall have segmental arches placed above (Fig. 138) to relieve them from the weight of the wall which otherwise would rest upon them. From the object of these additional arches, they receive the name of **relieving arches.**

The principles of the arch should be thoroughly understood by the engineer as well as the architect.

The form of the arch will depend upon the purposes which it has to serve, the locality, and the style of architecture

The full centre arch is the strongest, and should be used when great strength is required and no limit to the rise is imposed. The elliptical is regarded as the most graceful arch, the segmental as the most useful.

Pointed arches are used in buildings, especially those of the Gothic order, but are not as a rule used for bridges or similar structures.

369. **Origin and use of the arch.**—It is a matter in question, to what country or people the world is indebted for the arch. But there is no doubt that Europe is indebted to the Romans for the general use of the arch in building. The full centre and segmental arches especially were much used by them in the construction of both public and private works, as temples, palaces, private residences, baths, sewers, bridges, aqueducts, etc., whose remains are still to be seen. The Romans were the first to use the dome for covering temples.

Afterwards, the arch under various forms became an essential element in the construction of buildings throughout Europe. And still later it forms in the United States a prominent feature of all our constructions, although it has not by us been used to the same extent in bridges as by Europeans.

GENERAL RULES TO BE OBSERVED IN THE CONSTRUCTION OF MASONRY.

370. From what has preceded, the following general rules may be stated:

1. To build the masonry in a series of courses, which shall be perpendicular, or as nearly so as practicable, to the direction of the force which they have to resist.

2. To avoid the use of continuous joints parallel to the direction of the force.

3. To use the largest stones in the lower courses.

4. To lay the lower courses, the force acting vertically, on their **natural bed.** Where great strength is required in these courses, the beds should be dressed **square.**

5. To moisten all dry and porous stones before bedding

them in mortar, and to thoroughly cleanse from dust, etc., their lower surfaces, and the bed of the course on which the stones are to be laid.

6. To reduce the space between each stone as much as possible, and to completely fill the joint with mortar.

PRESERVATION OF MASONRY.

371. Masonry is frequently injured by the mortar being washed out of the joints by the weather, by unequal settling, or by the expansion and contraction of the material due to changes of temperature.

372. **Pointing.**—The washing out of the mortar from the joints may be prevented by means of **pointing**. This consists in cutting out the mortar at the edge of the joint to a depth of about an inch, brushing it clean, moistening it, and filling it with pointing mortar.

The pointing mortar is made of cement paste and clean, sharp sand, about one measure of paste to two and a half of sand; or if mixed dry, one of cement to three of sand by weight. It is made in small quantities at a time, the ingredients being mixed with a little water, and thoroughly incorporated by pounding with an iron pestle in an iron mortar.

The pointing mortar is then pressed into the joint and its surface rubbed smooth with an iron tool. The practice with the United States engineers is to calk the joints with a hammer and calking-iron and to rub the surface of the pointing with a steel polishing tool.

To obtain a good pointing is quite difficult, as the unequal amount of contraction and expansion of the stone and the pointing mortar causes the latter to crack, or to separate from the stone. Water getting into the cracks and freezing throws out the pointing. Some builders give the surface of the pointing such a shape that the water shall trickle over the pointing without entering the cracks usually found between the stone and the pointing.

The period at which pointing should be done is not fully agreed upon by builders, some preferring to point while the mortar in the joint is still fresh, or **green**, and others not until it has become hard. The latter is the better plan; the former is the cheaper, as the joints are more easily cleaned out.

The term **flash-pointing** is sometimes applied to a thin coating of hydraulic mortar, made with a large proportion of hydraulic cement, laid over the face or back of a wall to protect the joints or the stone itself from the action of moisture and the weather.

When used to protect the stone, the sand in the mortar should be coarse, and the mortar applied in a single uniform coat over the surface, which should be thoroughly cleansed from dust and loose mortar, and well moistened before the application is made.

373. **Precautions against unequal settling.**—A certain amount of settling always takes place in masonry, due to the shrinkage of the mortar and other causes, and the engineer must take every precaution to ensure that this settling shall be *equal* throughout. Otherwise, especially in parts sustaining unequal loads, and which are required to be firmly joined together, the unequal settling that takes place is accompanied by cracks and ruptures in the masonry.

To avoid this unequal settling, it is advised to use the same thickness of mortar throughout, to pay particular attention to the bond and correct fitting of the courses, and to carry up all parts of the wall *simultaneously*. If the walls are to be subjected to heavy vertical pressures, it is recommended to take the further precautions of using hydraulic instead of common mortar, of requiring the materials to be uniform in size and quality, and of delaying putting the permanent load on the walls until the season after the masonry is laid. It is also suggested to use a proof load, when practicable, before placing on the permanent one.

374. **Effects of temperature on masonry.**—Frost is the most powerful destructive agent against which the engineer has to guard in masonry constructions. During severe winters in the northern parts of our country, it has been ascertained, by observation, that the frost will penetrate earth in contact with walls to a depth of ten feet; it therefore becomes a matter of the first importance to use every practicable means to drain thoroughly all the ground in contact with masonry to whatever depth the foundations may be sunk below the surface; for if this precaution be not taken, accidents of the most serious nature may happen to the foundations from the action of the frost. If water is liable to collect in any quantity in the earth around the foundations, it may be necessary to make small covered drains under them to convey it off, and to place a stratum of loose stone between the sides of the foundations and the surrounding earth to give the water a free downward passage.

It may be laid down as a maxim in building, that mortar exposed to the action of frost before setting will be so much damaged as to impair materially its properties. This fact shows the necessity of using hydraulic mortar to a height of at least three feet above the ground when laying foundations and the structure resting on them; for although the mortar of the foundations might be protected from the action of the frost by the earth around them, the parts immediately above would be exposed, and would attract the moisture from the ground, so that the mortar, if of common lime, would not set in time to prevent the action of the frosts of winter.

In heavy walls the mortar in the interior will usually be secure against the action of the frost, and masonry of this character might be carried on until freezing weather commences; but in all important works it will be the safer course to suspend the construction of masonry several weeks before the ordinary period of frost.

During the heat of summer the mortar is apt to be injured by drying too rapidly. To prevent this the stone or brick should be thoroughly moistened before being laid; and afterwards, if the weather is very hot, the masonry should be kept wet until the mortar gives indications of setting. The top course should always be well moistened by the workmen when quitting their work for any short period during very warm weather.

The effects produced by a high or low temperature on mortar in a green state are similar. In the one case the freezing of the water prevents a union between the particles of the lime and sand; and in the other, the same result arises from the water being rapidly evaporated. In both cases the mortar is weak and pulverulent when it has set.

375. **Repairs of masonry.**—In repairing masonry it is necessary to connect the new work with the old. To do this, the surface of the old, where the junction is to be made, should be arranged in steps and the mortar along this surface be scraped and cleaned. The new work is then joined to the steps by a suitable bond, care being taken to have the surfaces fitted accurately, and to use the least amount of mortar that will effect the required object.

MENSURATION OF MASONRY.

376. Engineers, when measuring or estimating quantities of masonry, state them in cubic feet or yards. Builders and contractors often use other modes, as perches of stone, rods of brickwork, etc. To avoid misunderstanding, the engineer should inform himself of the modes used in the locality where his work is to be built.

PART V.

FOUNDATIONS.

CHAPTER XI.

377. The term, **foundation**, is used to designate the lowest portion or base of any structure.

This term is frequently applied to that portion of the solid material of the earth upon which the structure rests, and also to the artificial arrangements which may be made to support the base.

It is recommended to restrict the use of the term, **foundation**, to the lower courses of the structure, and to use the term, **bed of the foundation**, when either of the other two are meant.

378. In the preceding chapters, the foundations of the structures there considered have been regarded as secure. Since the permanence of structures depends greatly upon the safety of the foundations, it is plain that the importance attached by engineers to the proper construction of the latter cannot be over-estimated.

379. Foundations are liable to yield either by sliding on their beds or by turning over by rotation about one of the edges. In general, if care is taken to prevent rotation, there need be no fear of yielding by sliding, especially if the bed is a hard ground or other compact material.

If the bed is of a homogeneous material and the pressure borne by the foundations is uniformly distributed over it, there will be no tendency to overturn, and the settling, which always exists to a greater or less extent, will be uniform throughout.

If the material forming the natural bed is not homogeneous, or the centre of pressure does not coincide with the centre of figure of the base, unequal settling will take place, followed by cracks and ruptures in the masonry, and finally, under certain circumstances, by the destruction of the work.

The main objects to be attained, in preparing the bed and foundation of any structure, are to reduce the settling to the smallest possible amount, and to prevent this settling from being unequal.

380. The beds of foundations are divided into two classes:

1. **Natural beds**, or those prepared in soils sufficiently firm to bear the weight of the structure; and

2. **Artificial beds**, or those which require an artificial arrangement to be made to support the structure, in consequence of the softness or want of homogeneousness of the soil.

Before a selection of the kind of bed can be made, it is necessary to know the nature of the subsoil. If this is not already known, it is determined ordinarily by digging a trench or sinking a pit close to the site of the proposed work, to a depth sufficient to allow the different strata to be seen. For important structures, the kind of subsoil is frequently made known by **boring** with the tools usually employed for this purpose.

When this method is used, the different kinds and thicknesses of the strata are determined by examining the specimens brought up by the auger used in boring.

381. Soils are divided, with reference to foundations, into three classes:

1. Those composed of materials whose stability is not affected by saturation with water, and which are firm enough to support the weight of the structure.

2. Those firm enough, but whose stability is affected by the presence of water.

3. Compressible or soft soils.

Rock, compact stony earths, etc., are examples of the first class; clay, sand, fine gravel, etc., are examples of the second; and common earth, marshy soils, etc., are examples of the third.

The beds are prepared either on land or under the water.

FOUNDATIONS ON LAND.

There will be three cases, corresponding to the three kinds of soil in which the bed is to be prepared.

I. BEDS PREPARED IN SOILS OF THE FIRST CLASS.

382. **Rock.**—When rock forms the material in which the bed is to be made, it is only necessary to ascertain if the rock

has a sufficient area, is free from cavities, and sufficiently thick to support the structure without danger of breaking. If the rock be found too thin, the nature of the soil on which it rests must be determined. If there are any doubts on any of these points, a thorough examination into the thickness of the stratum and tests upon its strength should be made. It is also recommended, in case of important structures, to test further its strength by placing on it a trial weight, which should be at least twice as great as that of the proposed structure.

Having become satisfied with the strength of the rock, all the loose and decayed portions are removed and the surface levelled. If some parts are required to be at a lower level than others, the bed should be broken into steps. Fissures should be filled with concrete or rubble masonry. If this filling should be too expensive, arches should be used. In some cases, it is advisable to cover the whole surface of the rock with a layer of concrete.

The load placed on the rock should not exceed the limit of safety. This limit is taken usually at one-tenth of the load necessary to crush the rock.

A bed in solid rock is unyielding, and appears at first sight to offer all the advantages of a secure foundation. It is found in practice, that in large buildings some portions will not rest on the rock, but on some adjacent material, as clay or gravel. Irregularity of settlement will in such cases almost invariably follow, and give great trouble.

383. **Compact stony earths, etc.**—The bed is prepared in soils of this kind by digging a trench deep enough to place the foundation below the reach of the disintegrating effects of frost. A depth of from four to six feet will generally be sufficient.

The bottom of the trench is made level, both transversely as well as longitudinally, and if parts of it are required to be at different levels, it is broken into steps. Care should be taken to keep the surface water out of the trench, and, if necessary, to have drains made at the bottom to carry away the water.

The weight resting on the bottom of the trench should be proportioned to the resistance of the material forming the bed. The limit for a firm soil of this class is about twenty-five pounds per square inch.

It is usual, in order to distribute the pressure arising from the weight of the structure over a greater surface, to give additional breadth to the foundation courses; this increase of breadth is called the **footing** or **spread.** In compact soils, the spread is made once and a half the thickness of the wall, and in ordinary earth or sand, twice that thickness.

II. BEDS IN SOILS OF THE SECOND CLASS.

384. The bed is prepared in a soil of this kind by digging a trench, as in the previous case, deep enough to place the foundation of the structure bel w the injurious effects of frost. Since the soil is effected by saturation with water, the ground should be well drained before the work is begun, and the trenches so arranged that the water shall not remain in them. And in general, the less a soil of this kind is exposed to the air and weather, and the sooner it is protected from exposure, the better for the work.

In this case, as well as in the preceding, it was supposed that the layer of loose and decayed materials resting on the soil in which the bed is to be prepared was of moderate depth, and that the thickness of the stratum in which the bed is made was sufficient to support the weight of the structure.

It sometimes happens that this firm soil in which the bed is to be made rests upon another which is compressible, or which is liable to yield laterally. In such situations, the weight of the structure should be reduced to its minimum, and should be distributed over a bearing surface sufficiently large to keep the pressure on any portion of the bed within certain limits. If there is any danger from lateral yielding, the bed must be secured by confining the compressible or yielding soil so it cannot spread out. This may be done by using sheeting piles, or other suitable contrivance.

III. BEDS IN SOILS OF THE THIRD CLASS.

385. **In soft earths.**—The bed is prepared, as in the other cases, by digging a trench sufficiently deep to place the foundation courses below the action of frost and rain.

Greater caution, however, must be observed in a case of this kind than in any of the preceeding, to prevent any unequal settling.

The bottom of the trench should be made level and covered with a bed of stones, sand, or concrete.

If stone be used, it is the practice to pave the bottom of the trench with rubble or cobble stones, which are well settled in place by ramming, and on this paving lay a bed of concrete.

If sand is used, the sand is spread in layers of about nine inches in thickness, and each layer well rammed before the next one is spread. The total depth of sand used should be sufficient to admit of the pressure on the upper surface of the sand being distributed over the entire bottom of the trench. (Fig. 139.)

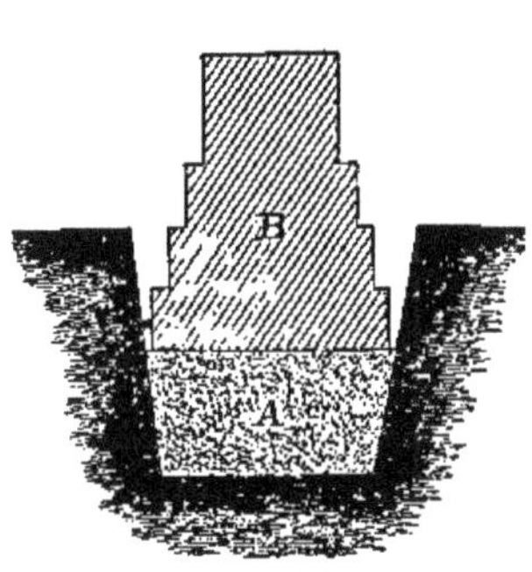

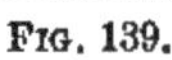
FIG. 139.

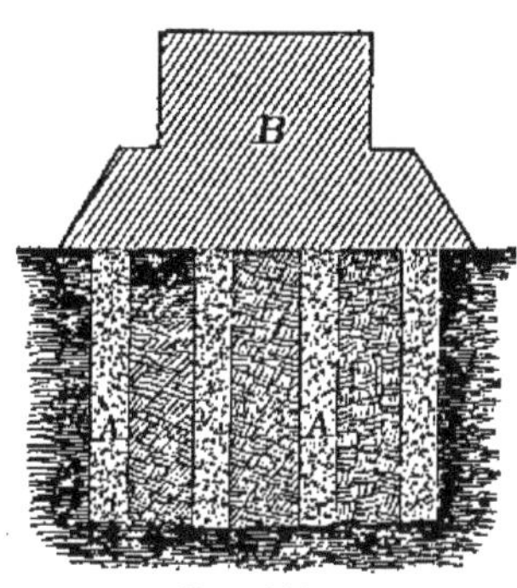

FIG. 140.

Another method of using sand for this purpose is to make holes in the soil or in the bottom of the trench (Fig. 140), and fill them with moist, well packed sand. The holes are about six inches in diameter and five or six feet deep.

Concrete may be used alone in the trench, or spread over a layer of stones well rammed in place. In either case, the concrete is spread in layers and rammed to form one compact mass. The upper surface is levelled off, and the foundation courses begun as soon as the concrete has *set*.

A concrete bed is also used when the soil is all sand; a trench is dug and the concrete laid as just described.

The pressure allowed on a concrete bed should not exceed one tenth part of its resistance to crushing.

By distributing the weight as nearly as possible uniformly over the foundation courses, the dangers of unequal settling may be avoided. If the structure rests on piers or other separate supports, these supports should be connected by inverted arches, and in this way the weight is distributed over the whole bed. If the weight of the structure varies in its different parts the surfaces of the bed should be proportioned accordingly, so as to have on each unit of surface the same amount of pressure

386. **In compressible soil.**—The principal difficulty met with in forming a sufficiently firm bed in a compressible soil arises from the nature of the soil and its yielding in all directions under pressure. There are several methods which have been used successfully in soils of this kind.

One method, when the compressible material is of moderate depth, is to excavate until a firm soil is reached, and then prepare the bed as described in the previous examples. The great objection to this method is the expense of excavation, especially when the depth of excavation is considerable.

A **second** method is to drive piles through the soft soil and into the firm soil beneath it. The piles are then cut off at a given level, fastened firmly together by heavy timbers, and a platform laid upon the top of the piles. On this platform the foundation courses of the structure rest.

A **third** is to use a modification of the last method. Instead of the piles reaching the firm soil, they are only driven in the compressible one. The platform is made to extend over so large an area that the pressure on the unit of surface produced by the weight of the structure is less than the limit allowed for this particular soil.

A **fourth** is also a modification of the second method, and differs from the last one in using piles of only five or six inches in diameter and five or six feet long. These piles are placed as close together as they can be driven, and support a platform, as in the second method. The object of the short piles is to compress the soil and make it firmer.

A **fifth** is to enclose the area to be covered by the structure by sheet-piles. The piles are driven to the firm soil, but not necessarily into it. The enclosed area is then covered with brush, fascines, or other similar materials, which are pressed down into the soft soil. When this upper layer is sufficiently firm, the foundation is begun.

This last method can only be used for small structures of a temporary nature. The stability of the construction depends almost entirely upon the power of the sheet-piles to resist the pressure transmitted to them by the compressible soil.

In general, if the firm stratum beneath the compressible soil can be reached by piles of ordinary dimensions, the second method is the one preferred, especially in those situations in which there is no danger of the piles rotting.

PILES.

387. **A pile** is a large piece of iron or timber, pointed at

one end, and driven or forced into the earth to be used generally as a support for some structure. Piles are classified, from the material of which they are made, into **wooden** and **iron**; from their length, into **short** and **long**; from the form of construction, into **round, square,** and **sheet-piles**; and from the method used to force them into the earth, into **common, screw,** and **pneumatic** piles.

388. **Short piles.**—These piles are usually round, from six to nine inches in diameter, and from six to twelve feet long, and made of timber, which may be oak, elm, pine, or other suitable wood, the particular kind depending upon the abundance of the wood in the vicinity of the work and the particular use to which the pile is to be placed. Their cross-section is sometimes a square. Their most general use is to compress and make firmer the soil in which they are driven.

389. **Long piles.**—These are either round or square in cross-section, and have a length of about twenty times their mean diameter of cross-section. The diameter of the small end should not be less than nine inches.

They are generally made of timber, the particular kind depending upon circumstances similar to those given for the short pile.

The long wooden pile is prepared for driving by having all knots and rough projections trimmed off, and having the end which is to enter the earth sharpened to a point.

This point should be kept on the axis of the pile, and the sharpening, which should extend for a distance equal to once and a half or twice the diameter should also be symmetrical with respect to the same line.

If the ground into which the pile is to be forced is stony or very hard, the lower extremity of the pile should be protected by an iron shoe. The shoe should be pointed, and may be made of cast iron.

The head of the pile should be protected from the blows used to force it down. This is usually effected by banding the head with a wrought-iron hoop, which is afterwards removed. Major Whistler's plan was to hollow out the head of the pile with an adze, the concavity in the head of the pile being made about one inch deep, and then to cover the head of the pile with a thin piece of sheet iron. By this means the piles were driven without injury.

As a rule, long piles are used to support a weight placed upon them. There are two cases, one in which the pile transmits the load to a firm soil, thus acting as a pillar; the other is where the pile and the load are wholly supported by the friction of the earth on the sides of the pile.

390. **Sheet-piles.**—These are flat piles of rectangular cross-section, driven side by side in a vertical position, or one that is nearly so, to form a **sheet.** The use of this sheet is either to prevent the materials enclosed by it from spreading out, or to protect them from the undermining action of water.

Sheet-piles are prepared for driving by having their edges fitted, so as to ensure a close contact. Sometimes each pile is "tongued and grooved," but this method is hardly ever necessary, for if the sides of the piles in contact are parallel and the piles well driven, the swelling of the wood by the water will ensure a sufficiently tight joint.

The sheet-piles are kept in position while they are being driven by resting them against horizontal pieces firmly bolted to guide-piles. The lower end of the sheet-pile is cut with an inclined edge for the purpose of giving the pile a **drift** towards the one next to it.

391. **Iron piles.**—Short, long, and sheet-piles are frequently made of iron. In many situations, iron piles can be used to advantage; it is not probable, however, that they will ever supersede those made of wood.

The long iron pile, when solid, is made of wrought iron. The best form for those of cast iron is tubular. The iron pile is forced into the earth either by means of a screw or by the pneumatic process. If a cast iron pile is to be forced down by blows on the head, a wooden punch must be used to avoid the danger of the breaking of the cast iron from the blows.

Sheet-piles of cast iron have been frequently used, especially in coffer-dams. They are from fifteen inches to two feet wide, half an inch thick, and generally strengthened by flanges or vertical ribs. The joints are made tight by making each pile overlap the two adjacent ones.

The difficulty of driving iron piles so that all their heads shall be on the same level is a serious objection to their use in many cases. This objection does not apply to their use in a coffer-dam, as it is of no consequence about having the heads of the piles on the same level.

392. **Screw piles.**—They are either of wood or iron. Generally they are made of iron. The screw blade is ordinarily of cast iron, fixed on the foot of the pile, and seldom consists of more than one turn. The diameter and the pitch of the screw vary with the nature of the soil and the load to be supported.

The piles are made either hollow or solid. The hollow

piles are of cast iron, from one to three feet in diameter, and generally cast in convenient lengths, which are afterwards connected together. Fig. 141 shows a cast-iron pile of the ordinary kind; it is about two feet and six inches in diameter. Solid piles are made of wrought iron, and are from four to nine inches in diameter. Fig. 142 shows one with a cast-iron screw.

Screw piles are applicable for use in sand, gravel, clay, soft rock, and alluvial soils. They can be forced into very hard soils, even into brickwork. To force them into the earth, it is usual to fix upon the top of the pile a capstan, and to apply the power to the levers which turn it. A strong frame-work is needed to hold the pile in its place while it is being screwed down.

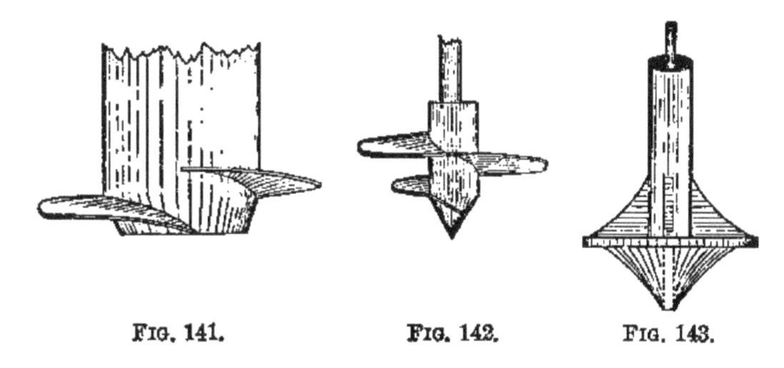

FIG. 141. FIG. 142. FIG. 143.

393. **Disk piles.**—These are hollow iron piles with the base enlarged by a broad disk attached to the foot (Fig. 143). They have been used successfully in light sand.

To sink them, the top is closed except where a tube of small diameter is inserted. Through this small tube, water is forced at high pressure by a force-pump, and as the water rushes out at the base of the pile, the sand is disturbed and the pile descends by its own weight. When it has descended far enough, the pumps are stopped, and the sand settling around the pile holds it firmly in position. Great caution should be observed to settle the foot of the pile some distance below the **scour**, or that point where there is danger of the sand being afterwards disturbed by water or any other cause.

394. **Pneumatic piles.**—These are iron cylinders often used instead of common piles to reach a firm stratum which lies below both water and a bed of soft material, as in the case of a bottom of a river.

The piles are sunk through this soft material in two ways,

either by exhausting the air from the interior of the cylinder, thus producing a pressure on the head of the pile; or by forcing air into the tube, thus driving the water out, so that workmen are able to descend to the bottom of the pile and remove any obstructions to its settling. The details of these methods will be given in another article.

395. **Means used to force common piles into the earth.**—Short, long, and sheet-piles of wood are forced into the earth most generally by blows delivered on the heads of the piles. The machines used for this purpose are called "**pile-drivers,**" and are of various kinds. The most common of these consists essentially of a large block of iron which slides between two uprights, termed **guides** or **leaders**. This block, called the **ram** or **monkey**, having been drawn to the top of the guides, is let fall and comes down on the head of the pile with a violent blow, forcing the pile into the soil.

The pile-driver may be worked by hand, horse, or steam power.

The simplest form of pile-driver is the **ringing engine.** In this machine the ram is attached to one end of the rope; the rope passes over a pulley, and its other end branches out into a number of smaller ropes, each held by a man. The men, all pulling together, lift the ram a few feet; then at a given signal all let go, and the ram falls on the pile. The number of men required will depend upon the weight of the ram. It is usual to allow about forty pounds to each man.

In the machine commonly used, the ram is raised by the power being applied to a windlass. The ram is held while being hoisted by tongs or **nippers,** the handles of which, when the ram has been raised to the proper height, come in contact with two inclined planes on the guides; these surfaces press the handles of the tongs together, open the tongs and let the ram fall. The tongs are so arranged that upon being lowered they catch hold of the ram by a staple or other contrivance on its upper surface.

If the piles are to be driven in an inclined position, it is only necessary to incline the guides. As a rule, the direction of the pile should be parallel to the pressure it has to support.

396. Other machines are frequently used to drive piles. The most important one is an application of the steam hammer. In this driver, the hammer is attached to a piston-rod which moves in a cylinder fixed on the top of a wrought-iron case between the guides.

The steam hammer is well adapted for continuous rows of piles, and can be economically used where there are a great number of piles to be driven, and where they are near each other.

In the ordinary pile-driver, the pile is driven by a comparatively small mass descending from a considerable height. But with the steam hammer, the pile is forced into the earth by the rapid blows of a heavy mass, delivered upon a block weighing several tons, placed directly over the head of the pile. The blows are given at the rate of one a second, and the hammer is raised each time only to a height equal to the stroke of the piston.

Various methods have been used in different machines for raising the ram. In some cases the pressure of the atmosphere has been tried with success. In one machine the explosive properties of gunpowder are the means used.

397. If the head of a pile has to be driven below the level to which the ram descends, another pile, termed a **punch**, is used for the purpose. A cast-iron socket of a suitable form embraces the head of the pile and the foot of the punch, and the effect of the blow is thus transmitted through the punch to the pile.

The manner of driving piles, and the extent to which they may be forced into the subsoil, will depend on local circumstances. It sometimes happens that a heavy blow will effect less than several lighter blows, and that piles, after an interval between successive volleys of blows, can with difficulty be started. Piles may be driven in rocky soils and even in rock itself, if holes are first made whose diameters are a little less than those of the piles. In this case the piles should be shod with an iron shoe. Careful attention is required in driving, for a pile has been known to break below the surface and to continue to yield under the blows of the ram by the crushing of the fibres of the lower end.

The test of a pile having been sufficiently driven, according to the best authorities, is that it shall not sink more than one-fifth of an inch under thirty blows of a ram weighing 800 pounds, falling five feet at each blow. A more common rule is to consider the pile fully driven when it does not sink more than one-fourth of an inch at the last blow of a ram weighing 2,500 pounds, falling 30 feet.

The least distance apart at which piles can be driven with ease is about two and one-half feet between their centres. If the piles are nearer than this, they force each other up during the driving. The average distance is generally about three feet.

If a pile has to be drawn out, as is often the case, a lever fastened by a chain to the head of the pile may be used. Where the pile is only partially driven, it may sometimes be drawn by fastening a chain around the head of the pile and attaching it to the nippers.

398. **Load on piles.**—Col. Mason's formula is

$$W = \frac{R^2}{R+p} \times \frac{h}{d},$$

in which W is the greatest load, R and p the weight respectively of the ram and pile, all in pounds; h the fall of the ram, and d the penetration of the pile at the last blow, both in feet. He used a factor of safety of 4.

Capt. Sanders' formula is $W = \frac{R}{8} \times \frac{h}{d}$; the quantities being the same excepting that W is the **safe** load.

The rule used by builders is to limit the load to 1,000 pounds on the square inch of the head when the pile transmits the weight to firm soil; and to 200 pounds when it resists by friction only.

399. **Preparation of bed in compressible soil, using common wooden piles.**—The piles having been driven to the firm soil beneath, their heads are sawed off at a given level and the whole system is firmly connected together by longitudinal and cross pieces notched into each other and bolted to the piles. On these piles a platform is laid; or the soft earth around the top of the piles is scooped out for five or six feet in depth, and this space filled with concrete.

If a platform is to be used, it is constructed as follows: A large beam, called a **capping,** is first placed on the heads of the outside rows of piles and is fastened to them by iron bolts, or wooden pins termed **treenails.** Sometimes an occasional tenon is made on the piles, fitting into a corresponding mortise in the capping. Other beams are then laid resting on the heads of the intermediate piles, with their extremities on the cappings, and are then bolted firmly to the piles and cappings. Another set of beams are laid at right angles to these, and are bolted to the piles. Where the beams cross each other, they are both notched so as to have their upper surfaces in the same plane. The beams which have their lengths in the direction of the longer sides of the structure are known as **string** pieces, and the other set are termed **cross** pieces.

A platform of thick planks is laid upon the upper surface of the beams and is spiked to them.

The cappings are sometimes of larger size than the other beams, in which case a **rabbet** is made in the inner edge so as to have the platform flush with the upper surface of the capping.

The whole construction is called a grillage and platform. (Fig. 144.)

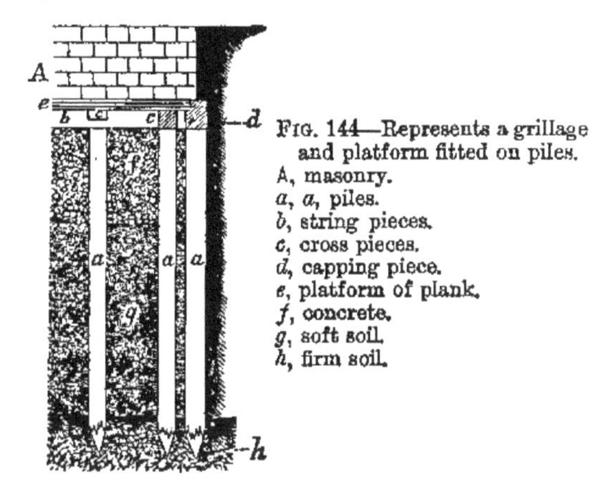

FIG. 144—Represents a grillage and platform fitted on piles.
A, masonry.
a, a, piles.
b, string pieces.
c, cross pieces.
d, capping piece.
e, platform of plank.
f, concrete.
g, soft soil.
h, firm soil.

400. When the firm stratum into which the piles have been driven underlies a soil so soft that there is doubt of the lateral stability of the piles, the soft soil should be scooped away and stones should be thrown between and around the piles to increase their stiffness and stability. (Fig. 145.)

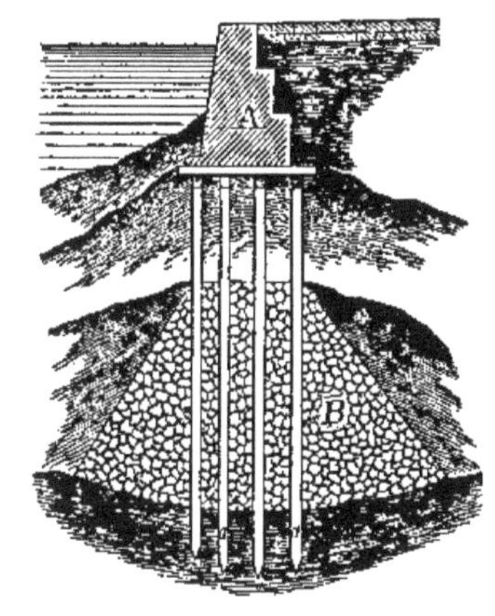

FIG. 145—Represents the manner of using loose stone to sustain piles and prevent them from yielding laterally.
A, section of the masonry.
B, loose stone thrown around the piles.

401. If the situation be such that decay in the timber is to be expected, the more costly method of excavation must be adopted.

The practical difficulty met when trenching in such cases, is the presence of water in such quantities as to seriously impede the work, even to the extent often of failure.

Pumps are used to keep the water out, and it may even be necessary to enclose the entire area by a sheet-piling. In this case, two rows of sheet-piles are driven on each side of the space to be enclosed, through the soft material and into the firm stratum beneath. The soft material between the rows is then scooped out, and its place filled with a clay puddling, forming a water-tight dam around the space enclosed. If the water comes from springs beneath the dam or from within the area enclosed, this method will fail, and it may be necessary to resort to some of the methods used for laying foundations under water.

CHAPTER XII.

FOUNDATIONS IN WATER.

402. Two practical difficulties meet the engineer in preparing beds of foundations under water. *One* is to make the necessary arrangements to enable the workmen to prepare the bed; and the *second*, having prepared the bed, to secure it against the deteriorating effects of the water and to preserve its stability.

Preparation of the bed.—The situation in which the bed is to be prepared may be either of two kinds: one is where it may be prepared without excluding the water from the place; and the other is where the water must be excluded from the area to be occupied before the bed can be made.

PREPARATION OF BED WITHOUT EXCLUDING THE WATER.

403. **Concrete beds.**—A bed of concrete is frequently used in water. To prepare the bed, the upper layer of loose, soft soil is removed by a dredging-machine or by other means, and the site is made practically level. The concrete is laid within this excavation. A conduit made of wood or iron, or a box or contrivance which opens at the bottom when lowered in position, may be used in laying the concrete.

A cylindrical conduit of boiler iron, made in sections of suitable lengths which can be successfully fastened on or detached as the case requires, has been used with success. The lower end of the conduit has the form of a frustum of a cone. The whole arrangement is lowered or raised and moved about at pleasure by means of a crane. The concrete is placed in the conduit at the upper end, and by a proper motion of the crane is spread in layers as it escapes from the lower end. By lifting and dropping the apparatus the layers can be compressed.

Bags filled with concrete have been used, with a moderate degree of success, for the same purpose.

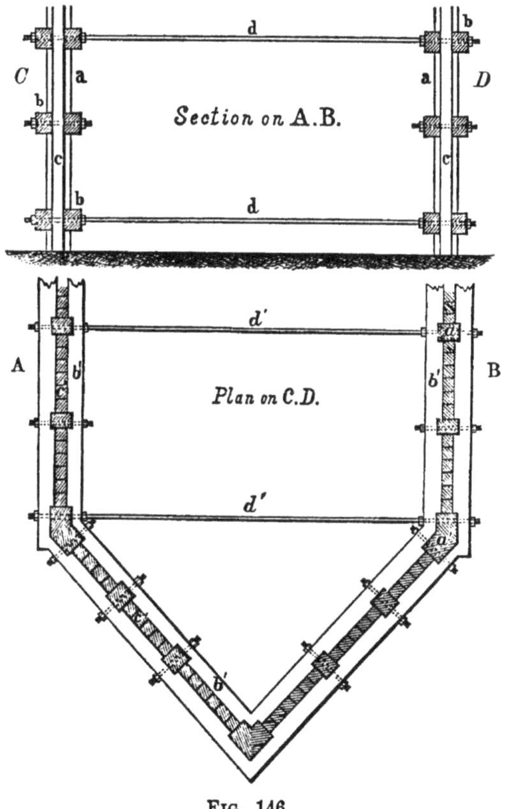

FIG. 146.

The object to be attained is to get the concrete placed in position in as nearly as possible the same condition as when it is made. If it be allowed to fall some distance through water, or be placed in a strong current, the ingredients of the concrete are liable to be separated.

Where the site is in flowing water, it is often necessary to provide some arrangement which, by enclosing the area of the site, will calm the water within the enclosure, and will thus prevent its injurious effect upon the fresh concrete before it has set.

404. The arrangement shown in Figure 146 was used for this purpose. It consisted of a framework composed of uprights connected together by longitudinal pieces in pairs; each pair being notched on and bolted to the uprights, leaving an interval through which sheet-piles were inserted. The sheet-piles were driven into close contact with the bottom, which was rock. The frame was put together on the shore and then floated to its place. It was secured in position by inserting the uprights in holes drilled in the rock. The sheet-piles c, c', were then inserted between the horizontal pieces b, b', and rested on the bottom. The whole area was thus enclosed by a wooden dam, within which the water was quiet. The concrete was then laid on the bottom of the enclosed space. To prevent the sides of the dam from spreading out iron rods d, d, d', d', were used to connect them.

405. **Beds made of piles.**—Common wooden piles are frequently used to form a bed for the foundation courses of a structure. They are driven through the soft soil into the firm stratum beneath, and are then sawed off on a level at or near the bottom. On these are laid a grillage and platform or other suitable arrangement to receive the lower courses. Where the bottom is suitable for driving piles, and there is no danger of scour to injure their stability, this method is economical and efficient. The foundation courses must be placed in position by some submarine process, as by the use of a diving-bell, or by means of a caisson.

406. **Common caisson.**—This caisson (Fig. 147) is a water-tight box, whose sides are ordinarily vertical, and which are capable of being detached after the caisson has been sunk in position. The bottom of the caisson, as it is to form a part of the foundation of the structure, is made of heavy timbers, and conforms in its construction to that of a grillage and platform.

The size of the timbers for the bottom is determined by the weight of the structure which is to rest on them, and for the sides, upon the amount of pressure from the water when the caisson rests on its bed.

The sides are generally made of scantling, covered with thick plank. The lower ends of the scantling or uprights fit into shallow mortises made in the cap pieces of the grillage. Beams are laid across the top of the caisson, notched upon the sides and projecting beyond them. These cross pieces are connected with the lower beams of the grillage by long iron bolts, which have a hook and eye joint at the lower end and a nut and screw at the upper. After the bolts are unscrewed at the top, they can be unhooked at the bottom, the cross beams raised, and the sides of the caisson detached.

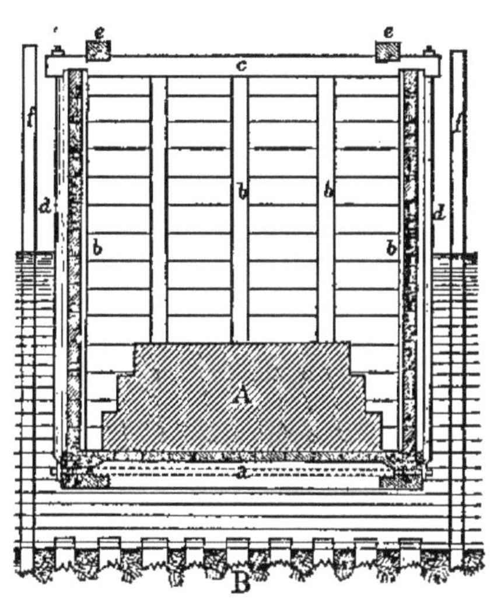

Fig. 147—Represents a cross-section and interior end view of a caisson. The boards are let into grooves in the vertical pieces instead of being nailed to them on the exterior.
a, bottom beams let into grooves in the capping.
b, square uprights to sustain the boards.
c, cross pieces resting on b.
d, iron rods fitted to hooks at bottom and nuts at top.
e, longitudinal beams to stay the cross pieces c.
A, section of the masonry.
B, bed made of piles.
f, guide piles.

In a caisson which was used in building a bridge pier, the exterior dimensions of the principal parts were nearly as follows:

The caisson was 63 feet long, 21 feet wide, and 15 feet deep. The cross beams on top were made 10 inches square in cross-section, and were placed about three feet apart; the uprights were of the same size as the cross pieces, and were placed about six feet apart.

Much larger caissons have been used, especially in some of the engineering constructions in England.

The caisson is built at some convenient place where it can be launched and towed to the position it has to occupy. The bed having been prepared by levelling off the bottom or by driving piles, the caisson is floated to and moored over the spot. The masonry courses are then laid on the bottom of the caisson, and are built up until the caisson rests on its bed. Just before it reaches the bed, it is sometimes settled in place, by admitting water into the interior, and an examination made as to its proper position. If it does not occupy its proper place, and there is a desire to change the position of the caisson, the gates by which the water was admitted are shut and the water is pumped out. The removal of the water will allow it to float and a rectification of its position may then be effected.

The caisson having been satisfactorily settled in position, the masonry is built above the surface of the water, and the sides are then detached and removed.

Caissons are frequently used whose sides are not detached. This is especially the case where the sides are of a permanent character. These might be termed permanent caissons.

407. **Permanent caissons.**—Caissons built with brick sides and timber bottoms were used to construct the sea-wall at Sheerness, in England, in 1811–12. After being sunk, they were filled with concrete.

Rankine mentions a kind that are built wholly of bricks and cement, and which are filled with concrete after being sunk in place.

408. **Diving-apparatus.**—The bed may be prepared as on dry land, provided some apparatus be used which will admit of the workmen executing their labors notwithstanding the presence of the water. Submarine or diving armor and diving-bells are devices which are frequently used for this purpose.

I. **Submarine armor.**—This is an apparatus to be used by a single person, and consists essentially of a metallic helmet from which the water is excluded by atmospheric pressure. The helmet encloses the man's head; rests upon his shoulders and is connected with an air and water-tight dress which he wears. He is supplied with fresh air forced through a flexible tube entering at the back of the helmet; a valve opening outwards allows the foul air to escape. To enable him to see, the helmet is provided with eye-holes protected by strong glass.

II. **Diving-bell.**—The form of diving-bell, commonly used, is that of a rectangular box with rounded corners. Holes protected by strong glass about two inches thick are made in the top to admit light into the interior. Fresh air is forced through a flexible tube into the bell by means of air-pumps. The bell is raised and lowered by means of a crane and windlass.

A bell, whose dimensions are four feet wide, six feet long, and five feet high on the inside, is of convenient size for laying masonry under water.

The diving-bell has been much used in laying submarine

foundations where there was no scour and where the bed was easily prepared.

409. **Pierre perdue.**—The methods just given are applicable to structures of moderate dimensions, but when the area occupied by the bed is very considerable, these methods are either inapplicable or require modifications. One known by the French as **pierre perdue** has been frequently used. It consists in forming an artificial island of masses of loose stone thrown into the water, and allowing the stone to arrange themselves. This island is carried up several feet above the surface of the water and the foundations are built upon it.

The structure should not be commenced until the bed has fully settled. If there is any doubt about this, the bed should be loaded with a trial weight, at least twice as great as that of the proposed structure.

This method can not be used in navigable rivers or other situations where it is of greater importance not to contract the water-way.

410. **Screw piles.**—Iron screw piles have been used with success for foundations in localities where the methods already mentioned were not practicable. They do not differ, in principle, from the common wooden pile. Iron piles last well both in fresh and salt water; whereas wooden piles can not be relied upon at all in salt water, and they will not last in fresh water unless entirely submerged.

Iron screw piles have been much used, in the United States, in the construction of light-houses on or near sandspits at the entrance of our harbors and on shoal spots off the coast, where it would be almost impossible to prepare the beds by any of the other more usual methods.

411. **Well foundations.**—In India, a method known as **well** or **block** foundations has been quite extensively used, especially in deep sandy soils. The method consists in sinking a number of wells close together, filling them with masonry, and connecting them together at top.

The method of sinking one of these wells is to construct a wooden curb about a foot in thickness; its cross-section being the same as that of the well, and to place it in position on the proposed site. On this curb a cylinder of brickwork is built to a height of about four feet. As soon as the mortar has set, the sand is scooped out from under the curb, and it descends, carrying with it the masonry. When the curb has settled about four feet, another *block* or height of masonry is added, and again the sand is scooped out from under the curb, and the whole mass descends as before. This process is then repeated and carried on until the curb has reached the required depth. Care must be taken to regulate the excavation so that the cylinder shall sink vertically.

From the very nature of the soil, water is soon met. As long as the water can be kept out either by bailing or by pumping, the work proceeds with rapidity. If the water comes in so fast that it cannot be exhausted by these means, the sand must be scooped out by means of divers or by some other method. Under these circumstances the excavation proceeds slowly and with difficulty.

When the curb reaches a firm stratum, or a depth where there is no danger of the foundations being affected by the water, the bottom is levelled, a concrete bed made, and the interior of the cylinder filled in solid with masonry. If the concrete bed is made without exhausting the water, the latter is pumped out as soon as the concrete sets, and the masonry is then built in the usual manner.

Cylinders of boiler iron have been used in the same way as the masonry curbs, and are an improvement upon them.

412. **Iron tubular foundations.**—This is a general name applied to large iron cylinders which are sunk through water and a soft bottom to a firm soil, and used to support a given structure in the same manner as common piles. The number and size of the tubes depend upon the weight to be supported and the means adopted to sink them.

The method just described for the well is frequently used for the iron tubes. Brunel, the English engineer, in building the Windsor Bridge, on the Windsor branch of the Great Western Railway, employed this method in constructing the abutments of the bridge. There were in each abutment six cast-iron cylinders, each six feet in diameter, and they were sunk to the proper depth by excavating the earth and gravel for the interior with dredges and by forcing the cylinders down by weights placed on the top of each one.

The concrete bed in the bottom was made by lowering the concrete in bags, which were arranged so that by pulling a rope the bags were emptied under the water in the proper place. When a sufficient quantity had been put in and had hardened, the water was pumped out and the cylinders filled in the usual manner.

This method does not differ in principle from a foundation on piles, and the same general rules apply as to the amount of load to be supported and the depth to which the pile is to be driven.

In some cases a clump of common piles was driven within

the cylinder at the bottom, and the spaces filled with concrete. In some of the recent constructions the piles extend to the top of the cylinder.

PREPARATION OF BED, THE WATER BEING EXCLUDED.

413. There are two cases: where the water is excluded by means of a dam, and where it is excluded by atmospheric pressure.

I. EXCLUSION OF WATER BY DAMS.

The dams used are the common **earthen** or **clay** dam, the common **coffer-dam**, and modified forms of the coffer-dam.

414. **Earthen dam.**—In still water not more than four feet deep, a dam made of earth or ordinary clay is usually adopted to enclose the given area and to keep out the surrounding water. This dam is made by digging a trench around the area to be enclosed and removing the soft material taken out; the earth or clay is then dumped along the line of this trench until it rises one or two feet above the surface of the water; as the earth is dumped in place it should be firmly pressed down, and when practicable, rammed in layers. Any good binding earth or loam will be a suitable material for the dam.

The dam being finished, the water within the enclosed area is pumped out, and the bed and foundations constructed as already prescribed for those "on land."

415. **Coffer-dam.**—Where the water is more than four feet deep, and especially if in running water, the common earthen dam would be generally too expensive a structure, even if it could be built. In a case of this kind, and where the water does not exceed twenty-five feet in depth, the common coffer-dam is usually employed.

The common coffer-dam (Fig. 148) is essentially a clay dam, whose sides are vertical and retained in position by two rows of piling.

The common method of constructing the coffer-dam is to drive two parallel rows of common piles around the area to be enclosed; the distance between the rows being equal to the required thickness of the dam, and the piles in each row being placed from four to six feet apart.

The piles of each row are then connected by horizontal

beams, called **string** or **wale** pieces, which are notched on and bolted to the piles on the outside of each row, about one foot above the highest water mark. On the inside of the rows, and nearly opposite to the wale pieces, are placed string pieces of about half the size, to serve as guides and supports to the sheet-piles.

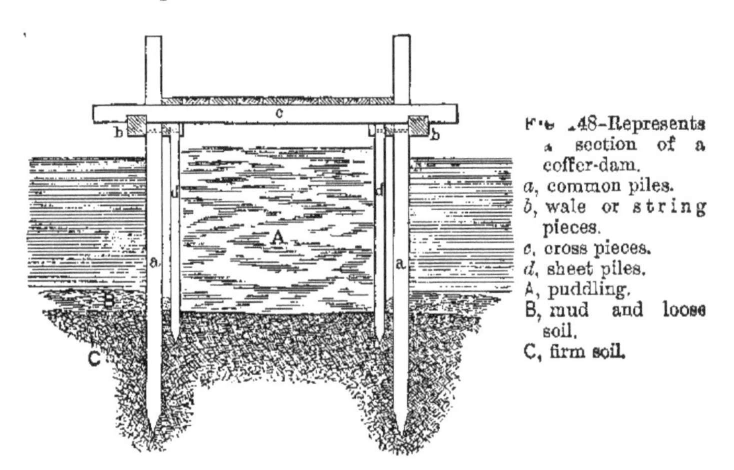

Fig. 148—Represents a section of a coffer-dam.
a, common piles.
b, wale or string pieces.
c, cross pieces.
d, sheet piles.
A, puddling.
B, mud and loose soil.
C, firm soil.

The two rows of piles are tied together by cross pieces notched on and bolted to the outer wale pieces. Upon these cross pieces are laid planks to form a scaffolding for the workmen and their tools, etc.

The sheet-piles are driven in juxtaposition through the soft soil and in contact with the firm soil beneath. They are about four inches thick and nine inches wide, and are spiked to the inner string pieces. Sometimes an additional piece known as **a ribbon** piece, is spiked over the sheet-piles.

These rows of sheet-piles form a **coffer** for the puddling whence the name of the construction. The sheet-piles having been driven and secured to the string pieces, the mud and soft material between the rows are scooped or dredged out.

The puddling which forms the dam is then thrown in and pressed compactly in place, care being taken to disturb the water as little as possible during the operation. When the top of the puddling rises to its required height, pumps are used to exhaust the water from the enclosed area. The interior space being free from water, the bed of the foundation is prepared as on dry land.

The puddling is composed of clay mixed with sand or

gravel, or of fine gravel alone, freed from all large stones, roots, or foreign material which may be mixed with it. The clay is worked into a plastic condition with a moderate amount of water, and then mixed thoroughly with a given quantity of sand or fine gravel. Care is taken that there are no lumps in the puddling after the mixing.

The dam is given the required strength ordinarily by making the thickness equal to the height of the dam above the ground or bottom on which it is to rest, when this height does not exceed ten feet. For greater heights the thickness is increased one foot for every additional height of three feet.

This rule gives a greater thickness than is necessary to make the dam water-tight, but adds to its stability. The stability of the dam is sometimes still further increased by supporting the sides of the dam by inclined struts, the upper ends of which abut against the inner row of common piles, and the lower ends against piles driven for that purpose into the ground.

416. The principal difficulties met with in constructing a coffer-dam are as follows:

First, To obtain a firm hold for the common piles; a difficult thing to do in deep muddy or rocky bottoms;

Second, To prevent leakage between the surface of the ground and the bottom of the puddling;

Third, To prevent leakage through the puddling;

Fourth, To exhaust the water from the enclosed area after the dam is finished.

These difficulties and the expense of construction of the dam, increase very greatly with the depth of the water. In deep water, the size and length of the piles and the amount of bracing required to resist the pressure of the water render the expense very great.

Common piles can not be efficiently used where the bottom is rocky. In a case of this kind, the following construction was successfully used:

Instead of the common piles, two rows of iron rods were used. These rods were "jumped" into the rock, a depth of fifteen inches. The sheet-piles were replaced by heavy planks which were laid in a horizontal position and fastened to the rods by iron rings. This method of fastening allowed the planks to be pushed down until each one rested on the one below it; the plank resting on the bottom being cut to fit the surface of the rock.

The frame was strengthened by bolting string pieces of

timber in pairs on both of its sides and by using inclined struts upon the interior.

The puddling was of the usual kind and was put in the dam in the way already described.

417. It will be very difficult to avoid leakage between the bottom of the puddling and the soil on which it rests unless the stratum of overlying soft soil be removed. It is therefore recommended for important works that a part of the dredging for this purpose be done before the common piles are driven.

Leakage through the puddling is mostly due to poor workmanship. If the sheet-piles are fitted and carefully driven, and the puddling is free from lumps and thoroughly mixed, leakage through the dam should not occur. It is not advisable to have bolts or rods passing through the dam, as leakage almost invariably takes place through the holes thus made. Fine gravel alone has been proved in some cases to be a better material for the filling than ordinary puddling.

Leakage due to springs in the bottom of an enclosed area is the great source of trouble, and in some soils is stopped with much difficulty. It may be necessary to fill in the whole area with a bed of concrete, and after it has set to pump out the water.

418. The water having been pumped out, the enclosed space is drained into some convenient spot in the enclosure, and arrangements are made to keep the interior dry. The bed having been prepared, the masonry is then built to the proper height. When it is above the surface of the water, the dam may be removed, and as there is danger of disturbing the bed if the piles were drawn out, it is customary to cut them off at some point below the water line, letting the lower ends remain as driven.

419. **Caisson dams.**—This name was given to a coffer-dam in which the outer row of common piles was replaced by structures resembling caissons, which were sunk and ballasted to keep them in position along the line which would have been occupied by the common piles.

The character of the bottom and the nature of the stream were such that common piles could not be used for the dam.

The caisson (Fig. 149) was a flat-bottomed boat, which having been floated to its place was sunk gradually, by the admission of water, until it rested on the bottom. A row of common piles was then placed in a vertical position against each side of the caisson and lowered until they rested on the bottom. They were then bolted in that position to the sides of the caisson. The caisson was then heavily loaded with stones and other weighty materials, until a considerable weight rested on the piles. It is observed, that instead of the piles being held fast by being driven into the ground, they are held in place by the sunken boat, and the whole arrangement takes the place of the outer row of piles in the common coffer-dam.

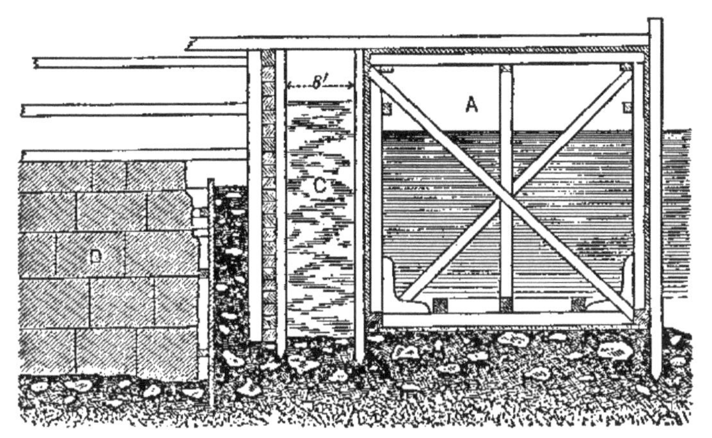

FIG. 149—Represents a cross-section of a caisson dam.
A, cross-section of caisson. C, puddling.
D, foundation courses of the pier.

To complete the dam, a row of posts, parallel to the inner row of piles, resting on the bottom and connected by a framework with the caissons, took the place of the inner row of piles in the common coffer-dam.

The sheet-piles were required only on the one side, the sides of the caissons being sufficient on the other. They were laid in a horizontal position, as shown in the figure. The puddling was in all respects the same as that described in the previous cases.

The masonry being finished, the loads were removed from the caissons. They were then pumped dry and the dam removed.

420. **Crib-work dam.**—A dam in which a crib ballasted with stone takes the place of the common piles, has been used with success.

In the example (Fig. 150), the cribs were built by laying the logs alternately lengthwise and crosswise, and fastening them together at their intersections by notching one into the other and pinning them.

On each crib a platform was laid about midway between the top and bottom, on which the stone was placed to sink the crib. The cribs were floated to the place they were to occupy and sunk gradually by loading stone on the platform. After they had been fully settled in their place, more stones were piled on until the required stability was secured.

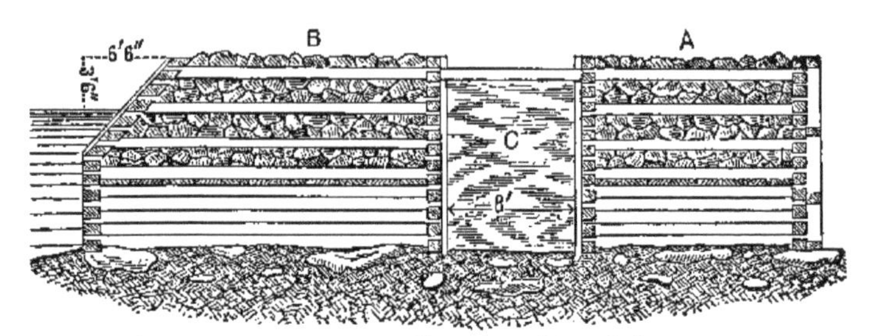

FIG. 150—Represents a cross-section of a crib-work dam. A, inner row of cribs. B, outer row of cribs. C, puddling.

Both of the preceding methods were used in constructing the piers and abutments of the Victoria Bridge, over the Saint Lawrence, at Montreal. A rocky bottom, covered with boulders, prevented the driving and the use of the common pile as in the ordinary method. There was also in the river a swift current, which in the spring of the year brought down large quantities of ice, the effect of which would have been to have destroyed any ordinary caisson or common coffer-dam.

It is seen that these dams do not differ in principle from the common coffer-dam, and that the modifications in each case consisted in finding for the common pile a substitute which would be stronger and equally effective.

II. EXCLUSION OF WATER FROM THE SITE BY ATMOSPHERIC PRESSURE.

421. In recent years, the use of compressed air has been extensively adopted as a means for excluding the water from the site of a proposed work, while the bed was being prepared.

There are two general methods of its application: in the pneumatic pile and in the pneumatic caisson.

422. **Pneumatic piles.**—Pneumatic piles are hollow vertical cylinders of cast iron, from six to ten feet in diameter, intended to be forced through soft and compressible materials to a firm soil beneath, and to be then entirely filled with

masonry or concrete or other solid material. Rankine classes them under the head of iron tubular foundations.

Their general construction and the mode of sinking them in the soil are shown in Fig. 151.

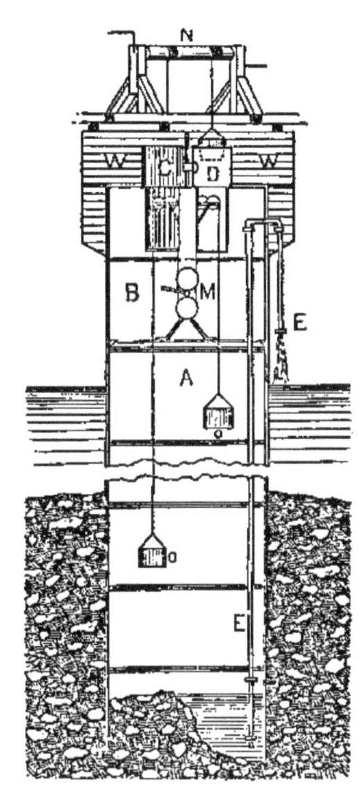

FIG. 151—Represents vertical section of a pneumatic pile.
A, body of cylinder.
B, the bell.
C, elevation of air-lock.
D, vertical section of air-lock.
E, water discharge pipe.
M, windlass on inside.
N, windlass on the top.
O, O, buckets ascending and descending.
W, W, iron weights.

In this example, shown in the figure, the cylinders were cast in lengths of nine or ten feet, with flanges on the interior at each end. These pieces were united by screw bolts passing through holes in the flanges, the joints being made water-tight either by an india-rubber packing or by a cement made of iron turnings.

To sink a pile of this kind, a strong scaffolding is erected over the site, and from which the lengths of the cylinders can be lowered and placed in position. On this scaffold a steam-engine is ordinarily placed, and furnishes the power required during the operation.

The lower edge of the lowest section of the cylinder is sharpened so that it may sink more easily through the soil.

The upper section, termed the "bell," is usually made of boiler iron, with a dome-shaped or flat top. An "air-lock" is used to pass the men and materials in and out of the cylinder. In this example there were two air-locks, which were placed in the top of the bell, as shown in the figure. Each lock had at the top a trap door which opened downwards, and at the side a door which opened into the interior of the pile. Stop-cocks were provided in each, communicating with the external air and the interior of the pile, respectively; they could be opened or closed by persons inside the tube, within the lock, or on the outside.

The bell was provided with a supply pipe for admission of compressed air, a pressure gauge, a safety valve, a large escape valve for discharging the compressed air suddenly when necessary, and a water-discharge pipe about two or three inches in diameter.

Windlasses placed within the cylinder and on the outside, as seen in the figure, were used to hoist the buckets employed in the excavation

The first operation in sinking the pile was to lower the lowest section, with as many additional lengths united to it as were necessary to keep the top of the cylinder two or three feet above the surface of the water, until it rested on the bottom. The bell and one additional length were then bolted to the top of the pile.

The weight of the mass forced it into the soil at the bottom of the river a certain distance, dependent upon the nature of the soil. As soon as the pile stopped sinking, the air was forced in by means of air-pumps worked by the steam-engine, until all the water in the tube was expelled. Workmen, with the proper tools, then entered the cylinder by means of the air-locks.

To get into the pile, the men entered the lock, closed all communications with the external air, and then opened the stop-cock communicating with the interior of the pile; in a few minutes the compressed air filled the lock, the men opened the side door and thus effected an entrance into the interior. To pass out it was only necessary to reverse this operation.

The gearing of the hoisting apparatus was so arranged that the buckets, when filled, were delivered alternately into the locks, and were then hoisted out by the windlass on the outside.

Care was taken to guard against the uplifting force of the compressed air within the pile. In the above example, a heavy weight, composed of cast-iron bars resting on brackets attached to the outside of the bell, was used to resist this action.

The workmen having descended to the bottom of the pile, excavated the material to the lower edge; they then took off the lowest joint of the water discharge pipe and carried it and their tools to the bell, and passed out of the lock. The valve for admitting compressed air was then closed and the large escape valve opened, allowing the compressed air to escape. The cylinder being deprived of the support arising from the compressed air, sank several feet into the soil, the distance depending on the resistance offered by the soil.

When the pile had stopped sinking, the escape valve was closed, the air forced in, and the operations just described continued. Great care was taken to keep the pile in a vertical position while sinking.

The pile, having reached the required depth, was then filled with concrete.

The usual method of filling the pile is to perform about one-half of the work in the compressed air and then remove the bell and complete the rest in the open air. In filling with concrete, it should be well rammed under the flanges and around the joints.

423. This description of a pneumatic pile, just given, is that of one of the piles used in the construction of a bridge over the river Theiss, at Szegedin, in Hungary.

The river, at this point, has a sluggish current with a gradual rise and fall of the water, the difference between the highest and lowest stages of water being about twenty-six feet. The soil of the bottom is alluvial, composed to a great depth of alternate strata of compact clay and sand.

The piles were sunk to about thirty feet below the bottom of the river, which latter was about ten feet deep at low water.

The excavation was carried down to within six feet of the bottom of the pile. Twelve common piles of pine were then driven within the cylinder, extending to a depth of twenty feet below it. The concrete was then thrown in and rammed in layers until its upper surface was on a level with that of ordinary low water.

The air-locks were about six feet and a half high and two and three-quarters in diameter.

424. In the first uses of the pneumatic piles, the cylinders were of small size, as many being sunk as were required to support the load, as in the use of common piles.

They were sunk into the soil by exhausting the air from the interior. The result following this removal of air was that the earth immediately under the pile was forced together with water, into the inside of the cylinder, and the pile sank into the opening thus made, both under its own weight and the pressure of the atmosphere.

This process is known as Dr. Pott's, and is well adapted to soft or sandy soils, when free from stones, roots, pieces of timber, etc. The presence in the soil of any obstacle which the edge of the tube cannot cut through or force aside, renders this method impracticable.

The next step was to increase the size of the pile. and instead of exhausting the air, to fill it with compressed air. The top being closed and the bottom open, all fluid matter was driven from the interior of the pile by the compressed air. By means of air-locks on the top of the cylinder, workmen were enabled to descend and remove the soil and such obstructions as prevented the pile from sinking. This process is generally known as "Triger's."

The air being compressed in the interior of the pile, the weight or the pressure downward was much lessened. To increase the pressure a weight was placed on the pile.

Although many improvements have been made in the details, the arrangements just described illustrate the general outline of all the pneumatic methods in use.

425. **Pneumatic method used by Mr. Brunel.**—The first improvement in the pneumatic method was that used by Mr. Brunel in preparing the bed for the centre pier of the Royal Albert Bridge, at Saltash, England.

This improvement consisted in confining the compressed air to a chamber at the bottom of a cylinder, the rest of the space inside of the cylinder being open to the air. The air chamber communicated with the outside air by means of a tube, six feet in diameter, with air-locks at the upper end. Outside of this tube, was another tube, ten feet in diameter, connecting the dome with the outside air. (Fig. 152.)

A dome, about 25 feet high, was built in the lower portion, so arranged that the top of the dome should be above the mud when the cylinder rested on the rock.

The chamber for the compressed air was annular, four feet wide, twenty feet high, was built around the inner circumference of the lower edge and was divided into eleven compartments by vertical and radial partitions; apertures in the partitions afforded communications from one to the other. An air passage at the top of the compartments connected them with each other, and with the vertical tube of six feet diameter before alluded to.

The cylinder was lowered into the water exactly over the place it was to occupy. As soon as it stopped sinking, the annular chamber was shut off from the rest of the dome, the air forced in, the water driven out, the workmen descended and dug out the mud and loose soil under the edge.

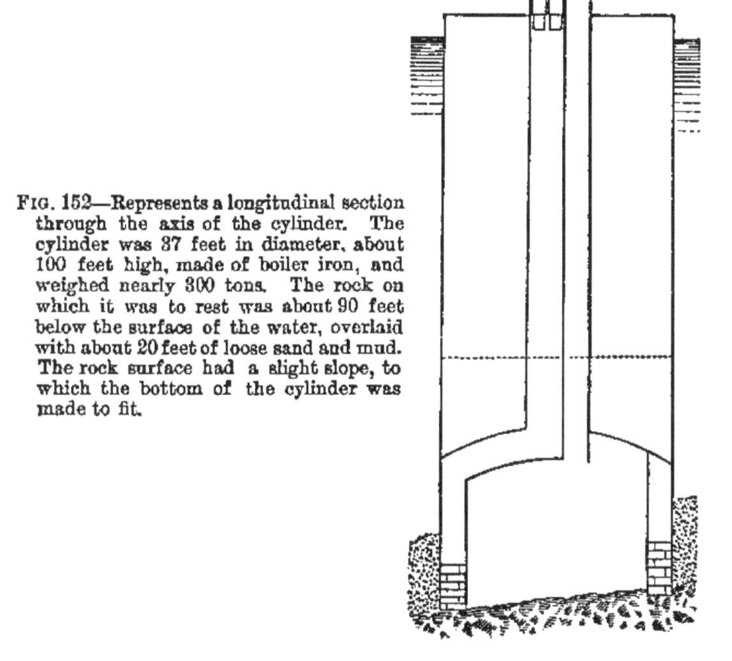

FIG. 152—Represents a longitudinal section through the axis of the cylinder. The cylinder was 37 feet in diameter, about 100 feet high, made of boiler iron, and weighed nearly 300 tons. The rock on which it was to rest was about 90 feet below the surface of the water, overlaid with about 20 feet of loose sand and mud. The rock surface had a slight slope, to which the bottom of the cylinder was made to fit.

When the rock was reached, a level bed was cut in its surface and a ring of masonry built. The water was then pumped out of the main tube and the masonry begun on the inside. As the masonry rose, the partitions, shaft, and the dome were removed. When the pier was above the surface of the water, the upper part of the cylinder, about fifty feet in length, was unbolted and taken away, it having been made in two sections for this purpose.

As the volume of the annular chamber in which the compressed air was used was small in comparison with the volume of the main cylinder, no extra weight was needed to balance the upward pressure.

The above is a good example of the pneumatic process combined with the principle of the coffer-dam.

426. **Pneumatic caisson.**—The next important modification in the pneumatic method was to combine the principle of the diving-bell with that of the common caisson. This combination is known as the **pneumatic caisson** and furnishes the means now most commonly used in situations like that at the Saltash bridge, and especially where the foundations have to support a great pressure.

It consists essentially of three parts: **1st, The caisson; 2d, The working chamber**; and **3d, The pneumatic apparatus** and its communications with the working chamber.

Caisson.—This does not differ in its principles of construction from the common caisson already described. The bottom is of wood or iron, made strong enough to support the structure with its load, and forms the roof of the working chamber. The sides are generally of wrought iron, and are not usually detached from the bottom when the structure is finished.

Working chamber.—This is below the caisson, and as just stated, the bottom of the caisson is the roof of the chamber. Its sides are firmly braced to enable it to resist the pressure from both the earth and water as it sinks into the ground. The chamber is made air and water tight.

Pneumatic apparatus and communications.—Vertical shafts, either of iron or masonry, passing through the roof of the chamber furnish the means of communication between the working chamber and the top of the caisson. The air-locks may be placed in the upper end of the shaft, as in the pneumatic pile, or at the lower end of the shaft where it connects with the working chamber.

The usual supply pipes, air-pumps, discharge pipes, etc., are required as in the other pneumatic methods.

Sinking the caisson.—It is moored over the place it is to occupy and is sunk gradually to the bottom as an ordinary caisson. Air is then forced into the working chamber, driving out the fluid matter; the earth and loose material are then dug out, while the caisson settles slowly under its own weight and that of the masonry until it rests on the firm soil or solid rock.

An outline description of some of the caissons recently used will more fully illustrate their construction and the method of sinking them.

427. **Pneumatic caissons used at L'Orient, France.**—These were used in laying the foundations of two of the piers of a railroad bridge over the river Scorff, at L'Orient, in France. The river bed consisted of mud from 25 to 45 feet deep, lying upon a hard rock. The surface of the water was about 60 feet above the rock at mean tide, and 70 feet at

high tide. It was essential for the stability of the piers that they should rest on the rock.

The caissons used were 40 feet long, 12 feet wide, and made of boiler iron.

The thickness of the iron forming the sides of the caisson varied according to the depth in the water, being greater for the lower than for the middle and upper parts. The ratio of the thickness was for the upper, middle, and lower, as 3, 4, and 5.

The working chamber was ten feet high and communicated with the upper chamber or bells, where the air-locks were placed, by two tubes for each bell; these tubes were each two feet and three-quarters in diameter. Each bell was ten feet high and eight feet in diameter, and contained two air-locks and the necessary hoisting gear; the full buckets ascended through one tube and descended through the other.

Fig. 153 shows the caisson used for the pier on the right bank.

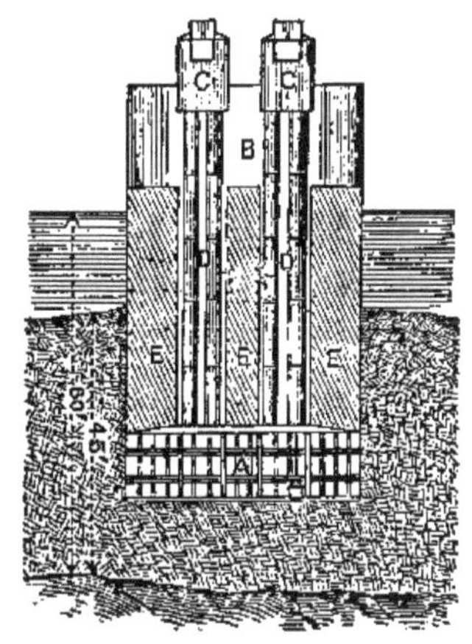

FIG. 153—Represents a vertical section of caisson and masonry of pier during the process of sinking.
A, the working chamber.
B, interior elevation of caisson.
C, C, elevation of the bells.
D, D, the communicating tubes.
E, E, masonry of pier, built as the caison was sinking.

When the rock was reached, its surface was cleaned off and a level bed made under the edges of the caisson. The working chamber was then filled up to the roof with masonry.

The pier was of concrete with a facing of stone masonry, and built up as the caisson was sinking to its place.

The working chamber being filled, the tubes were withdrawn and the spaces occupied by them filled with concrete.

Pneumatic Caissons at St. Louis, Mo.

428 At the time the foundations of the piers of the bridge over the Mississippi River, at St. Louis, were laid, the caissons there used were the largest that had ever been employed for such a purpose.

This bridge consists of three spans, supported on two piers

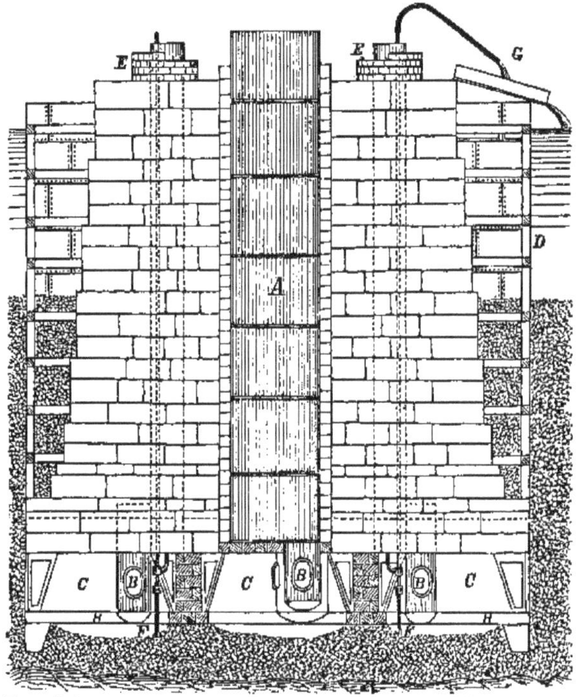

FIG. 154—Represents a section of the caisson used in construction of east pier of the bridge over the Mississippi River, at St. Louis, Mo.
A, main shaft. B, air-locks. C, working chamber. D, sides of caisson. E, side shafts. F, sand pumps. G, discharge of sand.

and the abutments. The river at this point is 2,200 feet wide at high water, with a bed of sand over rock. The rock slopes from the west to the east, the upper surface of the sand being practically level. The depth of the sand on the western shore was about 15 feet, and on the eastern nearly 100 feet.

As the scour on the bottom is very great in the Mississippi River, it was regarded as essential that the piers should rest on the rock. To penetrate this sand and lay the foundations on the rock, the pneumatic caisson was used.

Fig. 154 represents a section of the one used for the east pier. There the rock was 128 feet below the high-water mark. When the caisson was moored in position there was above the rock 35 feet of water and 68 feet of sand.

The plan of the caisson was hexagonal, the long sides being 50 feet each, and the short ones 35 feet each. The sides of the caisson were made of plate iron, three-eighths of an inch in thickness, and built up as the caisson sank.

The bottom, which was to support the masonry, was composed of iron girders, placed 5½ feet apart. Iron plates, ½ inch thick, were riveted to the under side of these girders to form the roof of the working chamber. The sides of the caisson, prolonged below the girders, formed the sides of the chamber, and were strongly braced with iron plates and stiffened by angle irons. The chamber, thus formed, was 80 feet long, 60 feet wide, and had an interior height of 9 feet. The interior space was divided into three, nearly equal, parts by two heavy girders of timber placed at right angles to those of iron, and intended to rest on the sand and assist in supporting the roof of the chamber. Openings made through the girders allowed free communication between the divisions.

Access to the top of the caisson was obtained by vertical shafts lined with brick masonry, and passing through the roof of the chamber. The air-locks were at the lower end of the shafts and within the chamber.

As the caisson descended, the masonry pier was built up in the usual manner, its foundation resting on the iron girders.

In the chamber were workmen who excavated the sand, and shovelled it under the sand-pumps. (Fig. 154.) A pump of 3½ inches diameter, working under a pressure of 150 pounds on the square inch, was capable of raising 20 cubic yards of sand 125 feet per hour.

When the caisson reached the rock, the latter was cleared of sand and the entire chamber then filled with concrete.

The experience acquired in sinking this caisson enabled the engineer to make material modifications in the details of the caissons subsequently used.

The health of the workmen was greatly affected by the high degree of compression of the air in which they had to work. In some cases the pressure was as high as fifty pounds on the square inch, and several lost their lives in consequence.

In the second pier, instead of filling the chamber entirely with concrete when the rock was reached, the space around the edges was only closed with concrete and the chamber was then filled with clean sand.

Pneumatic Caisson at St. Joseph, Mo.

429. This was used in 1871–2 in laying the foundations of the piers for a railroad bridge over the Missouri River, at St. Joseph, Mo.

For a reason similar to that given in the last case, it was decided to rest the piers on the rock below the bottom of the river. The rock was about sixty-seven feet below the level of high water, and was overlaid with mud and sand to depths varying from forty to the whole distance of sixty-seven feet. Six piers were used and were placed in depths of water varying at the low stage from zero to twenty-five feet; the difference between high and low water being twenty-two feet. Pockets of clay, with occasionally snags and boulders, were met with in the sand and mud.

The caisson used for pier No. 4 was made of twelve-inch square timber, and was at the bottom fifty-six feet long, and twenty-four feet wide. The sides of the working chamber were three feet thick, sloping inwards with a batter of $\frac{12}{1}$. It was built by placing a row of timbers in a vertical position, side by side, for the outside; then, inside of this, a second row was laid horizontally; and then, for the inside, a third row in a vertical position. The outer row extended one foot below the middle row, and the latter one foot below the third. A horizontal beam extending entirely around the interior was bolted to the sides of the chamber, one foot above the bottom of the inside row. A set of inclined struts rested on this beam, and abutted against straining beams framed into the roof of the chamber. The roof was solid timber, four feet thick, on which rested the grillage for the masonry of the pier. The grillage was made of timber, seven courses thick, each course being laid at right angles to the one below it. The timbers of each course were separated by a space of six inches, excepting the top course, which was solid.

All the timber work was accurately fitted, and the whole

bolted together so as to form one unyielding mass. The interior of the working chamber was calked, and was practically air-tight. The dimensions of the chamber were, on the inside, twenty-two feet wide and fifty-four long at the bottom: five feet wide and seven feet long at the top; and nine feet high at the centre. The grillage was drawn in so that its top was of the same dimensions as the base of the pier, being nine feet wide and twenty long, with curved starlings at each end.

The air-lock was four feet in diameter and seven high, made of plate iron, and placed in the middle of the top of the chamber. A door in the top of the air-lock opening downwards communicated with a vertical iron shaft three feet in diameter; the shaft extended above the top of the masonry and allowed access to the top of the caisson. An iron ladder in the shaft was used for ascent and descent. The usual supply and discharge pipes passed through the grillage to the working chamber.

The caisson was sunk by the process previously described. The arrangement of the lower bearing surfaces of the caisson are regarded as worthy of notice. The lower edge of the outside row of timbers was sharpened; as soon as it had sunk one foot, the under surface of the second or horizontal row came into play, adding a foot of bearing surface. When the caisson had descended two feet, the bottom of the inside or third row pressed on the soil, thus giving three feet of bearing surface. By this arrangement the amount of bearing surface was under the control of the engineer. If the soil through which the caisson was sinking was variable in its nature, that is, if on one side of the caisson it was soft, and on the other it was hard, the bearing surface could be increased on the soft side and diminished on the other. In this way the caisson could be kept vertical while sinking.

The greater part of the material excavated was mud or sand, and was discharged easily and rapidly by means of sand pumps. The clay, boulders, and snags were discharged through the air-lock.

The caisson was sunk at the rate of from five to seven feet in twenty-four hours.

When the caisson reached the bed rock, a wall of concrete, six feet thick, was built on the rock under the edges, and was solidly rammed under the three rows of timbers and up to and including the horizontal beam supporting the struts. Strong vertical posts were placed under the roof to assist in supporting it. The sand pumps were then reversed, and the chamber was filled with clean sand and gravel. A tube was so placed as to allow the escape of the water in the sand, so that the whole interior was compactly filled with solid material. The sand pumps were then withdrawn, and the shafts themselves were filled.

Caissons of the East River Bridge at New York.

430. The caissons used for the foundations of the piers in this bridge were rectangular in form, and made of timber.

The exterior of the bottom of the chamber in the Brooklyn caisson was 168 feet long and 102 wide. In the one on the New York side the width was the same, but the length was four feet greater.

Both were nine and a half feet high on the inside. The roof of the Brooklyn caisson was a solid mass of timber, fifteen feet thick (Fig. 155), and of the New York caisson, twenty-two feet thick.

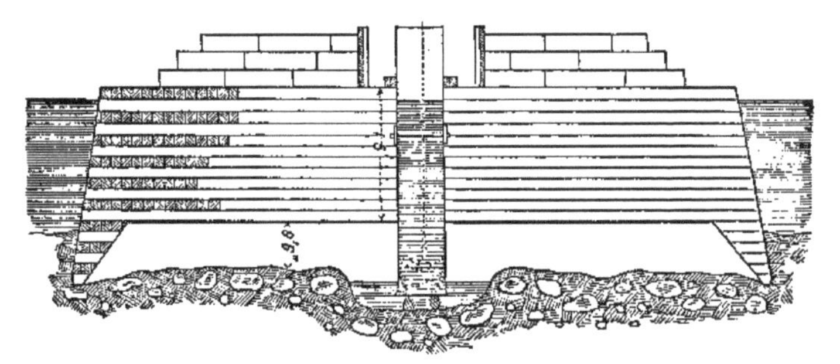

FIG. 155—Represents section through water shaft of the Brooklyn caisson, showing method of removing boulders or other heavy materials.

The sides of the caisson had a slope of $\frac{10}{1}$ for the outer face, and of $\frac{1}{1}$ for the inner, as shown in the figure. The outer slope was for the purpose of facilitating the descent of the caisson into the ground. The lower edge was of cast iron, protected by boiler iron, extending up the sides for three feet. The sides, where they joined the roof, were nine feet thick. The chambers were calked both on the outside and inside, to make them air-tight. As a farther security, an unbroken sheet of tin extended over the whole roof between the fourth and fifth courses, and down the sides to the iron edge. The New York chamber was, in addition, lined throughout on the inside with a light iron plate, to protect it from fire.

Each chamber was divided by five solid timber partitions into six compartments, each from twenty-five to thirty feet wide. Communication from one to the other was effected by doors cut through the partitions.

The air-locks were placed in the roof, projecting into the chamber four feet, and communicating at the top with vertical shafts of iron, built up as the caisson descended. The locks were eight feet high and six and a half feet in diameter.

The mud and sand were discharged through pipes by the compressed air. A pipe, three and a half inches in diameter, discharged sand from a depth of sixty feet at the rate of one cubic yard in two minutes, by the aid of the compressed air alone.

The heavy materials were removed through water shafts. These were seven and three-quarter feet in diameter, open at the top and at the lower end, the latter extending eighteen inches below the general level of the excavation. A column of water, in the shaft, prevented the compressed air from escaping.

The material to be removed through the water shaft was thrown into an excavation under the lower end of the shaft; it was there grasped by a "grapnel bucket," which was lowered through the shaft, and hoisted through the water to the top of the shaft, where it was removed.

After the caisson had reached the rock, the chamber was filled with concrete, in the usual manner.

The great thickness of the roof, and the moderate depth of water, enabled the engineer to dispense with the use of sides to the caisson, as the masonry could be kept always above the surface of the water.

Movable Pneumatic Caisson.

431. A pneumatic caisson has been successfully used in laying the foundations of piers of bridges, which differs from those already described, in its construction admitting of its being moved after completion of one pier, to another place for the same purpose. It was an iron cylinder, ten feet in diameter (Fig. 156), connected at its lower end with a working chamber, eight feet high and eighteen feet in diameter. On the roof of the latter was another chamber, annular in form, eighteen feet in diameter and about six feet high, so arranged as to allow of being filled with water when any additional weight was necessary, and being emptied of water and its

place supplied with compressed air when less weight was desired. On top of this annular chamber was a similar one arranged to be loaded with iron ballast. Strong chains attached to the roof of the working chamber and connected with a hoisting apparatus, placed on a strong scaffolding over the site of the pier, were used to lower and lift the cylinder, as necessity required.

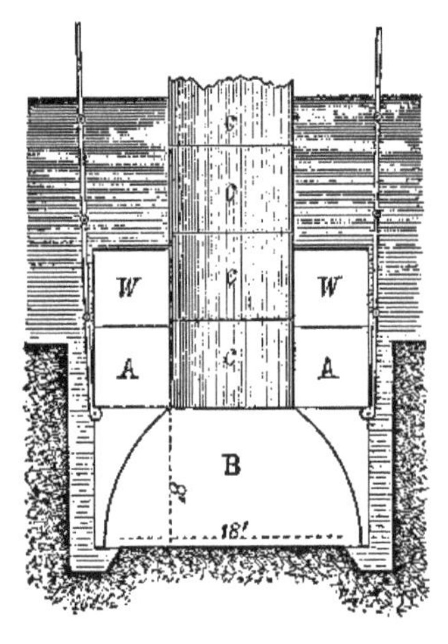

FIG. 156—Represents section of movable pneumatic caisson.
B, working chamber.
A, chamber for water, or for compressed air.
W, chamber for iron ballast.
c, c, elevation of lengths of the iron cylinder.

Air-locks, air-pumps, and all the necessary adjuncts of a pneumatic pile, were provided and used. Having reached the rock or firm soil, the bed and the foundations were constructed as already described. As the masonry of the pier rose, the whole apparatus was lifted by the chains and hoisting apparatus, the cylinder being lightened by expelling the water from the chamber, A, and filling the latter with compressed air. The masonry of the pier having risen above the surface of the water, the whole apparatus was removed and used in another place.

432. **Remark.**—It is seen that the pneumatic caisson, as before stated, is simply a combination of the diving-bell with the common caisson, the diving-bell being on a large scale, and its roof being intended to form a part of the bed of the foundation.

Experience has shown that the large caissons are more easily managed than the small ones. The circumstances of the case can only decide as to which is preferable, the caisson or the pneumatic pile. Either method is an expensive one, and is only employed in localities where the others are not applicable.

SECURING THE BED FROM THE INJURIOUS ACTION OF WATER.

433. The bed of a river composed of sand or gravel is liable to change from time to time, as these materials are moved by currents in the river. This change, when accompanied by an increase in depth of the river, is known as the "scour." Sometimes a scour will occur on one side of a structure and not on the other, producing an undermining threatening the stability of the masonry. Where common piles have been used, they have occasionally been washed out by this action. Even in rocky bottoms, when of loose texture, the rock will gradually wear away under the action of currents, unless protected

It therefore becomes an important point to provide security for the beds in all soils liable to any change. It is for this reason that in very important structures, the foundations are placed on the bed rock far below the possible action of currents, and so arranged that even if they should be exposed to a scour they would be safe. This requirement has caused the free use of the pneumatic methods.

Various expedients have been used to secure the beds where they do not rest on the rock or on a soil below the action of the water. A common method is to **rip-rap** the bed, that is, to cover the surface of the bottom, around the bed, with fragments of stone too large to be moved by the currents, and if the soil is a sand or loose gravel, to use clay in connection with the stone to bond the latter together.

Where the bed is made of piles, it is well to enclose the piles by a grating of heavy timber, before throwing in the stone. In some cases the foundations are **boxed**, that is, the piles are enclosed by a sheeting of planks, or by other device, so as to protect them from the scour.

PART VI.

BRIDGES.

CHAPTER XIII.

434. A **bridge** is a structure so erected over a water-course, or above the general surface of the ground, as to afford a continuous roadway between the opposite sides of the stream, or above the surface of the country, without obstructing those lines of communication lying beneath.

Such a structure, thrown over a depression in which there is ordinarily no water, is generally called a **viaduct.**

If the structure supports an artificial channel for conveying water, it is known as an **aqueduct;** and where it crosses a stream, it is frequently called an **aqueduct-bridge.**

Bridges may, for convenience of description, be classed either from the materials of which they are made: as **masonry** or **stone, iron, wooden** bridges, etc.; or from the character of the structure: as **permanent, movable, floating** bridges, etc.; or from the general mechanical principles employed in arranging its parts: as **arched, trussed, tubular** bridges, etc.

435. **Component parts.**—A bridge consists of **three** essential parts:

1st, The **piers** and **abutments** on which the superstructure rests; **2d,** the **frames** or other arrangements which support the roadway; and **3d,** the **roadway,** with the parts used in connection with it for its preservation or to increase its security, as the roof, parapets, etc.

Bridges are of various kinds, both in their general plan and dimensions. The latter are dependent upon the objects of and the circumstances requiring the erection of the bridge.

The simplest bridge is one in which the points of support

are so near together that two or more simple beams laid across the stream, or across an opening to be passed over, are sufficient for the frame; a few planks laid upon the beams may then form the roadway.

The supports being strong enough, the proper dimensions for the beams and for the planking are easily determined.

This calculation for the beams is made under the hypothesis that each is a simple beam, resting on two points of support at the extremities, strained by a load uniformly distributed over it, and also by a weight acting at the middle point.

The uniform load is the weight of the structure, ordinarily assumed to be uniformly distributed in the direction of its length. The weight at the middle represents the heavy body as it passes over; as, for example, a heavily loaded wagon for a common, and a locomotive for a railroad bridge. Having determined what this weight shall be, its equivalent uniform load may be obtained, and added to that already assumed; or if preferred, the uniform load may be replaced by its equivalent weight at the middle.

If the number of these beams be represented by n, and we suppose that they are at equal distances apart, then the total load on the bridge divided by n will give the load on each beam. Then by formulas already deduced we can, knowing the value for R, determine the proper breadth and thickness for each beam.

436. **Platform of roadway.**—In a common wooden bridge the roadway is generally of planks. These are of hard wood, from three to four inches thick, resting on longitudinal pieces placed from two to three feet apart from centre to centre. This thickness of plank is greater than is required for strength, but has been found necessary to enable the roadway to withstand the shocks, friction, and wear due to the travel over it.

If the longitudinal pieces which rest directly on the supports are too far apart to allow the plank to rest safely upon them, cross pieces, called **roadway bearers**, are placed upon the longitudinal pieces. On these cross pieces other longitudinal pieces, called **joists**, are placed close enough together, and the planking is laid upon the joists.

The particular kind and width of roadway will depend upon the character of the travel over the bridge. Knowing these, the weight per unit of length is quickly determined.

437. **Piers and abutments.**—Walls should be built to support the ends of the beams. These walls may be of stone, wood, or iron. Those placed at the ends of the bridge are

called **abutments**; the intermediate ones are termed **piers**; the distance or space between any two consecutive piers is called a **span**, and sometimes a **bay.**

If the frame of the bridge is of a form that exerts a lateral thrust, as, for instance, in an arch, the abutments and piers must be proportioned to resist this thrust.

As the foundations are exposed to the action of currents of water, precaution must be taken to secure them from any damage from this source. The piers and abutments must also be guarded against shocks from heavy bodies and against the damaging effects of floating ice.

438. **Wooden piers and abutments.**—Wooden abutments may be constructed of crib-work. The crib is ordinarily formed of square timber or logs hewn flat on two of their opposite sides. The logs are halved into each other at the angles, are fastened together by bolts or pins, and are sometimes further strengthened by diagonal ties. The rectangular space thus enclosed is filled with earth or loose stone. Very frequently the crib is built with three sides only. Another way of constructing the abutment is to make a retaining wall of timber by which the earth of the bank is held up.

The piers also are sometimes made of cribs. The cribs are floated to the spot, sunk in place, filled with stone, and built up to the proper height. There are serious objections to their use for piers, and they are recommended only where no injurious results will follow their adoption, and where it is not expedient to employ some one of the other methods.

The pier made of piles is the most common form of the wooden pier. It is constructed by driving piles from three to six feet apart, in a row, parallel to the direction of the current. The piles are then cut off at the proper distance above the surface of the water, and capped with a heavy piece of square timber. If the piles extend some distance above the water, they must be stiffened by diagonal braces.

In some cases the piles are cut off, at or just below the level of the water, so that the capping piece will always be kept wet. Mortises are made in this cap into which uprights are fitted; the uprights taking the place of the upper parts of the piles in the preceding case. Or, what is more common, a trestle made in the form of an inverted W is fitted on this cap, and the upper side of this trestle is capped with a square piece of timber.

Where the bottom is hard and not liable to "scour," the piles are dispensed with and the trestle alone is used. In this case the piece on which the trestle rests is laid flat on the bottom and is called the *mud-sill*. The upper part of the trestle is capped as before, and if necessary to get additional height another trestle is framed on top of this.

439. **Fenders and ice-breakers.**—Wooden piers are not constructed to resist heavy shocks from floating bodies. In positions exposed to such shocks, fenders should be built. A clump of piles driven on the exposed side of the pier, opposite to and some distance from it, will be a sufficient protection against ordinary floating bodies when the current is gentle. The piles should be bound together so as to increase their resistance; this may be done by wrapping a chain around their heads. If there is danger from floating ice, an inclined beam (Fig. 157), protected by iron, should be used to break up the ice as it moves towards the pier.

Elevation.

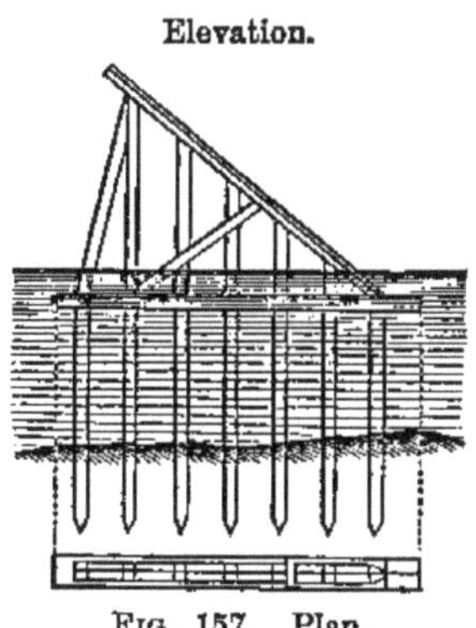

FIG. 157. Plan.

In rapid currents, where the ice is thick, a crib-work square in plan, with one of the angles up-stream, has been used. The crib was filled with heavy stone and the up-stream angle was given a slope and was protected by a covering of iron.

The construction shown in Fig. 158 is a good one. Its resisting power is increased by filling the interior with stone.

440. **Masonry piers and abutments.**—The methods, described in the chapters on masonry and foundations, are applicable to the construction of piers and abutments.

Since they are, from their position, especially liable to damage from the action of currents, both on the soil around them and on the materials of which they are made, particular attention should be paid to their construction.

In preparing the bed, a wide footing should be given to the foundation courses, if the soil is at all yielding, and whenever this footing does not rest on rock, means should be taken to secure the bed from any injurious action of the water.

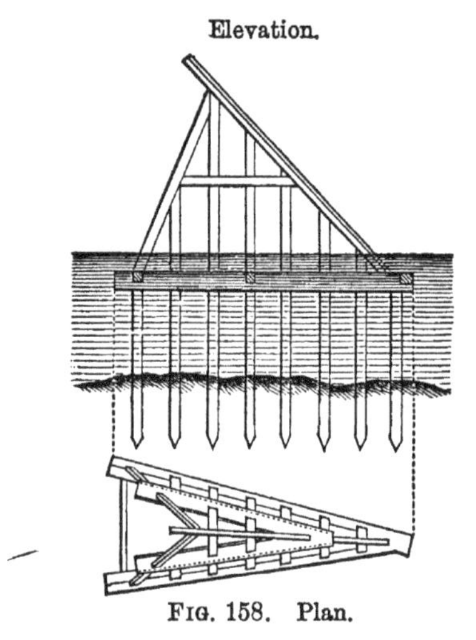

FIG. 158. Plan.

The piers, although they are generally built with a slight batter, may be built vertical. The thickness given them is greater than is necessary to support the load which is to be placed upon them, in order that they may better resist the shocks from heavy floating bodies and the action of the currents to which they are continually exposed.

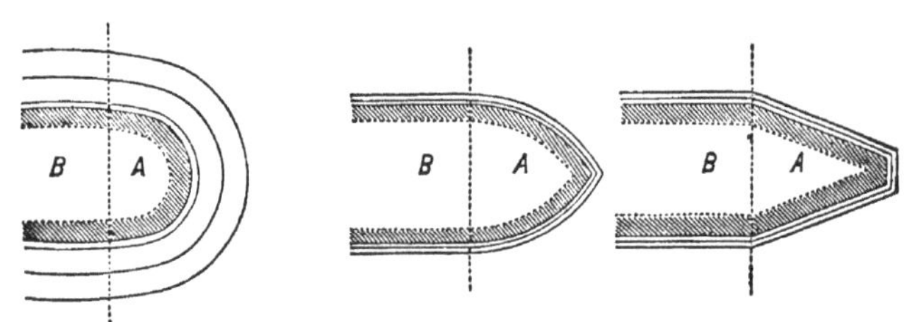

FIG. 159.—A, horizontal sections of starling.
B, same of pier.

They should be placed, if possible, so that their longest dimensions should be parallel to the direction of the current. They should have their up and down-stream faces either curved or pointed, to act as cut-waters turning the current aside, and preventing the formation of whirls, and to act as fenders.

These curved or pointed projections are called **starlings**. Of the different forms of horizontal section which have been given them (Fig. 159), the semi-ellipse appears to be the most satisfactory.

Their vertical outline may be either straight or slightly curved. They are built at least as high as the highest water line, and finished at the top with a coping stone called a **hood**.

In streams subject to freshets and to floating ice, the up-stream starlings are provided with an inclined ridge to facilitate the breaking of the ice as it floats against and by them. Where very large masses are swept against the piers,

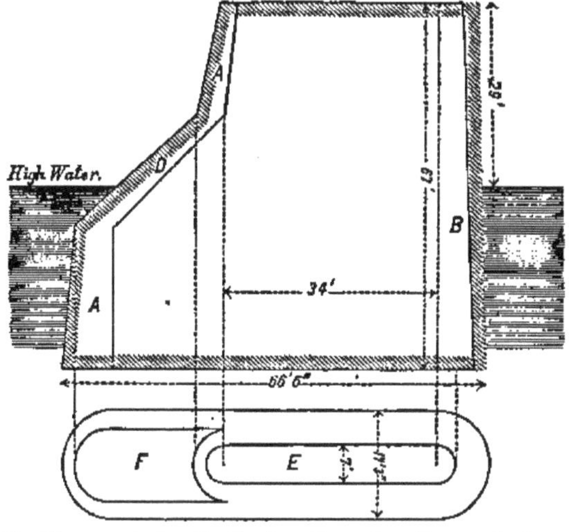

FIG. 160—Represents longitudinal section, elevation, and plan of a pier of the Potomac aqueduct bridge.
A, A, up-stream starling, with the inclined ice-breaker D, which rises from the low-water level above that of the highest freshets.
B, down-stream starling.
E, top of pier.
F, horizontal projection of ice-breaker.

it is not unusual to detach the ice-breakers and place them in front of the piers, as is generally done in the case of wooden piers.

Fig 160 represents the ice-breaker planned and constructed

by Colonel Turnbull, of the Topographical Engineers, United States Army, for the piers of the Potomac aqueduct bridge of the Alexandria Canal, at Georgetown, D. C.

The pier was at the bottom 66.6 feet long and 17.3 thick, and terminated by starlings whose horizontal cross-section was circular. The pier shown in the drawing was 61 feet high, and built with a batter of $\frac{12}{1}$.

The starlings were built up with the same batter, except that the up-stream one, when at the height of 5 feet below the level of high water, received an inclination of 45°, which it retained until 10 feet above it. From there to the top it had the same batter as the rest of the pier. The two lower courses of the ice-breaker were 22 inches thick, the rest being 18 inches. The stones were laid in cement, and no stone was allowed in the ice-breaker of a less volume than 20 cubic feet.

The ice brought down by the river at this point is often 16 inches thick, and the current is often six miles an hour. On such occasions the ice is forced up the ice-breakers to a height of 10 or 12 feet. The ice breaks by its own weight, and passes off between the piers without doing any harm.

Probably the ice-breakers of the International Bridge, over the Niagara River, at Buffalo, are more severely tested than any in our country. They are triangular in plan, have a slope of $\frac{1}{2}$, and are protected by iron plating.

441. **Iron piers and abutments.**—Until a very few years ago all piers were made either of masonry or timber. Where a solid bed could not be reached by excavation, piles were driven, their tops were sawed off, and on them a grillage and platform was placed to form the bed.

The substitution of iron for wood in many engineering structures, soon led to the use of iron in the above class of constructions.

Iron is used in the construction of piers and abutments in various forms as follows:

1. As piles or columns, wholly of iron; as screw piles.
2. As a hollow column, open at the bottom, and partly or entirely filled with concrete; the weight of the bridge resting on the iron casing.
3. As a cylinder, entirely filled with masonry or concrete; the weight of the bridge resting on the masonry, the iron casing serving to protect and to stiffen the column.
4. As a caisson; the sides being left standing.

The precautions recommended for stone and wooden piers are equally necessary for those made of iron.

442. **Approaches.**—The portions of the roadway, at each extremity of the bridge and leading to it, are termed the **approaches.**

These are to be arranged so that vehicles, using the bridge, may have an easy and safe access thereto.

The arrangement will depend upon the locality, upon the number and direction of the avenues leading to the bridge, upon the width of these avenues and upon their position, whether above or below the natural surface of the ground.

When the avenue to the bridge is in the same line as its axis, and the roadway of the avenue and of the bridge is of the same width, the abutment is generally made as shown in Fig. 161. The returns or short walls carried back parallel to

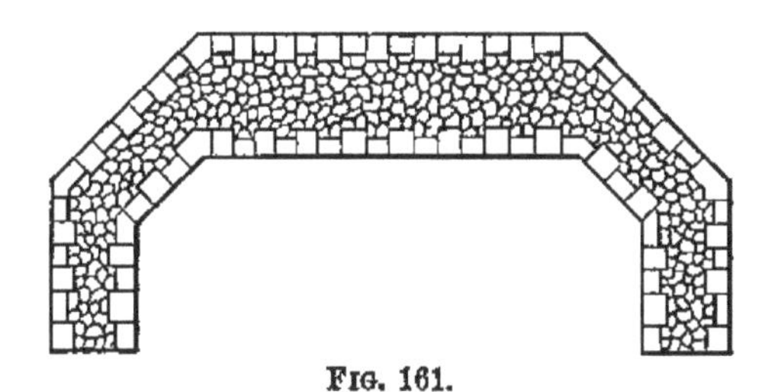

FIG. 161.

the axis of the road to flank the approach are called wing-walls, and are intended to sustain the embankment as well as to serve as a counterfort to the abutment.

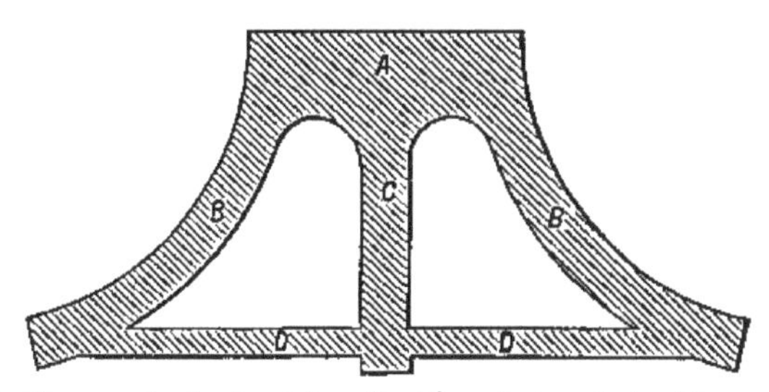

FIG. 162—Represents a horizontal section of an abutment, A, with curved wing-walls, B, B, connected with a central buttress, C, by a cross tie-wall, D.

When several avenues meet at the bridge, or it is necessary that the width of the approach shall be greater than the roadway of the bridge, the wing-walls may be given a curved shape, as shown in Fig. 162, in this way widening the approach.

When the soil of the river banks is bad, the foundation of the wing-walls should be laid at the same depth as that of the abutment. But if the soil is firm, they may be built in steps, and thus save considerable expense.

The rules for the dimensions of wing-walls are the same as for other retaining walls. A common rule is to make their length one and a half times the height of the roadway above the bed of the river, their thickness at bottom one-fourth their height, and to build them up in off-sets on the inside, reducing their thickness at the top to between 2 and 3 feet.

In some cases plane-faced wing-walls are arranged so that the faces make a given angle with the head of the bridge. The top of the wall is given a slope to suit the locality, and is covered by a coping of flat stones, to shelter the joints and to add a pleasing appearance to the wall (Fig. 163). The lower end of the coping is generally terminated by a **newel stone.**

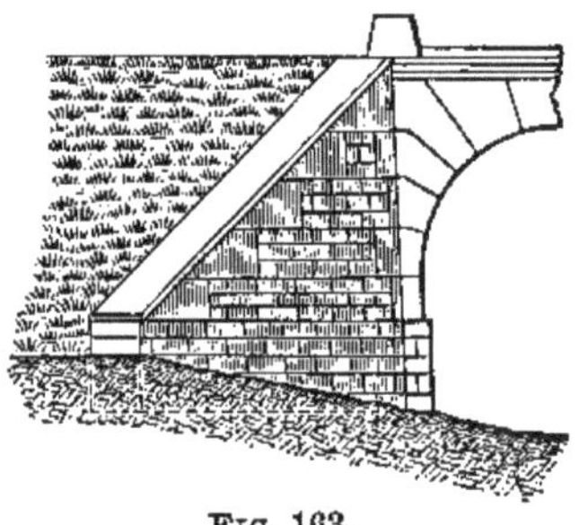

FIG. 163.

Instead of wing-walls, a single wall in the middle is used in many cases. The plan of the abutment in such a case is that of a T.

In case there are no wing-walls to retain the earth, the abutment wall must be sufficiently distant from the crest of the slope of the water-course to allow room for the slope of the embankment. This slope of the embankment may be the natural slope, or, if steeper, the embankment should be revetted with dry stone or sods, as shown in Fig. 164.

It may be necessary, to avoid obstructing the communications along the bank, to construct arched passage-ways under the roadway of the approaches.

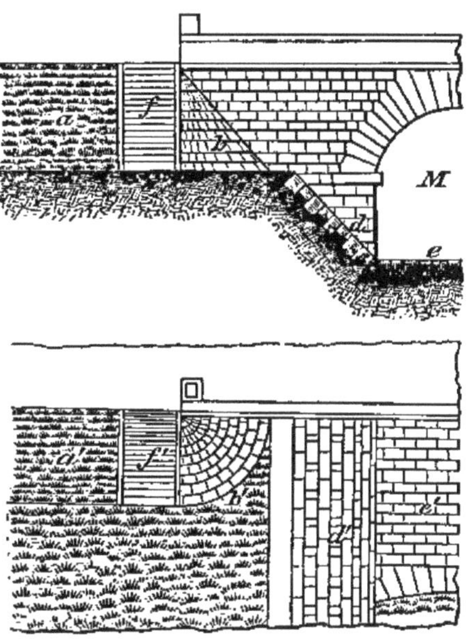

FIG. 164.—Plan and elevation showing a method of arranging the embankments where there are no wing-walls.
a, *a'*, side slopes of embankment of the approach.
b, *b'*, dry stone revetment of the slope towards the water-course.
d, *d'*, dry stone facing of the slope of the bank.
e, *e'*, paving used on the bottom of stream.
f, *f'*, stairs for foot passengers.

443. Water wings.—When the face of the abutment projects beyond the bank, an embankment faced with stone should connect it with points of the bank, both above and below the bridge. These are called **water-wings**, and serve to contract gradually the water-way of the stream at this point.

Where there is danger of the banks above and below the abutment being washed or worn away by the action of the current, it is advised to face the slope of the bank with dry stone or masonry, as shown in Fig. 164.

444. The frame.—It is evident that the arrangement used to support the roadway admits of the greatest differences in form. From these differences in the forms used, many classifications have been made.

According to the kind of frame, bridges may for analysis be classed as follows:

I. **Trussed Bridges**;
II. **Tubular Bridges**;
III. **Arched Bridges**; and
IV. **Suspension Bridges.**

Considering the simple bridge to belong to the first class, every bridge may be placed under the head of one or more of these divisions.

CHAPTER XIV.

I.—TRUSSED BRIDGES.

445. **A trussed bridge** is one in which the frame supporting the roadway is an open-built beam or truss.

A truss has been defined (Art. 252) to be a frame in which two beams either single or solid built, with openings between them, are connected by cross and diagonal pieces so that the whole arrangement acts as a single beam.

It generally has to sustain a transverse strain caused by a weight which it supports. To do this in the best manner, the axes of the pieces of which the truss is composed are kept in the same vertical plane with the axis of the truss, or are symmetrically disposed with reference to it.

Supposing the truss to rest on two or more points of support, in the same horizontal line, its upper and lower sides are called **chords.** In some cases the upper side has been called a **straining beam**, and the lower a **tie.** Sometimes both beams are designated as **stringers.** English writers call them **booms.**

Generally, both chords are straight and parallel to each other. Both may be and are sometimes curved; in some cases one is curved and the other is straight.

The secondary pieces, or those connecting the chords, are called **braces**, and are so arranged as to divide the frame into a series of triangular figures. The braces are known as **struts** or **ties**, depending upon the kind of strain they have to sustain. The triangles may be scalene, isosceles, equilateral, or right angled. They may be placed so as to form a system of single triangles, or by overlapping, form a lattice or trellis pattern.

446. Systems.—Trussed bridges are divided into three general systems:

1, The **triangular** system; 2, The **panel** system; 3, The **bowstring** system.

Other subdivisions are frequently made, based upon the particular arrangement adopted for the braces and upon the form given to the chords.

Special cases belonging to the systems are generally known by the name of the inventor: as Long's truss, Howe's, Fink's, etc.

The essential qualities in a truss are those already given for a frame (Art. 231), viz., **strength, stiffness, lightness,** and **economy of material.**

These qualities are dependent upon the kind of material used in its construction, the size of the pieces, and the method of arranging them in the frame. The latter gives rise to the variety of trusses met with in practice.

METHODS OF CALCULATING STRAINS ON THE DIFFERENT PARTS OF A TRUSS.

447. External forces acting on a truss.—It is necessary to know all the external forces which act on a truss, in order to determine the strains on its different parts.

The external forces which are considered, are:

1, The **weight** of the bridge;

2, The **moving** or **live** load;

3, The **reactions** at the points of support;

4, The **horizontal and twisting forces** which tend to push the frame in a lateral direction or around some line in the direction of its length.

1. The **weight of the bridge.**—Previous to the calculation of the strains, the weight is not known, since it is dependent upon the thing which we seek, viz., the dimensions of the parts of the bridge. An approximate weight is therefore assumed, being taken by comparison with that of some similar structure already built. The strains are then determined under the supposition that this is the weight of the bridge and the dimensions of its parts are computed. The weight is then calculated from these dimensions, and if the assumed weight does not exceed very greatly that of the one computed, the latter, and also the strains deduced therefrom, are assumed to be correct.

2. The **moving load.**—This is any load which may pass over the bridge, and when calculating the strains, should be assumed at its maximum; that is, as equal to or exceeding slightly the greatest load which will ever be placed on the structure. This load should be considered as occupying various positions on the bridge, and the greatest strains in these positions determined.

For a common road bridge, the load is assumed to be a maximum when the bridge is covered completely with men. This load is estimated at 120 pounds to the square foot, and must be added to the weight of the bridge.

For a railroad bridge, the load is assumed a maximum when a train of locomotives extends from one end of the bridge to the other. This load is assumed at one ton (2,240 lbs.) to the running foot.

Sometimes, common road bridges are liable to be crossed by elephants, in which case it is assumed that the maximum load is equivalent to that of 7,000 pounds supported on two points, six feet apart.

A load applied suddenly produces on the parts of a bridge double the strain which the same load would produce if it were applied gradually, beginning at zero and increasing gradually until the whole load rested on the bridge. A load moving swiftly on the bridge approximates in its effect to that of one applied suddenly.

Therefore, the action of a **live** load may be considered to be the same as that of a load of double its weight placed carefully on the bridge. The latter may then be treated as any stationary load added to the weight of the bridge; the strains can be determined in the usual manner.

To distinguish between these loads, it is usual to call the weight of the bridge the **permanent** or **dead** load, and that caused by bodies crossing the bridge the **moving**, the **rolling**, or the **live** load.

3. **Reactions of the points of support.**—The applied forces cause reactions at the points of support, which must be considered in the calculations, as external forces acting on the bridge; their value, therefore, must be determined. No sensible error is committed by regarding the reactions as vertical for trusses whose chords are straight and parallel to each other.

4. **Forces producing lateral displacement or twisting.**—The action of the wind on the sides of the truss tends to push the bridge in a horizontal direction. This pressure may be regarded as uniform over the entire extent of the

surface exposed. The best authorities assume this pressure ordinarily at forty pounds per square foot. The locality will decide as to the exact amount, since the force of the wind is greater in one place than in another. The wind gauge has recorded as high as sixty pounds in this locality.

Care is taken to guard against any forces which might produce a twisting strain, and to reduce their effect to a minimum. If there be any such forces acting, their effect on the bridge must be provided for.

The King-post Truss.

448. Excepting the triangular frame (Art. 256), the **king-post truss** is the simplest of the trusses belonging to the triangular system.

It is frequently employed in bridges of short span, and where the span is so small that the beam requires support only at its middle point.

For a single roadway, two of these frames are placed side by side, and far enough apart to allow room for the roadway between them. Roadway bearers are placed on the beams, or are suspended from them, to support the joists and flooring. Each truss will therefore be loaded with its own weight, one-half that of the roadway, and one-half of the live load. Knowing these weights, the strains on the different parts are easily determined, and the dimensions of the parts calculated.

To determine the amount and kind of strains on the parts, consider the load resting on the beams as uniformly distributed over them, and represent (Fig. 68), by w, the load on a unit of length of the beam, C; $2l$, the distance between the points of support.

The load on the beam, C, will be $2wl$. The post, g, is so framed upon the inclined braces, e, e, and into the beam, C, that the middle point of the beam is kept in the same straight line with its ends. C is therefore in the condition of a beam resting on three points of support in a right line. Five-eighths of the load $2wl$, is therefore held up by the king-post (Art. 186), and by it transmitted to the apex of the frame, the king-post sustaining a tensile stress. The amount of this stress being known, the dimensions required for the king-post are easily calculated.

The stresses developed and the dimensions of the braces are determined as in Art. 256.

If the middle point of a beam, as A B (Fig. 165), is supported at C by inclined braces resting against the abutments, the amount and kind of stresses in the braces are the same as those in the king-post truss; in this case the horizontal thrust at the lower ends of the braces, instead of being taken up by a tie beam, will act directly against the abutments.

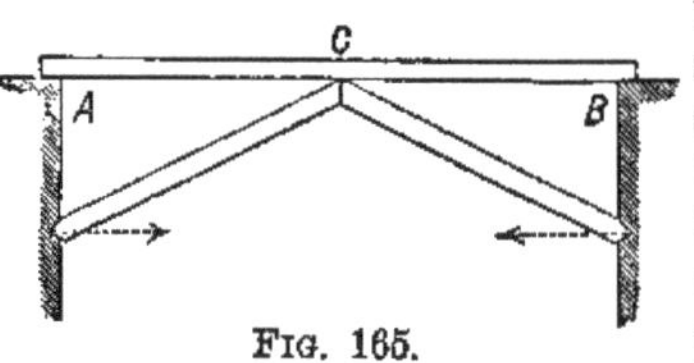

FIG. 165.

Inverted king-post.—If the king-post truss be inverted, and supported at the extremities (Fig. 166), the amount of stress in each piece will be the same as before. The strains, however, will be reversed in kind; that on the beam, and that on the king-post, being compression, and those on the braces being tension.

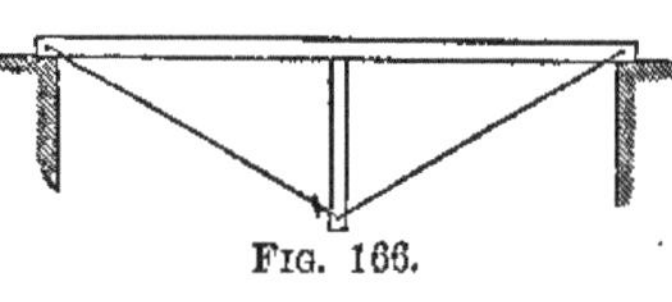
FIG. 166.

FINK TRUSS.

449. **Fink truss.**—This is the name by which a truss devised by Mr. Albert Fink, civil engineer, is generally known (Fig. 167). It consists of a combination of inverted king-post trusses, as shown in the figure. There is a primary truss, A O B; two secondary ones, A K C and C L B; four tertiary ones, A P D, D M C, etc.

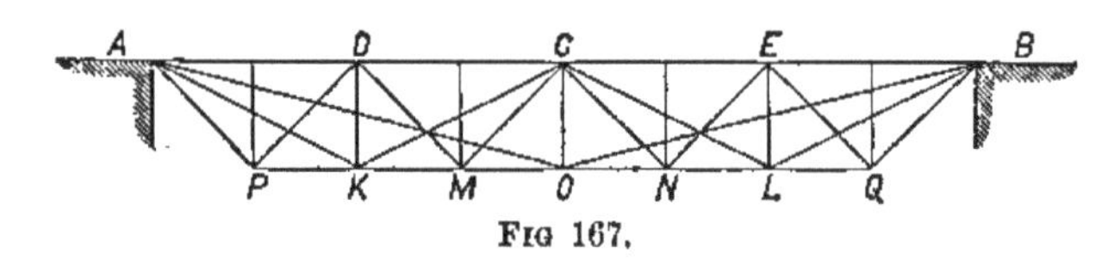

FIG 167.

The load may be upon the upper or the lower chord, as the circumstances may require. The strains on the different parts are easily determined, when the weights to be placed upon the bridge are known.

If the load should be on the upper chord, there would be no necessity for a lower chord, so far as strength is concerned.

450. **Bollman truss.**—If the braces all pass from the foot

of the posts to the ends of the chord, as in Fig. 168, the truss thus formed is known as Bollman's truss.

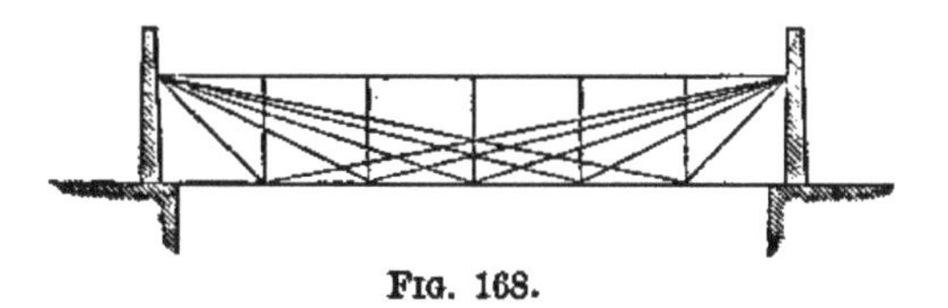

FIG. 168.

The calculations for the strains do not differ in principle from those in the preceding case. It is observed that the ties are of unequal length in each of the triangular frames of this truss; excepting the one at the middle post.

There is no necessity, as in Fink's, for a lower chord if the load is placed above the truss.

From the fact that they need but one chord, both of these constructions are frequently called "**trussed girders,**" to distinguish them from the ordinary bridge-truss, which, by the definition given for a truss, requires two chords.

I. The Triangular System.

451. The term, **triangular truss**, is ordinarily used to designate a truss whose chords are connected by inclined braces, so arranged as to divide the space between them into isosceles or equilateral triangles, as shown in Fig. 169.

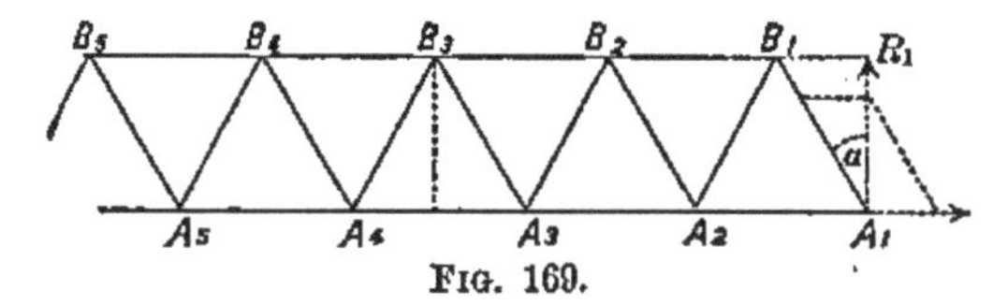

FIG. 169.

In the isosceles bracing the braces are generally arranged so as to make an angle of forty-five degrees with the vertical, although sometimes other angles are used.

When the triangles are equilateral, the truss is known in England and in the United States as the "**Warren girder,**" and in other countries as the "**Neville.**"

The strains on this truss may be determined by the methods given in Arts. 260–1–2, or they may be determined by using the reactions of the points of support when these reactions are known. The following is an example of the latter method.

452. *Let it be required to determine the strains produced upon the different parts of a triangular truss by a weight supported at the middle point of the truss.*

The truss is supposed to be resting on firm points of support at its ends, these supports being in the same horizontal line.

Represent by (Fig. 169),

2W, the weight resting on the truss at the middle;

R_1 R_2, the reactions at the points of support;

a, the angle $R_1 A_1 B_1$ between the brace and a vertical line.

Since the load is at the middle, the reactions due to it are $R_1 = W$, and $R_2 = W$.

The strains in one half will be equal to the corresponding strains in the other half. Take the right half, as shown in the figure, and on $R_1 = W$, as a resultant, construct a parallelogram of forces, the components of which are in the directions of the pieces, $A_1 B_1$ and $A_1 A_2$. These components will be respectively equal to $\frac{W}{\cos a}$ and $W \tan a$. Going to B_1, and resolving $\frac{W}{\cos a}$ into two components, one in the direction of $B_1 B_2$, and the other in the direction of $B_1 A_2$, their values will be $2W \tan a$ and $\frac{W}{\cos a}$. Performing the same operation at A_2, for the components in the directions of $A_2 A_3$ and $A_2 B_2$, there are found the same values just determined, $2W \tan a$ and $\frac{W}{\cos a}$. At B_2, A_3, B_3, etc., until the point of application of the force is reached, similar expressions for the stresses will be found.

Hence the stress in $B_1 B_2$ is equal to $2W \tan \alpha$; in $B_2 B_3$, this same amount is increased by that in $B_1 B_2$, or $4W \tan \alpha$ for the stress in $B_2 B_3$; on $B_3 B_4$, $6W \tan \alpha$., etc. The stress in $A_1 A_2$ is $W \tan \alpha$; in $A_2 A_3$, it is $2W \tan \alpha$ increased by that in $A_1 A_2$, or in all, $3W \tan \alpha$; in $A_3 A_4$, $5W \tan \alpha$, etc.

There is no increase of the stress as we pass from one brace to another, the intensity being the same for each, viz., $\frac{W}{\cos \alpha}$.

An examination of the forces acting will show the nature of the strain in each piece. The direction of the component of the reaction along the axis of the chord is towards its centre or middle point; the strain is therefore one of compression, and increases from each end toward the middle.

On the lower chord, the strain is in the opposite direction, and is therefore tensile, increasing in amount towards the centre, as already shown.

In the right half of the truss, the strain on the brace A_1 B_1 and on those parallel to it, is compressive; on those not parallel to it, the strain is tensile.

The pieces of the other half are strained in a similar manner; on the corresponding pieces, the strains are equal in amount, and they are also of the same kind on the braces, being tensile in those parallel to the brace A_1 B_1, and compressive in the others.

It will be noticed that these results are identical with those already obtained in Art. 260.

In the above example, the load was placed at the middle point of the truss, but if the load had been placed at any other point, the process used to obtain the strains would be the same; it would only be necessary to find the corresponding values for R_1 and R_2, and substitute them in the foregoing expressions.

453. *Let it be required, to determine the strains produced upon the parts of this truss by a uniform load distributed over the lower chord.*

The effect of the uniform load upon the truss may, without material error, be considered to be the same as that produced by a series of weights acting at the points A_1, A_2, A_3, A_4, etc., each weight being equal to that part of the uniform load resting on the adjacent half segments.

Denote by n the number of these points thus loaded, and by $2w$, the load at each point.

Their total weight on the chord will be $2nw$, and the reactions at the points of support due to them will be, at each support, equal to nw.

To determine the strains, proceed as before. Construct the parallelogram on $R_1 = nw$, and determine the stresses in A_1 A_2 and A_1 B_1, which are found to be $nw \tan \alpha$, and $\frac{nw}{\cos \alpha}$.

Going to B_1, the stress in B_1 B_2 is $2nw \tan \alpha$, and that in B_1 A_2 is $\frac{nw}{\cos \alpha}$. At A_2 the components of $2w$, acting at this point in the direction of A_2 A_3 and A_2 B_2 must be subtracted from those of the transmitted forces along these lines. The stress in A_2 A_3 will therefore be $2nw \tan \alpha - 2w \tan \alpha = 2(n-1) w \tan \alpha$. To this must be added the stress already determined in A_1 A_2, which gives the total stress in A_2 A_3 to be $w [n + 2(n - 1)] \tan \alpha$.

The stress in $A_2 B_2$ is $\frac{nw}{\cos \alpha} - \frac{2w}{\cos \alpha}$, which may be written $\frac{(n-2)w}{\cos \alpha}$. Going to B_2 the stress in $B_2 B_3$, produced by the strain on the brace $A_2 B_2$, is $2(n-2)\,w \tan \alpha$, to which the stress in $B_1 B_2$ is to be added, making the total stress $2(n-2)w \tan \alpha + 2nw \tan \alpha$, which may be written $4(n-1)\,w \tan \alpha$. The stress in $B_2 A_3$ is the same as that in $A_2 B_2$, or is equal to $\frac{(n-2)w}{\cos \alpha}$.

It is plain that the stress in any segment of the upper chord is obtained by adding to the stress transmitted to it by the brace with which it is connected, the respective stresses in each of the segments preceding it; and, that the same law obtains for the stresses in the lower chord.

It is to be noticed that the stresses in the first pair of braces are the same in intensity but different in kind, being compressive for the first and tensile for the second, as in the last case; that in the next pair the intensity differs from that in the first by $\frac{2w}{\cos \alpha}$; that the stresses in the third pair differ from the second by the same quantity; and hence, that the stress in any pair may be obtained when that in the preceding one is known by subtracting $\frac{2w}{\cos \alpha}$ from it. It is noticed that those braces whose tops incline towards the middle point of the truss are compressed, while those that incline from it are extended.

It is seen that while the strains on the braces decrease from the ends towards the middle, that it is the reverse for the chords; in both the upper and lower, the strains increase from the ends to the middle.

The stresses thus determined may now be written out, as follows:

1. The compressions on the braces, $A_1 B_1$, $A_2 B_2$, $A_3 B_3$, etc., are

$$\frac{nw}{\cos a}, \frac{(n-2)w}{\cos a}, \frac{(n-4)w}{\cos a}, \frac{(n-6)w}{\cos a}, \text{ etc.}$$

2. The tensions on the braces, $B_1 A_2$, $B_2 A_3$, $B_3 A_4$, etc., are the same in amount, viz.,

$$\frac{nw}{\cos a}, \frac{(n-2)w}{\cos a}, \frac{(n-4)w}{\cos a}, \text{ etc.}$$

3. The compressions on the segments of the upper chord are, for $B_1 B_2$, $B_2 B_3$, $B_3 B_4$, etc.,

$$2nw \tan a,\ 4(n-1)\,w \tan a,\ 6(n-2)\,w \tan a,\ 8(n-3)\,w \tan a, \text{ etc.}$$

4. The tensions on the segments of the lower chord are, for $A_1 A_2$, $A_2 A_3$, $A_3 A_4$, etc.,

$$nw \tan a,\ [n+2\,(n-1)]\,w \tan a,\ [n+4\,(n-2)\,w \tan a,\ [n+6\,(n-3)]\,w \tan a, \text{ etc.}$$

General term.—By examining the expressions just obtained for compression on the segments of the upper chord, it is seen that a general term may be formed, from which any one of these may be deduced upon making the proper substitution. Let the segments be numbered from the ends to the middle, by the consecutive whole numbers, 1, 2, 3, 4, etc., and represent the number of any segment by m. Then,

$$2m\,(n-m+1)\,w \tan a,$$

will be the general term expressing the intensity of the stress in the mth segment.

It is seen that the term,

$$[n+2\,(m-1)\,(n-m+1)]\,w \tan a,$$

will represent the amount of tension on the mth segment of the lower chord.

The value of $m = \frac{n+1}{2}$, corresponds to a maximum in the first expression, and upon substitution gives $\frac{1}{2}(n+1)^2\,w \tan a$ for the maximum compression. The value of $m = \frac{n+2}{2}$, corresponds to a maximum in the second, and upon being substitituted in it gives $\frac{1}{2}[\,(n+1)^2-1]\,w \tan a$ for the maximum tension. The quantity, $n+1$, denotes the number of bays in the lower chord, which if we represent by N, the expression,

$$\tfrac{1}{2}N^2\,w \tan a,$$

will very nearly correspond to the maximum tension or compression upon the chords.

Strains on the chords.—The strains on the chords vary from segment to segment, but are uniform throughout any one segment. If the segments were infinitely short, the strains in that case would be a continuous function of the abscissa, and the rate of increase could be represented by the ordinates of a parabola. Suppose a vertical section made, cutting the truss between A_4 and B_4, and A_4 taken as the centre of moments.

From the principle of moments, there must be for equilibrium,

$$C_1 \times d = \tfrac{1}{2}wx^2 - R_1 x,$$

or

$$C_1 = \frac{wx^2 - 2R_1 x}{2d},$$

in which x is the distance of the centre of moments from A_1; C_1 is the stress in the upper piece $B_3 B_4$; d the distance between the axes of the chords; w the uniform load on the unit of length; and R_1 the reaction at the point of support A_1. This is the equation of a parabola whose axis is vertical and whose vertex is over the middle of the truss.

Remark.—The usual method of computing the strains upon the pieces of a truss is that of adding and subtracting for each consecutive piece, as shown in the previous methods for calculating strains. General formulas are used in connection with these methods to check the accuracy of the computations.

II. The Panel System.

454. If the ties of the triangular truss be pushed around until they are vertical, we shall have the method of vertical and diagonal bracing referred to in Article 263, and the resulting truss will be a type of the system. In England this truss is frequently called the **trellis girder**, and in France the **American beam.** (Fig. 170.)

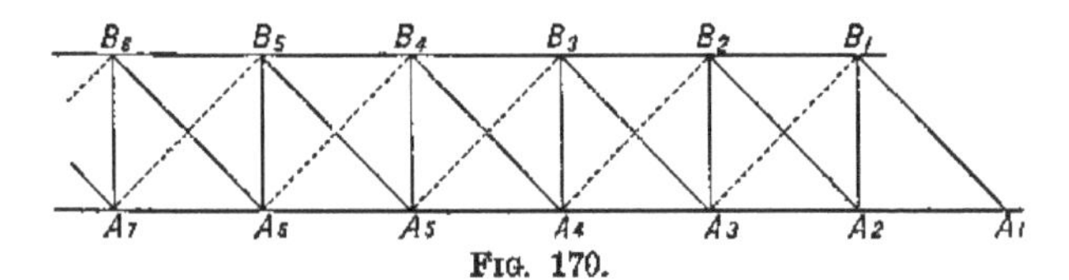

FIG. 170.

The methods already given for the determination of the strains on the parts of a Warren truss, and on a frame where vertical and diagonal bracing is used, can be applied to this truss.

The space included between any two consecutive verticals is known as a **panel**; hence the name of the system.

Diagonal pieces, as shown by the dotted lines in the figure, called counter-braces, are generally inserted in each panel. Their particular use will be alluded to in another article.

The Queen-post, or Trapezoidal Truss.

455. This is the simplest truss belonging to the panel system, and is much used in bridges where the span is not greater than forty or fifty feet. Its parts are most strained when the load extends entirely from one end to the other. Suppose this load to be uniformly distributed over the lower chord, $A_1 A_4$, and represent by (Fig. 171),

l, the length of the segment $A_1 A_2$;

w, the weight on the unit of length; and by

a, the angle of $R_1 A_1 B_1$.

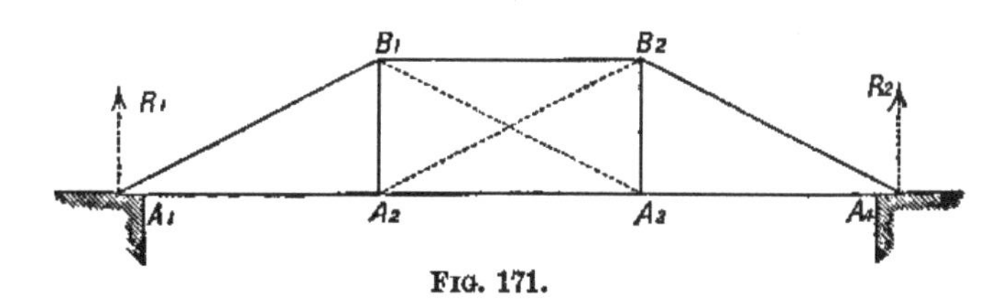

Fig. 171.

Since the segments $A_1 A_2$, $A_2 A_3$, $A_3 A_4$, are ordinarily equal to each other, $3l$ will be the length of the lower chord, and $3wl$ will be the total load on the truss. The queen-posts are framed into the lower chord; the latter, therefore, has four points of support. Supposing the lower chord to be a single beam, or so connected as to act like one piece, each post would sustain $\frac{11}{10}$ of wl. Each weight is transmitted to the upper end of its post, where it is held in equilibrium by two forces, one acting in the direction of the inclined brace, and the other in the direction of the chord $B_1 B_2$. The components along $B_1 A_1$ and $B_2 A_4$ are each equal to $\frac{11}{10} \frac{wl}{\cos a}$, and those along $B_1 B_2$ are equal to $\frac{11}{10} wl \tan a$. The two latter balance each other, producing a strain of compression on the upper chord. The other two produce compression on the braces, which, transmitted to the points of support, causes a strain of tension on the lower chord and a vertical pressure on the points of support. Knowing the amount and kind of strains, the dimensions of the pieces can be calculated.

Instead of considering the lower chord as a beam resting on four points of support, it is more usual to consider that one-third of the entire load is held up by each post, and one-sixth at each point of support at the ends; or, if the segments are unequal in length, to consider the weight held up by each post to have the same proportion to the whole load that the segments have to the entire length of the chord $A_1 A_4$. The remarks made upon the inverted king-post truss will apply to this frame, if inverted.

The queen-post truss, in its present shape, will not change its form under the action of a load uniformly distributed over it; when loaded in this manner, the truss is said to be **balanced.** If, however, the load be only partially distributed over it, so that the resultant acts through some other point than the middle of the truss, the truss may become distorted by a change of figure in the parallelogram $A_2B_1B_2A_3$. The truss is then said to be **unbalanced.**

Sometimes, a certain amount of stiffness in the joints and of resistance to bending in the pieces, give sufficient rigidity to the truss, and may be relied upon to prevent distortion under light loads.

As the load moves from one point to another, a change of form will generally take place, due to the elasticity of the materials of which the frame is made and to the imperfection of the joints. To prevent this change of form, diagonal pieces are inserted, as shown in the dotted lines of the figure. The truss is then said to be **thoroughly braced.**

A truss is said to be thoroughly braced when the parts are so arranged that no distortion takes place under the action of its usual load, whatever may be the position of the load.

A truss may be distorted and even broken, by an excessive load, notwithstanding the use of braces, but this distortion is excluded by the definition of a frame, given in Art. 230.

In the calculations to determine the strains, the joints of the truss are considered to be perfect.

III. The Bowstring System.

456. The **common bowstring girder** is one in which the upper chord is curved into either a circular or parabolic form and has its ends secured to the lower chord, which is straight (Fig. 107). The horizontal thrust of the upper beam is received by the lower chord; the latter therefore acts as a tie, and as a consequence, the reactions at the points of support are vertical. The intermediate space between the **bow** and the **string** is filled with

a diagonal bracing, like that used in the triangular or panel systems, for the purpose of stiffening the truss.

The load straining the girder may rest directly upon the lower chord or be suspended by vertical ties from the upper one, and the greatest stresses developed in the pieces of the girder are found by the usual methods.

Where the span is of considerable length, the usual practice is to form the upper chord of a number of straight pieces, the intersections of whose axes are in the curve of the bow. (Fig. 172.)

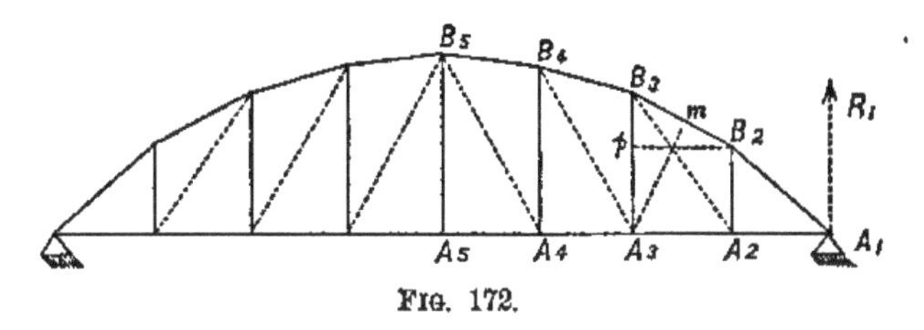

FIG. 172.

To find the strains produced upon the parts of a truss belonging to this system, by a uniform load resting on the lower chord, which is connected with the upper one by vertical ties dividing the truss into an even number of panels of equal horizontal length, represent by

$2a$, the length of the lower chord;

f, the rise of the curve, or depth of the truss at the centre;

w, the weight on the unit of length of the lower chord;

P_1, the stress in any piece of the upper chord; and

T_1, the stress in the lower chord.

Take the origin of the co-ordinates at A_1, the axis of X coinciding with the axis of the lower chord, and Y perpendicular to it.

Disregarding the braces, and supposing a vertical section made on the left of A_3, and very near to it, and B_3 taken as the centre of moments.

Taking the moments around this point, there results

$$T_1 \times A_3B_3 = R_1x - \frac{wx^2}{2} = \frac{wx}{2}(2a - x), \quad (155)$$

x, representing the distance A_1A_3.

Taking the curve containing the intersections B_1, B_2, B_3, etc., to be a parabola, its general equation when referred to the vertex and tangent at that point is

$$x^2 = 2py.$$

The vertex being the origin, the value of $y = f$ gives $x = \pm a$, or

$$a^2 = 2pf,$$

whence,

$$2p = \frac{a^2}{f},$$

which being substituted for $2p$ in the equation of the parabola, gives

$$x^2 = \frac{a^2}{f} y, \text{ or } y = \frac{f}{a^2} x^2, \quad . \quad . \quad . \quad (156)$$

Placing the origin at A_1, the equation of the curve will be

$$y = \frac{f}{a^2}(2ax - x^2). \quad . \quad . \quad . \quad (157)$$

Since A_3B_3 is equal to y, for the value of x equal to A_1A_3, there follows from the substitution of this value of A_3B_3, in equation (155),

$$T_1 = \frac{wx}{2} \frac{(2a - x)}{y} = \frac{w}{2} \frac{a^2}{f}, \quad . \quad . \quad (158)$$

Hence, the strain on the lower chord, produced by a uniform load, is constant throughout.

It is observed that this is the same value obtained for the horizontal component of the thrust in Art. 228.

In the same section, taking the moments around A_3, the lever arm of the strain on B_2B_3, is A_3m drawn perpendicular to the piece and through the centre of moments.

There results

$$P_1 \times A_3m = \frac{wx}{2}(2a - x). \quad . \quad . \quad (159)$$

Through B_2, draw a straight line parallel to the lower chord. From the triangles B_2B_3p and A_3B_3m, we have the proportion,

$$B_2B_3 : B_2p :: A_3B_3 : A_3m.$$

The first term of this proportion is the length of the piece of the upper chord in this panel, and varies in length for each panel from A_1 to the centre. The second term is the horizontal length of the panel and constant. Representing the former by v, and the latter by l, and substituting in the above proportion, we obtain

$$v : l :: y : A_3m. \quad \therefore A_3m = y \frac{l}{v},$$

or
$$A_2m = \frac{f}{a^2} x (2a - x) \frac{l}{v}.$$

Substituting which in equation (159), we get

$$P_1 = \frac{w}{2} \frac{a^2 v}{fl}. \quad . \quad . \quad . \quad . \quad . \quad (160)$$

This shows that the strain is independent of x and dependent upon v the only variable present, and that it increases as v increases, or is greatest at the points of support.

Suppose a brace to be inserted in this panel, joining A_2 and B_3, or B_2 and A_3. A section taken midway between A_2 and A_3 would cut the upper chord, the lower, and the brace. For an equilibrium, the algebraic sum of the horizontal components and of the vertical components of all the forces must be separately equal to zero.

Represent the strain on the brace by F, and the angles made by the brace and the piece $B_2 B_3$ of the upper chord with a vertical, by α and β, respectively.

The first of these conditions of equilibrium can be expressed analytically, as follows:

$$P_1 \sin \beta - F \sin \alpha - T = 0.$$

But $P_1 \sin \beta = T = \frac{w}{2} \frac{a^2}{f}$, hence

$$F \sin \alpha = 0, \quad \text{or} \quad F = 0.$$

That is, there is no strain on the brace produced by a load uniformly distributed over the truss.

If the load had been placed directly upon the upper chord, there would have been no strain on the verticals.

If the triangular instead of the panel system had been used for the bracing, its use would have been simply to transmit the loads on the lower to the upper chord. Knowing the angle of the bracing, the strain on any brace could be easily determined.

The vertical component of P_1 may be obtained as follows:

Let y' and y'' be the ordinates of the lower and upper extremities of any piece, as $B_2 B_3$, of the upper chord.

Let v, the length of the piece, denote the intensity of the strain on the piece, then $y'' - y'$ would represent its vertical component.

From the equation of the curve, we have

$$y'' = \frac{f}{a^2} x'' (2a - x''), \text{ and } y' = \frac{f}{a^2} x' (2a - x').$$

But $x'' = x' + l$, substituting which in the first of these equations for x'', and then from this result subtracting the second of the equations, we get

$$y'' - y' = \frac{fl}{a^2}(2a - 2x' - l).$$

Representing the vertical component by V, we may form the following proportion:

$$v : y'' - y' :: P_1 : V.$$

Substituting for P_1 and $y'' - y'$, the values just found, and solving, we find

$$V = \frac{w}{2}(2a - 2x' - l), \quad . \quad . \quad (161)$$

for the vertical component.

Other Forms of Bowstring Girders.

457. The common bowstring girder has been used in an inverted position by simply turning it over, so that the bow was below and the straight chord above. This inversion causes no difference in principle, the amount of strains on the different parts remains the same as before; the kinds of strains are changed, being compression on the straight chord and tension on the lower one.

By combining this inverted with the other truss, that is, by making both the upper and lower chords curved, another form is obtained. This arrangement was used by Brunel in the Saltash Bridge.

Where the amount of material forms an important item, both in the weight and cost of the structure, as in the case of very large spans, the last form can be more advantageously used than any of the other forms of bowstring girders.

The great objection to the bowstring girder, compared with the trusses of the other systems, is the inferior facilities it affords for lateral bracing.

Compound Systems.

458. If two or more of the trusses already described be combined, there is formed a class known as **compound trusses.** This term is sometimes limited to a com-

bination made of two or more of different systems, particular names being given to those made of the same system.

As they can be always resolved into their simple parts, there is no need of a separate classification except for descriptive purposes.

459. **Lattice truss.**—If the segments of the simple triangular bridge truss (Fig. 169) be bisected, and braces inserted in the intervals thus formed parallel to the braces already used, a truss similar to that shown in the Fig. 173 is formed.

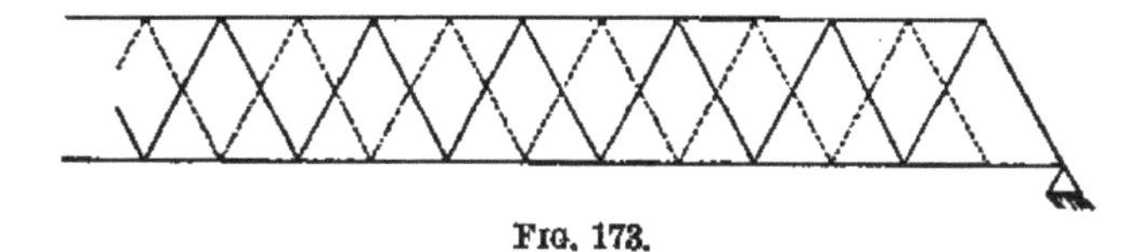

FIG. 173.

The dotted lines show the intermediate braces. This is called a **double triangular** truss, although sometimes it is known as the **half-lattice**.

By dividing the segments into three, four, or more equal parts, and inserting a corresponding number of braces, the triple, quadruple, etc., triangular trusses are formed. They are generally known as **lattice** trusses, or girders.

To determine the strains on a truss of this kind, it is usual to consider the truss as composed of two, three, four, or more simple triangular trusses, as the case may be, and find the strains on each one separately. These are then added and the strength of the truss considered as that of the whole combined. Under this supposition, the braces are regarded as separate from each other, and only fastened at their ends. In fact, they are generally fastened together at their intersections, which adds to the strength of the combination but complicates the problem of finding the amount of strain on each piece.

A subdivision of a truss of the panel system, and putting in another set of panels of the same size, will give a compound truss which has been much used. A calculation of the strains is made in the same way as that just described.

Strains Produced by a Moving Load.

460. Loads placed in particular positions, or stationary loads, have been the only forces considered in the previous examples. As a bridge affords continuous roadway between two points,

it is subjected to strains produced by loads which move over it, and it is essential that the action of the moving loads on the parts of the bridge be known.

With the exception of the shearing strain, it has already been shown that the strains produced by a moving load are the greatest when the centre of the load is at the centre of the bridge, and will be at the maximum when the moving load covers the entire structure.

If, then, the maximum moving load that will ever come upon the bridge be supposed to have its centre at the middle of the bridge, and the parts of the bridge determined under this supposition, the bridge will possess the requisite strength.

When the shearing strain enters as an important element, its maximum value should be obtained, and the parts of the bridge proportioned accordingly.

461. **Counter-braces.**—The dotted lines in Fig. 170 represent pieces of the truss known as counter-braces. If the truss supports only a load at the middle point, or a load uniformly distributed over the entire truss, these counter-braces are not necessary. In ordinary trusses they are needed to resist the action of moving loads.

Take the simple triangular bridge truss, and suppose it strained by a live load which is uniformly distributed over the lower chord. Let this live load extend from either end of the truss and for a distance equal to one-fourth of the span. The resultant of the load acts through its middle point, which is at a distance from the end of the truss equal to one-eighth of the span.

The nearest abutment, or point of support, will therefore support seven-eighths of this live load and the farthest abutment will support one-eighth. The strains on the chords and braces can be determined by the methods already explained.

The strains produced upon the diagonals between the end of the load and the middle of the truss, by the one-eighth of the live load going to the farthest point of support are of opposite kind to those which would be produced on the same pieces by the dead load. That is, the braces whose tops incline towards the middle of the truss are extended by the action of this eighth instead of being compressed, and the other braces are compressed by it instead of being extended. Some of the braces, therefore, are under certain circumstances, liable at one time to be extended and at another time to be compressed, and must, in consequence, be constructed to resist both kinds of strains. In the panel system, each brace is generally constructed to take only one kind of strain. Hence, in those panels where a change of strain is liable to take place, another brace must be inserted to take this new strain. The braces required by the dead load are called **main braces**; the extra braces inserted in those panels where a change of strain may occur, are called **counter-braces.** The main braces are necessary in every panel, and it has also been the custom to use counter-braces in every panel. The main and counter-braces generally cross each other in the middle of the panel, the angles which they make with a vertical being supplements of each other. It is evident that there is no necessity for counter-braces in any of the panels except those between the extreme positions of the points of "no shearing" strain and the middle of the truss.

Length and Depth of a Truss.

462. The **length** of a truss depends upon the span and whether the truss is to rest on two or more points of support. Assuming that the truss rests on two points of support, the length depends upon the span. The span depends upon several things: the navigability of the stream, character of the freshets, the movement of ice, the practicability of obtaining inexpensive and good foundations, etc.

Over wide river bottoms, marshes, etc., where good foundations are easily procured without much expense, the spans range from twenty-five to fifty feet. Over important rivers, from 150 to 250 feet.

Extra wide spans are frequently required for bridges over the main channel of very important rivers. The central span of the Victoria Bridge, over the St. Lawrence River, is 330 feet. The channel spans of the Louisville bridge, over the Ohio River, are 370 and 400 feet respectively. The central span of the St. Louis bridge, over the Mississippi River, is 515 feet.

The **depth** of the truss, in terms of its length, varies from one-tenth to one-fifteenth in England and from one-sixth to one-tenth in the United States.

The Graphical Method.

463. The **graphical method** is much used to determine the strains on the different parts of a bridge truss. This method possesses many advantages and grows in favor with engineers as it becomes better known.

By its use the engineer is enabled to make an independent investigation of the strains and to test the accuracy of his calculations by a comparison of the results obtained through two independent methods.

The graphical method is based on the simple principles much used in mechanics: that **a force** may be represented by **a straight line**; that the force **is completely given** when the **length** of the line, its **direction**, and **point of application** are known; and that if **two forces** having a common point of application are given, that a **third force** may be determined, which acting at the common point will produce the **same effect as the two acting simultaneously**. This third force is determined by the use of the principle of the "parallelogram of forces."

464. **Two forces having a common point of application.**—Suppose two forces, P_1 and P_2, acting at the point A_1 (Fig. 174).

From any assumed point, as O, draw a right line parallel to the direction of the force P_1, and lay off on this line, ac-

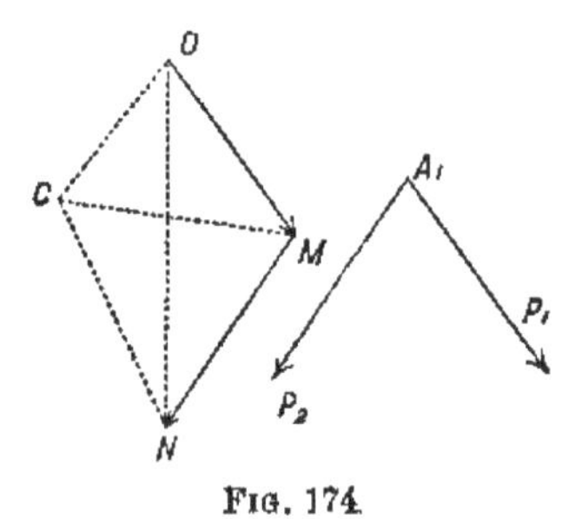

FIG. 174.

cording to some assumed scale, the distance O M, to represent its intensity. From the end, M, of the distance just drawn, draw the line M N parallel and equal to P_2. Join N and O by a straight line, and N O will be parallel and equal to the resultant of P_1 and P_2. Its intensity can be obtained by measuring the distance O N with the same scale used to lay off O M and M N.

If a force equal to and parallel to N O acts from A_1 upwards, there would be an equilibrium among the three forces at A_1. It therefore follows that if three forces at any point are in equilibrium, the three sides of a triangle, which are respectively parallel to the directions of these forces, may be taken to represent their intensities.

Assume any point, as C, and from it draw to the extremities O and N of O N, the right lines C N and C O. These distances, C N and C O, may be taken as the intensities of two components which, acting at A_1 in directions parallel to these lines respectively may be used to replace the resultant, O N.

And in general, any two right lines drawn from any assumed point, which may be called **the pole**, to the ends of the straight line representing a force, may represent the components of that force.

465. **Any number of forces in the same plane having a common point of application.**—Whatever be the number of forces acting at A_1, the right lines representing them in intensity, if drawn parallel to their directions and in order, either from the right to the left or the reverse, each from the end of the other, will form a polygon whose sides may be taken to represent the forces, acting at A_1.

If the last line drawn terminates at the starting point of the polygon, the forces are in equilibrium; if not, then the right line drawn, joining the extremity of the last side with this point, will represent the force, which, being added to those acting at A_1, will produce an equilibrium.

It is evident that if a diagonal be drawn in this polygon, it may be taken as the resultant of the forces on either side of it and may be used to replace those forces.

The polygon constructed by drawing these lines parallel to the forces is called the "**force polygon,**" and when it terminates at the point of beginning, the polygon is said to be "**closed.**"

If the forces act in the same straight line, the polygon becomes a right line.

466. **A system of forces in the same plane with different points of application.**—It will only be necessary, in this case, to produce the lines of direction until they intersect. It is then the case just considered. It may be that the point of intersection will not be found within the limits of the drawing. Under this supposition, a point of the resultant may be determined as follows:

Let P_1 and P_2 be any two forces which do not intersect within the limits of the drawing, their points of application being A_1 and A_2 respectively. (Fig. 175.)

Draw O M and M N, respectively, equal and parallel to P_1 and P_2. The line O N will give the direction and intensity of their resultant. From any point, as C, draw the right lines, O C and C N. These are the components which may be taken to replace O N. Assume any point on P_1, as *a* and draw through it the lines, *ac* and *ab* parallel to C O and M C, respectively. Where *ab* intersects P_2, as at *b*, draw *ba* and *bc* parallel to C M and N C. Produce the lines *ac* and *bc* until they intersect. Their point of intersection will be one point of the resultant, which can now be constructed. The same method holds good if the forces are parallel. If there were more than two forces the same method can be used.

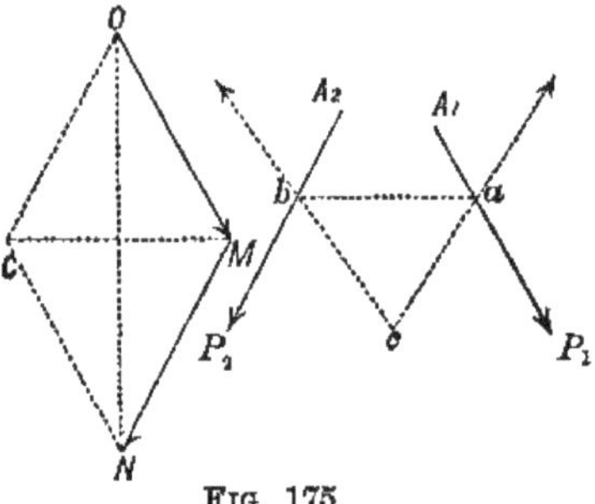

FIG. 175.

If perpendiculars are let fall from the point of intersection, *c*, upon the directions of the forces P_1 and P_2, it can be easily shown that they are to each other inversely as the forces. That is, if the perpendicular let fall on P_1 is represented by p', and that on P_2 by p'', that there is the following proportion :

$$p' : p'' :: P_2 : P_1.$$

This is also true for the perpendiculars let fall from any other point of the resultant.

467. **Parallel forces.**—The principal forces acting on engineering structures are due to the action of gravity, and in these discussions such forces are taken as parallel and vertical.

Let P_1, P_2, P_3, etc., be a system of parallel forces acting at the points A_1, A_2, A_3, etc., in the same plane. (Fig. 176.)

Lay off from O, on a straight line parallel to A_1, P_1, the distance O 1 equal to its intensity, and from 1 to 2, the intensity of P_2, and then from 2 to 3, the intensity of P_3, etc. The straight line of O 5 will be the force polygon, and in this case equal to the resultant, as all the forces are acting in the same direction. From any assumed point, *c*, as a pole, draw straight lines to O, 1, 2, 3, etc., or extremities of the forces just laid off on the line O 5. The perpendicular, C H, is called the "**pole distance.**" Assume a point on the the right of P_1, as *a*, and through it draw a straight line parallel to O C.

From the point b, where this line intersects P_1, or P_1 produced, draw a line parallel to C 1, and from the point where this intersects P_2 produced, draw one parallel to C 2, etc., until lines parallel to all the lines drawn from C have been drawn.

These forces P_1, P_2, etc., may be supposed to act at these points, b, c, d, etc. If the points a and g are fixed, and the others are all connected by flexible cords, the whole arrange-

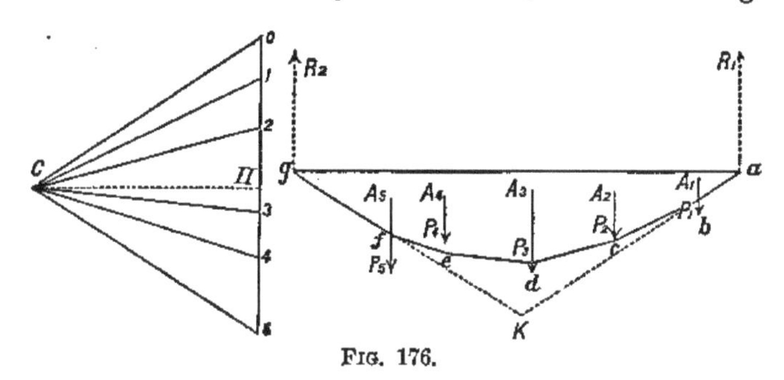

FIG. 176.

ment would form a **funicular** machine or polygon. The three forces acting at any one of these points are represented by the three sides of a triangle, and are therefore in equilibrium. The broken line, a, b, c, d, e, f, g, thus formed, is called the "**equilibrium polygon.**"

If ab and gf be produced until they intersect, their intersection will be one point of the resultant of the system of forces, and the resultant may at once be constructed.

468. Suppose ag to be the axis of a beam resting in a horizontal position upon two points of support at a and g, and acted upon by a system of forces whose resultants correspond in direction with those of the forces P_1, P_2, P_3, etc. In order that an equilibrium should exist, there must be vertical reactions acting upwards at these points, a and g, and their sum must be equal to the resultant. Represent these reactions by R_1 and R_2. If the resultant passes through the middle point of this line, ag, that is, if the forces are distributed symetrically with respect to the middle point, the reactions will be equal to each other.

Examining the equilibrium polygon, it is seen that the resultant of P_1 and P_2 must pass through the intersection of ab and cd; that the resultant of R_1 and P_1, through the intersection of ag and bc; of P_1, P_2, and P_3, through the intersection of ab and de; and so on. A simple inspection of the force polygon will give the direction and intensity of any of these resultants.

469. **Bending moment of any section, and the shearing strain.**—Let it be required to determine the bending moment and the shearing strain on any section of a beam resting on two points of support and holding up four unequal weights at unequal distances apart.

Theorem.—*The moment of a force, around any centre is equal to the "pole-distance" multiplied by a straight line drawn through this centre parallel to the force and limited by the components of the force.*

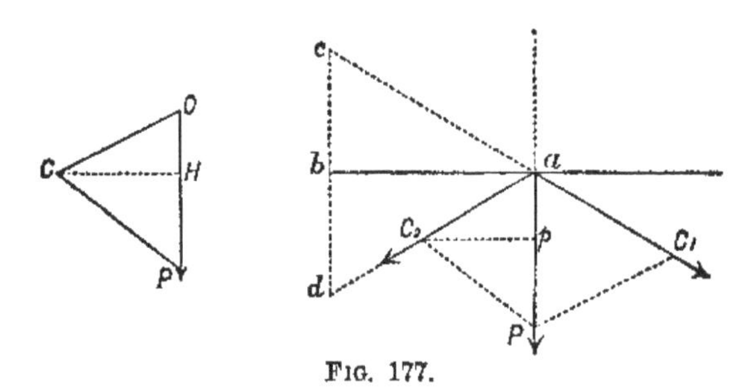

FIG. 177.

Let the force, P (Fig. 177), be resolved into any two components, aC_1 and aC_2, which are represented by the right lines C O, C P, drawn from the pole to the ends of the force in the force polygon. The moment of P with respect to any point, as b, is $P \times ab$. From C_2 draw the line C_2p, perpendicular to P. This is equal to the "pole distance," which represent by H. Through b draw cd parallel to P, and limited by C_1 and C_2 produced. From similar triangles, the following proportion is obtained:

$$P : H :: cd : ab, \quad \text{or} \quad P \times ab = H \times cd,$$

which was to be proved.

In Fig. 178 the bending moment at O′ is $R_1 \times AO'$, which, as has just been shown, is equal to $H \times pp_1$; at O″ the bending moment is $R_1 \times AO'' - W_1 \times A'O''$. The components of W_1 are ab and bc. Hence the moment of W_1 at O″ is $H \times p'_1 p_{\prime\prime}$, and the total moment is $H \times p'p'_1$.

And as this is true for any section, it is seen that the bending moments are proportional to the ordinates drawn from the closing line to the sides of the equilibrium polygon. And at any section, it is equal to the product of H and the ordinate of the equilibrium polygon corresponding to the section under consideration.

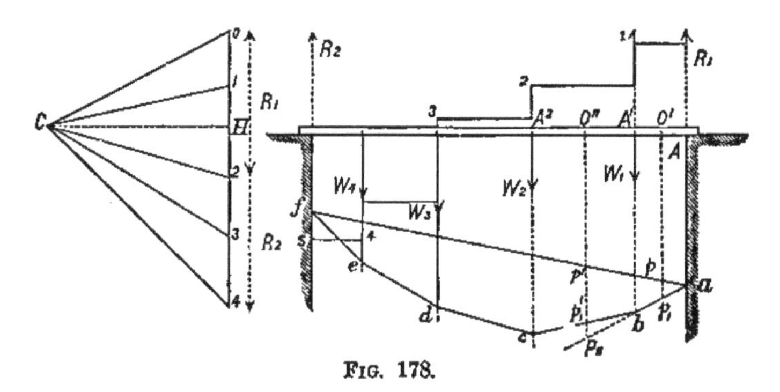

FIG. 178.

The ordinate is measured by the scale used for the equilibrium polygon, and the pole distance, H, by the scale for the force polygon. These may be drawn on the same or different scales, whichever is the most convenient.

Representation of the shearing strain.—The shearing force between R_1 and W_1, is R_1. At W_1, the shearing force is R_1-W_1; at W_2, it is $R_1-W_1-W_2$, etc. Hence, the line, R_1, 1, 2, 3, etc., represents graphically the shearing forces for all parts of the beam.

An examination of the figure shows that the shearing force is greatest where the bending moment is the least, and the reverse.

470. **Couples.**—It has been assumed, in the previous discussions and examples, that the forces were in equilibrium, or by the addition of a single force an equilibrium could be established.

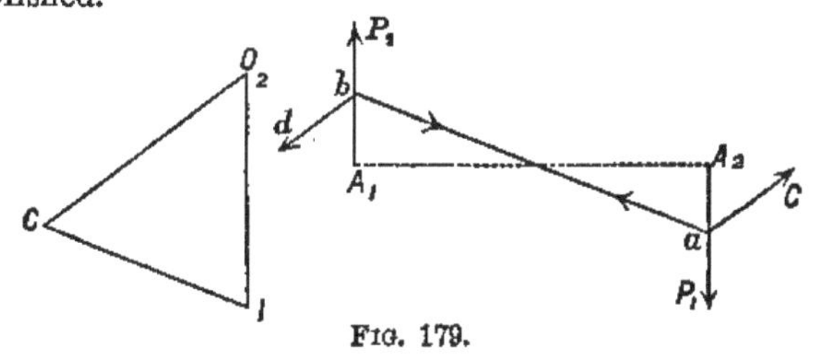

FIG. 179.

If two forces form a couple, they cannot be replaced by a single force. Let P_1 and P_2 be a couple (Fig. 179), and O 1 2 the force polygon.

It is seen that this force polygon closes, that is, the resultant is zero. From any point on P_1 draw *ac* and *ab* parallel to C O and 1 C. At *b*, where *ab* intersects P_2 or P_2 produced, draw lines parallel to C 1 and 2 C. The lines *ac* and *bd* are parallel. Therefore the equilibrium polygon will not close, or the lines will intersect at an infinite distance. A result which was to be expected. (Art. 98, Analytical Mechanics.)

The figure shows that the components of the forces P_1 and P_2, which act in the direction of the line *ab*, are equal and directly opposed to each other, and that the other two are parallel, forming a couple. Hence, it is concluded that a couple can be replaced by another without changing the action of the forces.

From what has been shown, it is evident that if both the force and equilibrium polygon close, that an equilibrium exists among the forces. But if the force polygon closes and the equilibrium does not, that the forces cannot be replaced by a single force, but only by a couple.

471. **Influence of a couple.**—Let A B (Fig. 180) be a beam fastened at its ends and acted upon by the couple P_1 P_2.

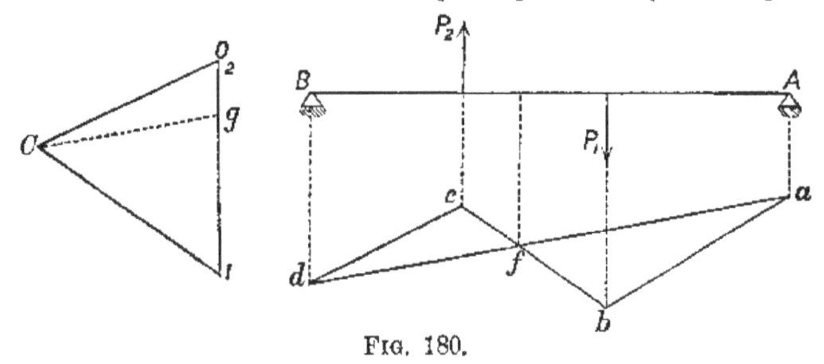

FIG. 180.

The beam being fastened, the reactions at A and B will keep the couple from moving and the four forces will be in equilibrium. Construct the force polygon, O 1 2, and from a pole, C, draw the lines C O, C 1. Form the equilibrium polygon, *a b c d*, of the forces P_1 P_2; produce *b a* and *c d* until they intersect the lines of direction of the reactions; join *a* and *d* and this will be the closing line of the polygon. Parallel to this line draw C*g* in the force polygon. An examination of the force polygon shows that O *g* is the vertical reaction acting downwards at B, and *g* O, the reaction at A, acting upwards, which with the couple P_1 P_2 form an equilibrium.

The ordinates drawn from the closing line, *ad*, upon the sides, *ab*, *bc*, and *ad*, multiplied by the pole distance give the bending moments for the corresponding sections of the beam.

In the preceding examples the force polygon has been given, and from it the equilibrium polygon has been constructed. Inversely, the equilibrium polygon being given, the force polygon is easily constructed.

472. From the preceding demonstrations, the following theorem may be enunciated:

Theorem.—*If straight lines be drawn through any assumed point parallel to the sides of a polygonal frame, then the sides of any polygon whose angles lie on these radiating lines may be taken to represent a system of forces which, if applied to the angular points of the frame, will be in equilibrium among themselves.* And the converse, *that if a system of external forces acting at the angles of a frame are in equilibrium,* that from an assumed point drawing straight lines parallel to the sides of the frame, and then parallel to the directions of these forces drawing straight lines whose successive intersections are on the successive radial lines, *the distances cut off by the second set will represent the strains on the corresponding sides of the frame.*

Let A B C (Fig. 181) be a triangular frame acted upon at the points A B C by a system of external forces which are in equilibrium. Let P_1 P_2 P_3 be the resultants of the forces acting at these points, and suppose that these resultants are in the plane A B C.

From an assumed point, P, draw the straight lines, P 1, P 2,

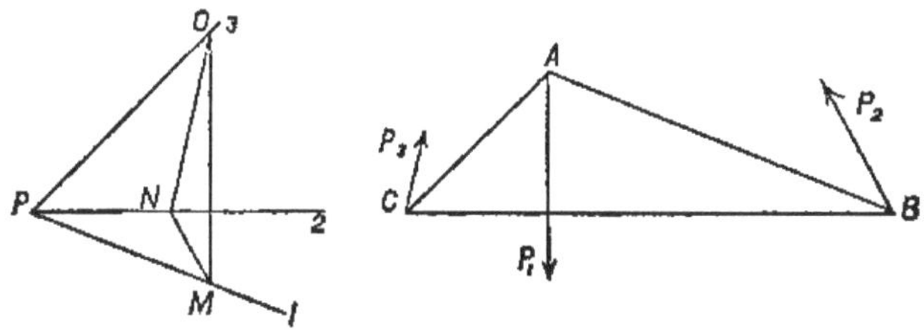

FIG. 181.

and P 3, respectively, parallel to the sides A B, B C, and C A. Through an assumed point, as O, on the line P 3, draw the line O M parallel to the direction of the force P_1, and from its point of intersection with P 1, draw the line M N parallel to the force P_2.

Join N and O by a straight line, and this will be parallel to the force P_3. The triangle, O M N, will be the force polygon.

The distance, P O, will measure the force acting along the piece A C; P M, that along A B; and P N, that along B C.

If the external forces are parallel the polygon becomes a straight line, which will be divided into segments by the lines drawn parallel to the sides of the frame. Each segment will represent the external force acting at one of the angles of the frame, and the distances cut off will represent the forces acting along the adjacent pieces.

An application of these principles will enable the student to determine graphically the strains on the different parts of a frame, and test the accuracy of calculations already made by other methods.

Working, Proof, and Breaking Loads.

473. **Ultimate strength** of a structure.—The object of the calculations made to determine the strength of a given structure is to find the load which, placed on the structure, will cause it to give way or break in some particular way. This load is called the **ultimate strength or breaking load** of the structure.

Working load.—As the bridge must not be liable to yield or give way under any load which it is expected to carry, it is made several times stronger than is actually necessary to sustain the greatest load which it will ever have to support. The greatest load thus assumed is called the **working load.**

The ratio of the breaking load to the working load, or "factor of safety," is assumed arbitrarily, limited by experience. It is usually taken from four to six for iron, and even as high as ten for wooden bridges. It should be large enough to ensure safety against all contingencies, as swift rolling loads, imperfect materials, and poor workmanship.

Proof load.—When the bridge is completed, it is usual to test the structure by placing on it a load greater than it will ever have to support in practice. A train of locomotives for a railroad bridge, and a crowd of men, closely packed, upon an ordinary road bridge, are examples. These loads are known as **proof loads.**

A proof load should remain on the bridge but for a short time, and should be removed carefully, avoiding all shocks. Excessive proof loads do harm by injuring the resisting properties of the materials of which the bridge is built.

Wooden Bridge-trusses.

474. Both the king and queen-post trusses, as stated in a

previous article, are frequently made entirely of wood, and are used in bridges of short spans.

A compound truss, entirely of wood, the outline of which is shown in Fig. 182, has been used in bridges for spans of considerable width.

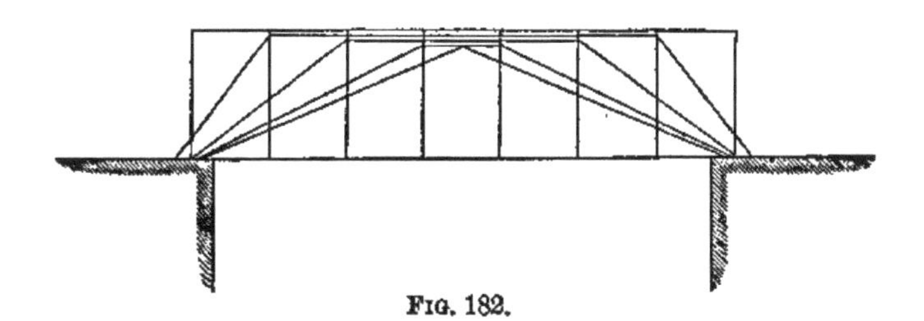

FIG. 182.

The celebrated bridge at Schaffhausen, which consisted of two spans, the widest being 193 feet, was built upon this principle.

475. **Town's lattice truss.**—This truss was made entirely of wood, and at one time was much used in bridge construction. It belongs to the triangular system. The chords (Fig. 183) were built of beams of timber, and frequently of plank of the same dimensions as that used for the lattice. They were in pairs, embracing the diagonals connecting the upper and lower chords. The diagonals were of plank, of a uniform thickness and width, equally inclined towards the vertical and placed at equal distances apart. They were fastened to the chords, and to each other at their intersections, by treenails, as shown in the figure.

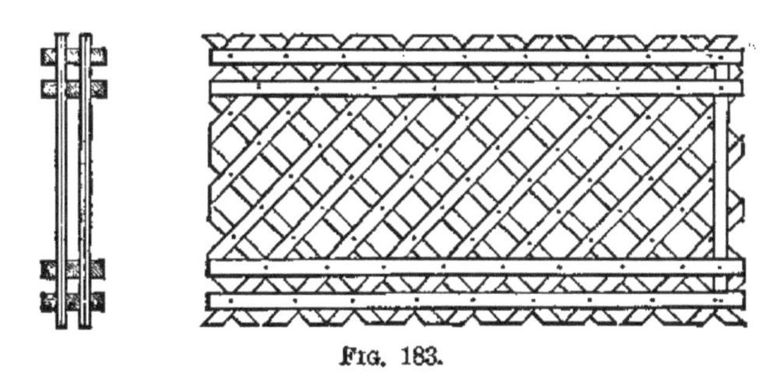

FIG. 183.

This truss was frequently made double. In case the lattices were separated by a middle beam, as shown in the cross-section in Fig. 183, the chords, instead of being in pairs, were made of three beams, placed side by side.

When the truss was of considerable depth, intermediate longitudinal beams were used to stiffen the combination, as shown in the figure.

This truss possessed the advantages of a simple arrangement of its parts and ease of construction. It also possessed the disadvantages of a waste of material and a faulty construction by which the strength of the truss depended upon the strength and the perfect fitting of the treenails.

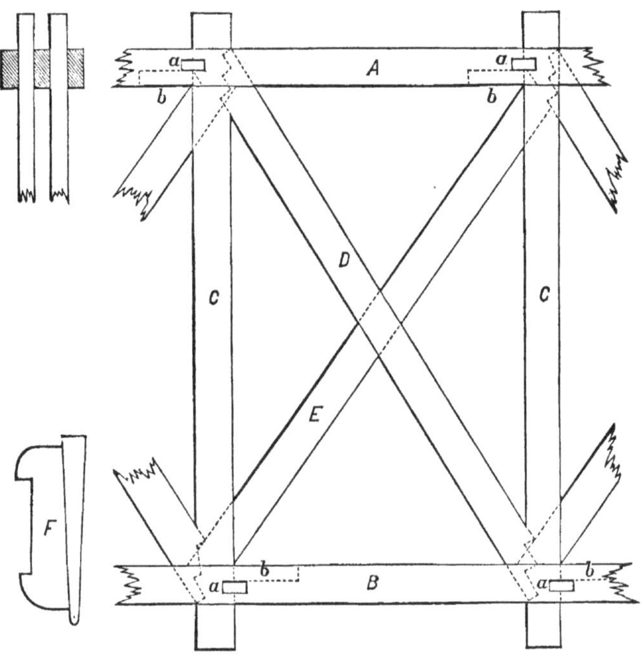

FIG. 184—Represents a panel of Long's truss.
A and B, upper and lower chords.
C, C, uprights, in pairs.
D, main braces, in pairs.
E, counter-brace, single.
a, *a*, mortises where gibs and keys are inserted.
b, *b*, blocks behind uprights, fastened to the chord.
F, gib and key of hard wood.

476. Long's truss.—This truss belongs to the panel system, and was built entirely of wood. It was one of the earlier trusses used in the United States, and takes its name from

Colonel Long, of the Corps of Engineers, United States Army, who invented it. It was one of the first trusses in which a scientific arrangement of the parts was observed. (Fig. 184.)

All the timber used in its construction had the same dimensions in cross-section.

Each chord was composed of three solid-built beams, placed side by side, with sufficient intervals between them to allow of the insertion of the uprights. The uprights which connected the chords were in pairs, and fastened to the chords by **gibs** and **keys**. These gibs were inserted in rectangular holes made in the chords, and fitted in shallow notches cut in the uprights. Pieces of wood wide enough to fill the space between the beams, about three or four inches thick and two feet long, were inserted between the beams of the chords, behind the uprights, and fastened to the beams by treenails. These were for the purpose of strengthening the uprights and preventing their yielding at the notches.

The main braces were in pairs, and were joined to the uprights, as shown in the figure. The counter-braces were single, and were placed between the main braces, abutting against or fastened upon the upper surface of the middle beam of the chords. Generally they were fastened to the main braces by treenails at their intersections.

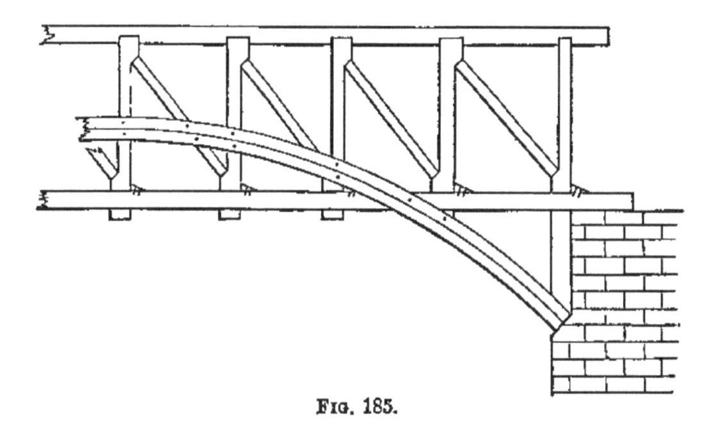

FIG. 185.

477. Burr's truss.—This is another of the earlier wooden trusses, much used at one time in the United States. This truss (Fig. 185) belongs to a compound system, being composed of a truss of the panel system, stiffened by solid-built curved beams, called **arch timbers**. These arch timbers were in pairs, embracing the truss and fastened to it at the different intersections of the pieces of the truss with the curved beams, as shown in the figure.

478. **Other forms of wooden trusses.**—The trusses already named may be considered as typical trusses. There are many others, all of which may be referred to one of the systems already given, or a combination of those systems. Haupt's lattice, Hall's lattice, McCallum's truss, etc., are examples of some of the different forms of wooden bridge-trusses.

Bridge-trusses of Wood and Iron.

479. **Canal bridge.**—A truss composed of wood and iron, which has been much used for common road bridges over the New York State canals, is shown in Fig. 186.

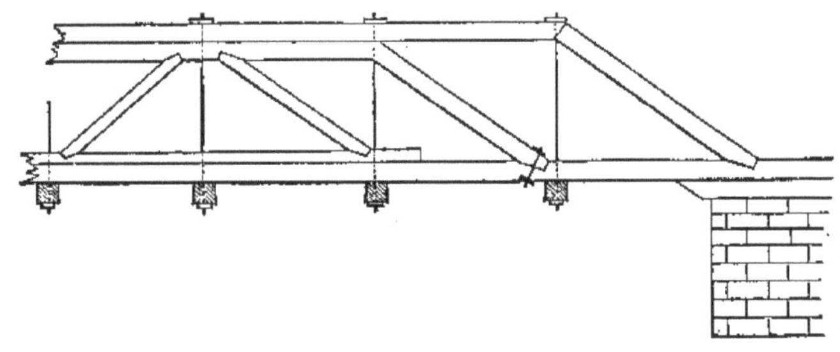

FIG. 186.

In this truss, the chords and diagonals are of wood, and the verticals of iron. In some cases, the lower chord is also of iron.

480. **Howe's truss.**—A popular truss for bridges, both common and railroad, and one which has probably been used more than any other, is known as the **Howe truss.** (Fig. 187.)

This truss belongs to the panel system. The chords and braces are made of wood, and the verticals of iron.

The chords are solid-built beams of uniform cross-section throughout.

The braces are also of uniform size, the main braces being in pairs, and the counter-braces single, and placed between the main braces, as in Long's truss. Between the ends of the braces and the chords, blocks of hard wood or of cast iron, inserted in shallow notches in the chords, are used as shown in the figure. The faces of the blocks should be at right angles to the axes of the braces.

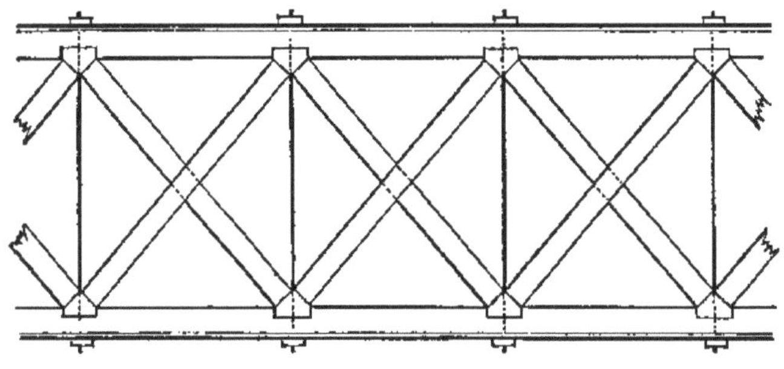

FIG. 187.

The verticals are in pairs, and pass through the blocks and chords, and are secured by nuts and screws at both ends, or by heads at the ends with a nut and screw arrangement at the middle. By tightening the screws, the chords are drawn towards each other, and the reverse. To prevent the edges of the nuts from pressing in and injuring the timber, washers, or iron plates, are placed between the nut and the wood.

Where the pressure on the block is great, an iron block or other arrangement is placed between the block on one side and the washer on the other, to prevent the block from crushing into the chord.

It is seen that there is an excess of material in some of the chords and the braces. The corresponding gain obtained in reducing the amount of material, by proportioning the pieces to the strains they would have to support, would not pay the cost of extra time and labor required; these pieces are there fore made, as a rule, with a uniform cross-section.

There would be a gain if the verticals were proportioned to the strains which they have to support, instead of being made of uniform size.

It is observed that the framing is such that the diagonals will only take a compressive strain, and the verticals a tensile one.

481. **Pratt's truss.**—If the framing of the Howe truss is changed so that the diagonals will only take a tensile strain, and the verticals a compressive one, there results the truss

known as Pratt's. The chords and verticals in this case are of wood, and the diagonals are of iron.

482. There are quite a number of trusses besides those just named, which are composed of wood and iron. Those mentioned are typical ones, and illustrate fully the method of combining the two materials in the same structure.

Iron Bridge-trusses.

483. Bridge-trusses made entirely of iron, or of iron and steel are much used at the present time.

Trusses of iron belong to the three systems already described, viz: the triangular, the panel, and the bow-string systems, and are generally known by the names of their inventors.

At one time, the use of cast-iron for the compressive members of a truss was much favored by builders in the United States. At the present time, wrought iron or steel is preferred to cast-iron for all the parts of a truss.

The trusses known as Fink's, Bollman's, Warren's. Jones, Whipple's, Murphy-Whipple, Linville, Post's, etc., are some of the trusses made entirely of iron which are most frequently seen in use in the United States.

Fink's truss.—The principles of Fink's truss are given in Art. 449. The arrangement of its parts enables the truss to resist in the best manner the effect produced by a moving load, or by changes of temperature. The lower extremities of the verticals being free to move, the verticals remain normal to the curve assumed by the chord under the straining force, and the distances of their lower ends from the connection of the ties with the chord remain relatively the same. None of its parts are therefore unequally strained by the force producing the deflection.

Bollman's truss.—The principle on which this truss is constructed is mentioned in Art. 450. In order to avoid the ill effects of unequal expansion or contraction of the ties produced by changes of temperature, a compensating link is used, by means of which the pin holding the ties is enabled to change its position as the ties contract or expand, without straining the verticals.

Warren's truss.—The principle of this truss is explained in Article 451. It is ordinarily made entirely of wrought iron. In some cases the braces are of cast iron, in the form of hollow pillars, with wrought-iron ties enclosed. The brace is thus composed of two distinct parts, and is better suited to resist the strains which it has to sustain.

Jones's truss.—This truss is the Howe truss in principle, all the parts being of iron.

Whipple's truss.—This truss is one of the first used in this country made entirely of iron. It is composed of cast and wrought iron; the former being used for the compression members, and the latter for the tension members.

This truss (Fig. 188) belongs to the panel system. The upper chord is usually made of hollow tubes of cast iron, in sections, whose lengths are each equal to a panel distance.

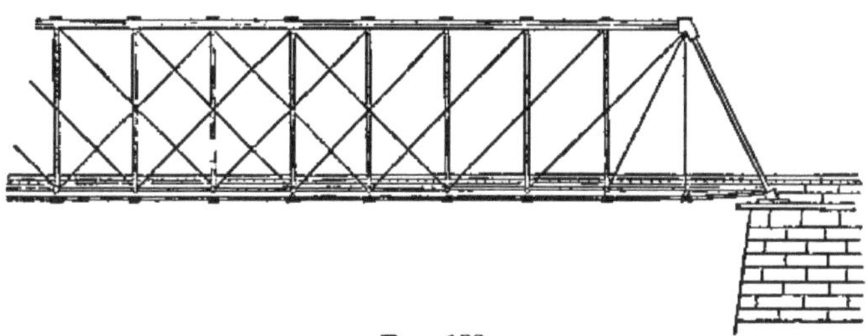

FIG. 188.

The lower chord is made of links, or eye-bars, of wrought iron, which fit upon cast-iron blocks. These blocks hold the lower ends of the vertical pieces.

The vertical pieces are of cast iron, and are so made that the inclined pieces can pass through the middle of them. The parts are frequently trussed by iron rods, to prevent bending.

The inclined pieces are wrought-iron rods, and it is seen that each of them, excepting those at the ends, crosses two panels.

An examination of this truss shows that the inventor has considered economy of material in making the verticals, struts, and the diagonals, ties. In principle it corresponds with the Pratt truss.

Murphy-Whipple truss.—This is the Pratt truss, entirely of iron, with some of the details of Whipple's.

Linville truss.—This is Whipple's truss made entirely of wrought iron, the verticals being wrought-iron tubular columns.

Post's truss.—This truss is composed of cast and wrought iron. Its peculiarity lies principally in its form (Fig. 189);

the struts, instead of being vertical, are inclined towards the centre of the bridge, making an angle of about 23° 30′ with the vertical, as shown in the figure. The ties cross two panels, and make an angle of 45° with the vertical. The counterties make the same angle, but cross only one panel.

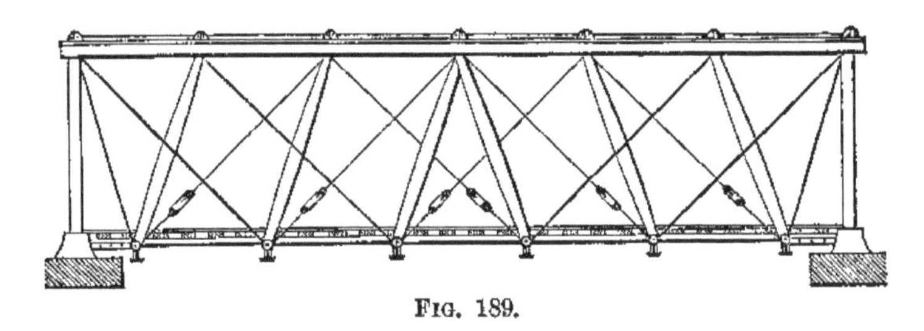

FIG. 189.

The inclination given to the struts was for the purpose of obtaining the same strength with a less amount of material than that obtained when the struts were vertical.

Lattice trusses.—Lattice trusses, made entirely of iron, are frequently used in railroad bridges. They do not differ in principle from the lattice truss made of wood.

Continuity of the Truss.

484. Various opinions have been held as to the advantages obtained in connecting the trusses over adjacent spans, so that the whole arrangement should act as a single beam.

If the load is permanent, or the weight of the structure is very great, compared with the moving load, it is advisable to connect the trusses, so that they shall act as a single continuous beam.

But when this is not the case, the effect of a heavy load is to reverse the strains on certain members of the trusses over the adjacent spans; a result which is to be avoided, and hence the trusses are ordinarily not rigidly connected.

CHAPTER XV

II.--TUBULAR AND IRON PLATE BRIDGES.

485. Bridges of this class are made entirely of wrought iron.

A **tubular girder** is one which is made of iron plates, so riveted together as to form a hollow beam. These girders may be placed side by side and a roadway built upon them, forming a simple bridge, which in principle, would not differ from the simple bridge described in Art. 435.

When the tube is made large enough to allow the roadway to pass through it, it is called a **tubular bridge.**

The difference in construction between the tubular bridges and the tubular girders consists in the arrangements made to stiffen the four sides of the tube.

The three great examples of tubular bridges are the Britannia Bridge, across the Menai Straits, in Wales; the Conway Bridge, over the Conway River, in Wales; and the Victoria Bridge, over the St. Lawrence River, at Montreal, Canada.

The Britannia Bridge consists of two continuous girders, each 1,487 feet long, resting on three piers and two abutments. Each tube is fixed to the central pier and is free to move on rollers placed on the other piers and abutments. The middle spans are 459 feet each, and the shore spans are 230 feet each. The bridge is 100 feet above the surface of the water.

The Conway Bridge consists of two tubes, separated by a few feet, over a span of 400 feet.

The Victoria Bridge is a single tube, 6,538 feet long, resting on piers, forming twenty-four spans of 242 feet each, and a centre span of 330 feet, or twenty-five spans in all. The tube is made continuous over each set of two openings, the middle of the tube being fixed at the centre pier of the opening and the extremities being free to move on rollers placed on the adjacent piers.

The centre span is level, and is about sixty feet above the surface of the water. From the centre span the bridge slopes downward at an inclination of $\frac{1}{132}$.

In the Conway and Britannia bridges, the tops and bottoms are made *cellular;* that is, the plates are so arranged as to form rows of rectangular cells (Fig. 190). The joints of the cells are connected and stiffened by covering plates on the outside and by angle-irons within.

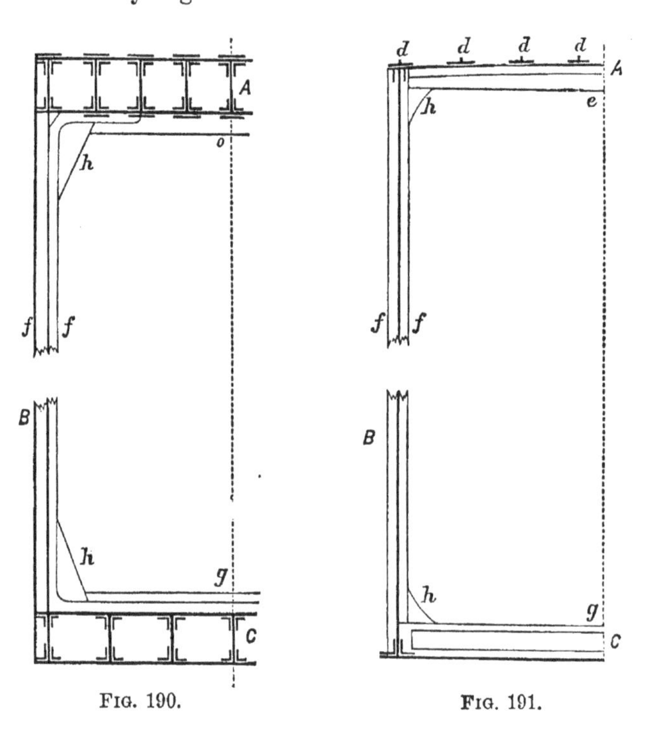

FIG. 190. FIG. 191.

The top, A, is composed of eight cells, each of which is one foot and nine inches wide, and one foot and nine inches high, interior dimensions The bottom, C, is divided into six cells, each of which is two feet and four inches in width, and one foot and nine inches high. These dimensions are sufficiently large to admit a man for painting the interior of the cells and for repairs.

The sides, B, are composed of plates set up on end (Fig. 192), their edges adjoining, and connected by means of vertical T-iron ribs, *f*, *f* (Fig. 190). The horizontal joints of the side plates are fastened by covering strips. The connection between the sides and top and bottom is strengthened by gussets, *h*, *h*, riveted to the interior T-irons.

In the Victoria Bridge the top and bottom, instead of being cellular, consist of layers of plates riveted together and stiffened by means of ribs (Fig. 191). The top, A, is slightly arched, and is stiffened by longitudinal T-irons, *d, d, d,* placed about two feet three inches apart, and by transverse ribs, *e,* about seven feet apart. The bottom, *c,* is stiffened by T-shaped beams, *g,* which form the cross-pieces of the roadway.

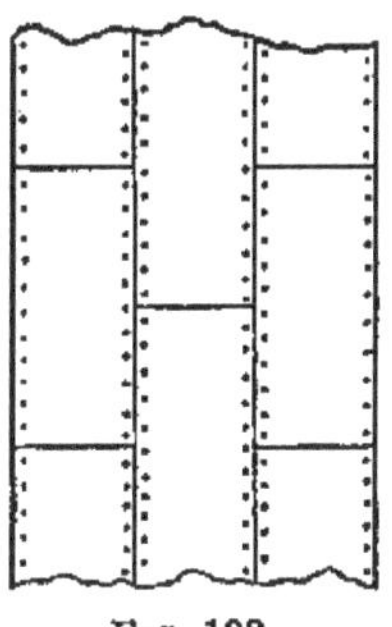

FIG. 192.

Erection.—There are three methods which have been used to place tubular bridges in position: 1, building the tube on the ground, and then lifting it into place; 2, constructing the tube, and moving it endwise upon rollers, on the piers; and 3, building it in position on a scaffold.

The first of these methods was adopted for the Britannia Bridge and the third for the Victoria Bridge.

Cambering.—If the top and bottom of the tube were made horizontal, the tube would when placed in position suffer deflection at the middle point from its own weight. In order that it may be horizontal after it has fully settled in position, the tube is made convex upwards. This convexity is called the **camber** of the tube or truss. The expression for maximum deflection of a beam in a horizontal position resting upon two points of support will give the amount of camber to give the tube. The camber given the Britannia Bridge was eighteen inches.

Remark.—Tubular bridges of these types are not now in much favor with the engineering profession, and few, if any, will ever be built in the future. The same amount of material in the form of a truss bridge will give a better bridge.

486. **Plate bridges.**—If we were to suppose the top removed from the tubular bridge, or to suppose the diagonals of the lattice truss to be multiplied until the side was a continuous piece, we would obtain the **plate girder.** In cross-section, the girder is ⌶-form (Fig. 193). Its general construction conforms to that given for the tubular bridge.

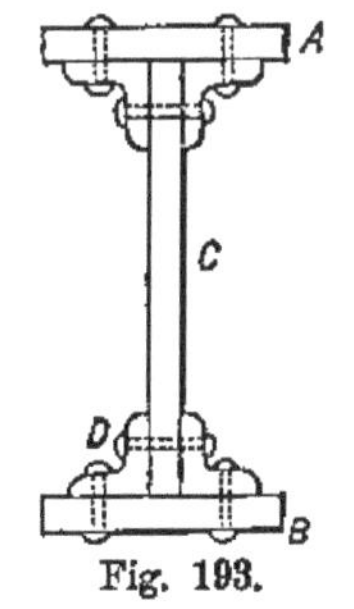

Fig. 193.

The joints of the flanges, A and B, are connected by covering plates; the web, C, is generally of thin plate. The web and flanges are fastened by angle-irons, D, riveted to both of them. The sides are stiffened by T-irons, as in the tubular bridges.

The advantages gained by using this class of bridges are confined to shallow bridges of moderate span. When the span exceeds sixty feet, it is more economical to use one of the iron trusses already named.

CHAPTER XVI.

III.—ARCHED BRIDGES.

487. **Arched bridges** are made either of masonry, of iron, or of steel.

The form of arch most generally used is the cylindrical. The form of soffit will be governed by the width of the span, the highest water level during the freshets, the approaches to the bridge, and the architectural effect which may be produced by the structure, as it is more or less exposed to view at the intermediate stages between high and low water.

Oval and segment arches are mostly preferred to the full centre arch, particularly for medium and wide bays, for the reasons that for the same level of roadway they afford a more ample water-way under them, and their heads and spandrels offer a smaller surface to the pressure of the water during freshets than the full centre arch under like circumstances.

The full centre arch, from the simplicity of its construction and its strength, is to be preferred to any other arch for bridges over water-courses of a uniformly moderate current, and which are not subjected to considerable changes in their water-levels, particularly when its adoption does not demand expensive embankments for the approaches.

If the spans are to be of the same width, the curves of the arches should be the same throughout. If the spans are to be of unequal width, the widest should occupy the centre of the structure, and those on each side of the centre should either be of equal width, or else decrease uniformly from the centre to each extremity of the bridge. In this case the curves of the arches should be similar, and the springing lines should be on the same level throughout the bridge.

The level of the springing lines will depend upon the rise of the arches, and the height of their crowns above the water-level of the highest freshets. The crown of the arches should not, as a general rule, be less than three feet above the highest known water-level, in order that a passage-way may be left for floating bodies descending during freshets. Between this, the lowest position of the crown, and any other, the rise should be so chosen that the approaches, on the one hand, may not be unnecessarily raised, nor, on the othe other, the springing lines be placed so low as to mar the architectural effect of the structure during the ordinary stages of the water.

488. **Masonry arches.**—These may be of stone, of brick, or of mixed masonry. The methods of construction, already described under the heads of Foundations and Masonry, are applicable to the construction of masonry arches used for bridges. As the foundations and beds of the piers and abutments are exposed to the action of the water, precaution should be taken to secure them. (Art. 440.)

Centres.—The centres used should be strong, so as to settle as little as possible during the construction of the arch, and for wide spans, should be so constructed that they can be removed without causing extra strains on the arch. This is effected by removing the centering from the entire arch at the same time. Removing the centering is termed **striking the centre.**

In wide spans, the centres are struck by means of an arrangement of wedge blocks, termed **striking plates.** This arrangement consists in forming steps upon the upper surface of the beam which forms the framed support for the centre. On this a wedge-shaped block is placed, on which rests another beam, having its under surface also arranged with steps. The struts of the rib of the centering either abut against the upper surface of the top beam, or else are inserted into cast-iron sockets, termed **shoe-plates**, fastened to this surface. The centre is struck by driving back the wedge block. When the struts rest upon intermediate supports between the abutments, folding wedges may be placed under the struts, or else upon the back pieces of the ribs under each bolster. The latter arrangement presents the advantage of allowing any part of the centre to be eased from the soffit, instead of detaching the whole at once, as in the other methods of striking wedges.

Another method of striking centres is by the use of sand. In this method, the centres rest upon cylinders filled with sand. These cylinders are arranged so that the sand can run out slowly near the bottom. When ready to strike the centre, the sand is allowed to run out of the cylinders, and all the ribs gradually and evenly settle down away from the soffit. The sand having run out, the centre can then be removed in the ordinary manner.

489. **Iron arched bridges.**—Next to masonry, cast iron is the material best suited for an arched bridge. It combines great resistance to compression or strength, with durability and economy; qualifications already given as requisite for an engineering structure.

Wrought iron is sometimes used for arched bridges. Where the bridge is liable to considerable transverse strains or shocks, wrought iron would be a better material than cast iron.

490. **Construction.**—Instead of the soffit being a continuous surface, as in the masonry arch, it is formed, in the iron arch, of curved iron beams placed side by side at suitable distances apart, and bound together by lateral bracing. This lateral bracing binding the ribs together, the proper abutting of the ends of the ribs, and the fastening of them upon the bed-plates or skew-backs of the abutments, form the most important part of the construction.

The ribs are generally made in segments, the joints being in the direction of the radii of curvature of the under surface of the rib. To guard against any possibility of accident, the segments are bolted together at the joints, forming in this way a continuous curved beam.

The form of the under surface of the rib is either parabolic or circular, more generally the latter. The depth of the rib is taken ordinarily at about $\frac{1}{60}$th of the span.

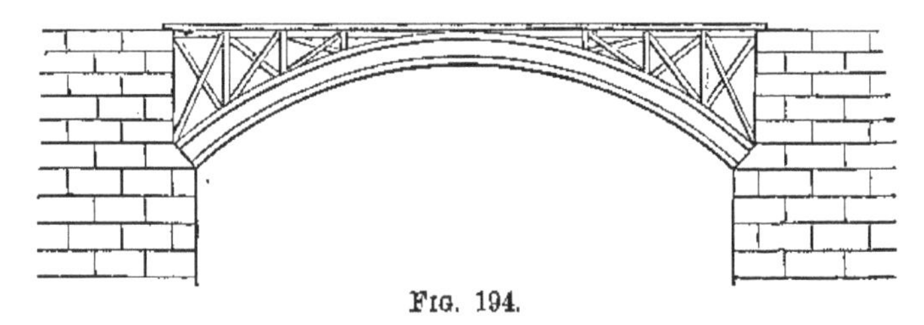

FIG. 194.

The rib may be solid, having a cross-section of the usual I-shape, the upper and lower flanges being equal; or it may be tubular; or it may be open-work, similar to a truss in which the chords are curved.

The first is the usual form. The other forms have been and are frequently used, but require no particular description.

Whatever be the form of cross-section of the rib, it is usual to place above the crown a horizontal beam, generally of wrought iron, suitably stiffened by covering plates and angle irons. (Fig. 194.)

The connection of this beam with the curved rib is made by a truss-work, called the spandrel filling, as shown in the figure.

On the horizontal beams the roadway is placed.

491. **Expansion and contraction.**—The rib is frequently hinged at the crown and ends, and sometimes at the ends only, to provide for the expansion and contraction of the metals produced by changes of temperature.

It is a matter of doubt whether anything is gained by this provision, as the friction arising from the great pressure on the joint probably prevents the motion of rotation necessary to relieve the arch from the increased strain.

492. **Arched bridges of steel.**—Bridges of this class, made of steel, do not differ in principle from those in iron. The most noted example of the steel arch is that used in the St. Louis and Illinois Bridge, across the Mississippi River, at St. Louis, Missouri.

In this bridge, the portion which corresponds, in the previous descriptions, to the rib, is composed of two tubular steel ribs placed directly one over the other and connected by a truss-work.

The segments of each of the tubular ribs are straight throughout their length, instead of being curved. The ends of each segment are planed off in the direction of the radius of curvature, and abut against the ends of the adjacent segment, to which they are joined and fastened. In this way the tube is made continuous; but instead of being curved, it is polygonal, as in the case of the bowstring girder. The tubes are connected by a truss-work, and the whole forms a rib of the third class.

493. **Eads' patent arch bridge.**—Captain Eads, the engineer of the St. Louis Bridge, has patented an arch bridge, the principle of which is shown in Fig. 195.

This arch is hinged at the crown, C, and springing lines, A and B, to provide for the expansion and contraction of the metal used in its construction. This arrangement of hinging the arch at the crown reduces the construction to that of two inclined beams resting against each other at C. Each beam is a truss belonging to the triangular system and having curved chords.

The line, A C B, is the arc of a parabola, whose vertex is

at C. The lines, A D C and C E B, are also arcs of parabolas. The maximum depth of either truss must not exceed one-half the rise of A C B.

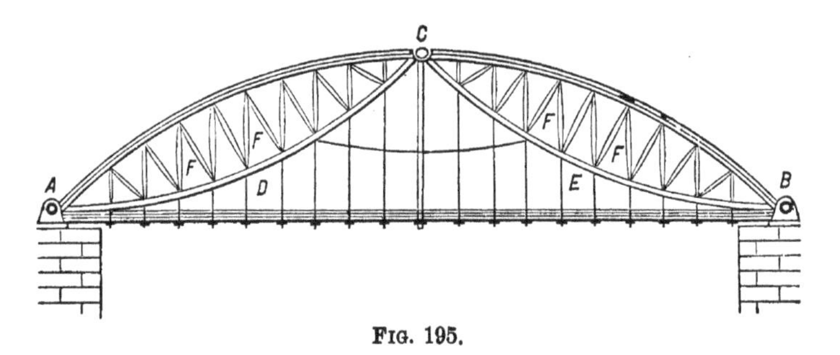

FIG. 195.

494. **Cases** in which the **arch may be preferred** to the truss.

The arch will usually be found to be a less expensive structure than the truss, when the banks are of rock forming good natural abutments.

It will oftentimes be more economically employed where a deep valley is to be spanned and where high arches can be used.

It is to be preferred when the roadway is a very heavy one, as in the case of a macadamized, or similar covering.

It is frequently selected in preference to a truss, from architectural considerations.

CHAPTER XVII.

IV. SUSPENSION BRIDGES.

495. A **suspension bridge** is one in which the roadway over the stream or space to be crossed is suspended from chains or wire ropes. The chains or wire ropes pass over towers, the ends of the chains being securely fastened or "**anchored**" in masonry at some distance behind and below the towers. The roadway, usually of wooden planking, is

supported by suspending rods placed at regular distances along the chains. (Fig. 196.)

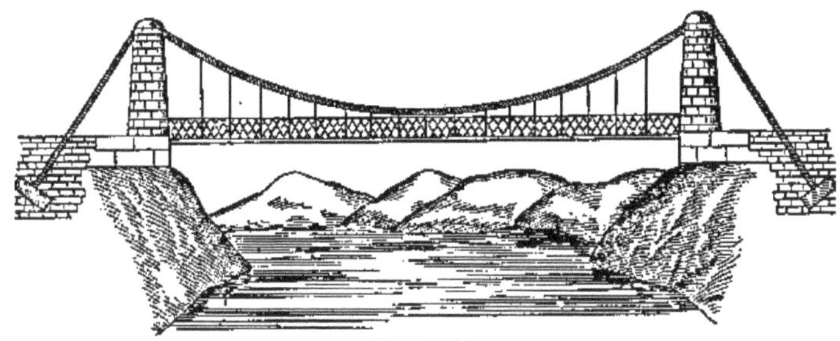

FIG. 196.

Suspension bridges are used principally for spans so great that they can not be crossed by arches or truss-work at a reasonable cost. Sometimes they are used, where the span is not very great, as a roadway only for foot passengers, especially over high-banked rivers, ravines, and similar places where the cost of a bridge of the other kinds would be out of proportion to the service required.

496. A suspension bridge consists of the **towers** or **piers**, over which the main chains or cables pass; the **anchorages**, to which the ends of the cables are attached; the **main chains** or **cables**, from which the roadway is suspended; the **suspending rods** or **chains**, which connect the roadway with the main chains; and the **roadway**.

497. **Towers.**—The towers, frequently termed piers, are made generally of masonry, although iron has sometimes been used. The particular form of the towers will depend in a measure upon the locality and the character of the surroundings. Their dimensions will depend upon their height and the amount of strains which they will have to resist.

Their construction will be governed by the rules already given for the careful construction of masonry.

A cast-iron saddle on rollers, to allow of free motion in the direction of the length of the main chains, is placed on each tower. (Fig. 197.)

FIG. 197.

The main chains may be fastened to these saddles, but they are generally passed over them.

The strains on the towers are produced by the vertical and horizontal components of the tensions in the cables.

The tower must be built expressly to resist the crushing forces due to this vertical component of the tension and the weight of the masonry.

If the saddle was not free to move, the horizontal force tending to push the tower over would be equal to the difference of the horizontal components of the tension in the two branches of the main chain. But since the saddle, by means of the rollers, is free to move, the horizontal force acting at the top of the tower must be less than the friction of the rollers.

498. **Anchorage.**—If the shore or bank be of rock, a vertical passage should be excavated and a strong iron plate placed in the bottom and firmly imbedded in the sides of the passage. Through this plate the ends of the main chains are passed and firmly secured on the under side. After the chains are put in place the passage should be filled with concrete and masonry.

If the rock is not suitable, a heavy mass of masonry should be built of large blocks of cut stone, well bonded together for this purpose. In this case it is advisable to construct a passage way, so that the chains and the fastenings may be examined at any time. This mass of masonry, or the natural rock to which the ends of the chain are fastened, is frequently called the **abutment.** Its stability must be greater than the tension of the chains. The principles of its stability are precisely the same as those for the abutment of an arch; its weight and thickness must be sufficient to prevent its being overturned; and its centre of resistance must be within safe limits.

499. **Main chains** or **cables.**—These may be made of iron bars, connected by eye-bar and pin joints; of iron links, as in common chains; of hoop or strap iron; of ropes or cables of wire, and in some cases of vegetable fibre, as hemp, flax, or bark. When of ropes or strap iron they are of uniform cross-section; when of links they may have variable cross-sections.

The smallest number of cables in a suspension bridge is two, one to support each side of the roadway. Generally more than two cables are used, since, for the same amount of material, they offer at least the same resistance, are more accurately manufactured, are liable to less danger of accident, and can be more easily put in place and replaced than a single chain of an equal amount of material.

Discussions have arisen as to the respective advantages possessed by the chain and wire cables, some engineers preferring the former to the latter, and the reverse. The wire cable is generally adopted in the United States.

The wire cable is composed of wires, generally from $\frac{1}{8}$th to $\frac{1}{6}$th of an inch in diameter, which are brought into a cylindrical shape by a spiral wrapping of wire. Great care is taken to give to each wire in the cable the same degree of tension.

The iron wires are coated with varnish before they are bound up into the cable, and when the cable is completed the usual precautions are taken, as in other iron-work, to protect it from rust and the action of the weather.

If the load placed on a cable be a direct function of its length, the curve assumed by the mean fibre of the cable will be a catenary. If it be a direct function of the span, it will be a parabola. But the weight resting on the main chains is neither a direct function of the length of the cable, nor of the span, but a function of both. The curve is therefore neither a catenary nor a parabola. But since the roadway, which forms the principal part of the load, is distributed very nearly uniformly over the span, the curve approaches more nearly the parabola and in practice is regarded as such a curve.

Knowing the horizontal distance between the tops of the towers and the deflection, the corresponding length of the cable between the two points of support may be obtained by the operation of rectifying the curve of a parabola. (Church's Integral Calculus, Art. 235.) The length obtained by this method will be expressed in terms containing logarithmic functions. For this reason approximate formulas are made which will give the length, in most cases, near enough for practical purposes. Rankine gives the following approximate value for the length of a parabolic arc:

$$s = x + \tfrac{2}{3}\frac{y^2}{x} \quad (\textit{nearly}). \quad . \quad . \quad (162).$$

Where the cable is to have a constant cross-section throughout, the area of this section must be proportioned to the greatest tension upon the cable. This tension is greatest at the points of support when they are of the same height, or at the highest point when the heights are unequal.

If the main chain is made of bars or links, it may be proportioned to form a chain of uniform strength, in which case the cross-sections will be made to vary from the lowest point to the highest, increasing in area of cross-section as the strain of tension increases. The horizontal component, or tension of the lowest point, is dependent upon the parameter of the curve. It therefore follows that for the *same curve* and the *same load* on the unit of length throughout, the horizontal component is the same for a bridge of a span of ten as for one of a thousand feet. And it is also plain that the wider the span, the deflection remaining constant, the greater will be the tension on the cable, and the reverse.

500. **Suspending chains.**—The roadway is suspended from the cables by wire ropes or iron rods, which are placed at equal distances along the cable, for the purpose of distributing the load as uniformly as possible over the cables.

If the cables are composed of links or bars, the suspending rods may be attached directly to them. If of rope, either of wire or of vegetable material, the suspension rod is attached to a collar of iron of suitable shape bent around the cable, or to a saddle-piece resting on it.

Where there are two cables, care must be taken to distribute the load upon the cables according to their degree of strength.

In the Hungerford Suspension Bridge the method adopted was as follows: The suspension rod, A (Fig. 198), was attached to a triangular plate, B, which hung by the rods, C and D, from the main chain, E and F. By this arrangement half of the load on the rod, A, was supported by each of the main chains, E and F.

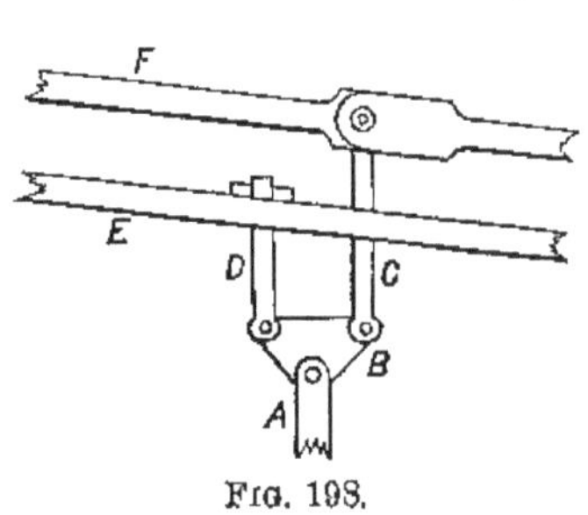

Fig. 198.

The suspending rods may be vertical or inclined. In recent constructions they are frequently inclined inwards, for the purpose of giving additional stiffness to the framing. The cross-section of the rod is constant, and is determined by the amount of strain on the upper section.

501. **Roadway.**—The roadway in its construction does not differ in principle from that used for other forms of bridges. The roadway bearers are supported by the suspension rods. On the bearers are laid longitudinal joists, and on them the planking, or the planking is laid directly on the roadway bearers. The latter are stiffened by diagonal ties of iron placed horizontally between each pair of roadway bearers.

502. **Oscillations.**—Suspension bridges, from the nature of their construction, are wanting in stiffness, and hence are peculiarly liable to both vertical and horizontal oscillations, caused by moving loads, action of winds, etc.

These oscillations cannot be entirely prevented, but their effect may be reduced so as to be almost harmless.

When the banks will admit of it, guy-ropes of wire may be attached to the roadway and fastened to points of the bank beneath the bridge. The guy-ropes directly under the bridge will be the most effective in resisting the vertical oscillations; those oblique to the bridge, for resisting the horizontal.

The elder Brunel fastened the roadway to a set of chains, whose curve was the reverse of that of the main chains. The reversed chains had a cross-section of about one-third of the main chains, and preserved the shape of the roadway under a movable load even better than the guys.

Engineers have made many efforts to provide for this want of stiffness in suspension bridges and to fit them for railroad uses.

A heavy moving load coming on a suspension bridge, when at a point, as M (Fig. 199), causes the roadway and cables to assume positions similar to those indicated by the

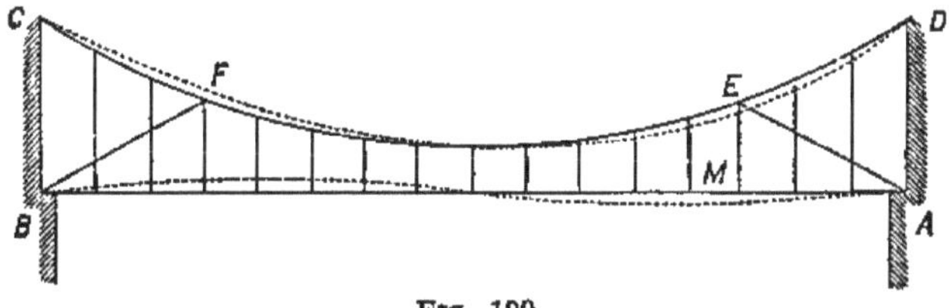

FIG. 199.

dotted lines in the figure. To prevent this deformation, the cables are fastened at the points of greatest change by chains, AE and BF, attached to the piers. These are known as Ordish's chains.

Roebling effected the same result by fastening these points of change in the roadway to the top of the towers, by the lines, Da, Db, etc., as shown in Fig. 200.

It is agreed at the present time that the best method of increasing the stiffness of a suspension bridge is to use, in addition to the chains just named, trussed parapets on each side of the roadway. These parapets form two open-built beams, strongly connected and braced by the roadway, and supported at intermediate points by the attachments to the main chains. Each end of the roadway is firmly secured to the base of the tower.

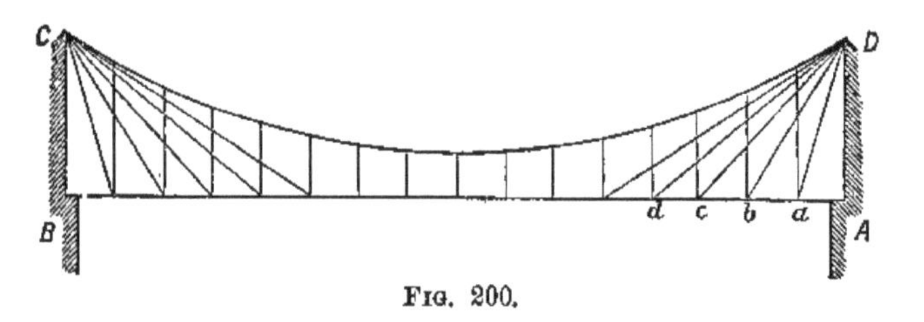

FIG. 200.

The objection to this method is the increase of the weight placed upon the main chains.

503. **Niagara Suspension Bridge.**—This bridge was planned and constructed by Roebling, and illustrates the method of stiffening just described.

The bridge affords a passage-way over the Niagara River, a short distance below the Falls, both for a railroad and a common road. It consists of two platforms (Fig. 201), one above

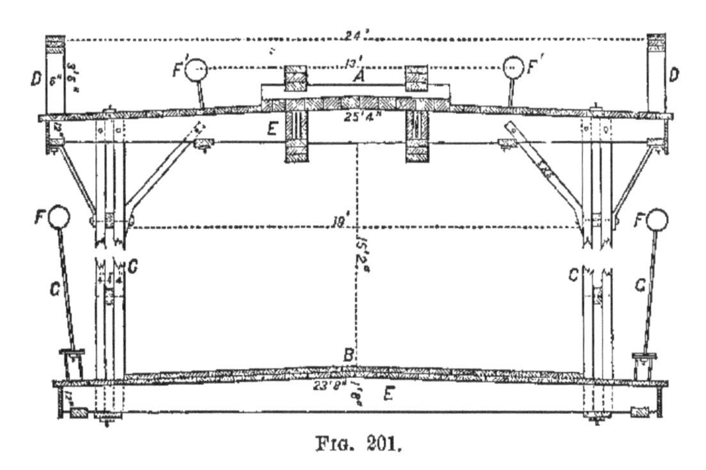

FIG. 201.

the other, and about fifteen feet apart; the upper is for the railroad track, and the lower, B, is for the common road. The platforms are connected by a lattice truss-work, C, C, on each side, which serves to increase its stiffness. The whole bridge is suspended by four main wire cables, F, F, F′, F′, the upper two being connected with the upper platform, and the lower two with the lower platform.

Each platform consists of a series of roadway bearers in pairs; the lower covered by two thicknesses of flooring-plank, the upper by one thickness; the portion of the latter immediately under the railroad track having a thickness of four inches, and the remainder on each side but two inches.

The roadway bearers and flooring of the upper platform are clamped between four solid-built beams; two above the flooring, which rest on cross supports; and two, corresponding to those above, below the roadway bearers; the upper and lower corresponding beams, with longitudinal braces in

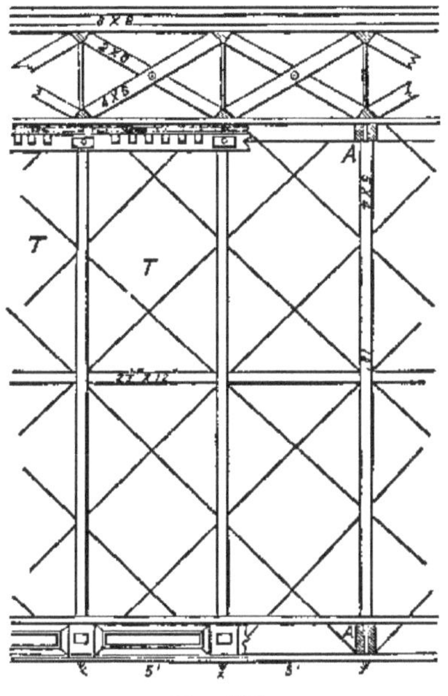

FIG. 202.

pairs between the roadway bearers and resting on the lower beams, being firmly connected by screw-bolts. The rails are laid upon the top beams, forming the railroad track, A. A parapet, D, D, of the form of the Howe truss is placed on each side.

The lattice-work, C, C, which connects the upper and lower platforms, consists of vertical posts in pairs (Figs. 202), and of diagonal wrought-iron rods, T, T. The rods pass through cast-iron plates fastened above the roadway bearers of the upper platform, and below those of the lower, and are brought to a proper bearing by nuts and screws on each end. A horizontal rail of timber is placed between the posts of the lattices at their middle, to prevent flexure.

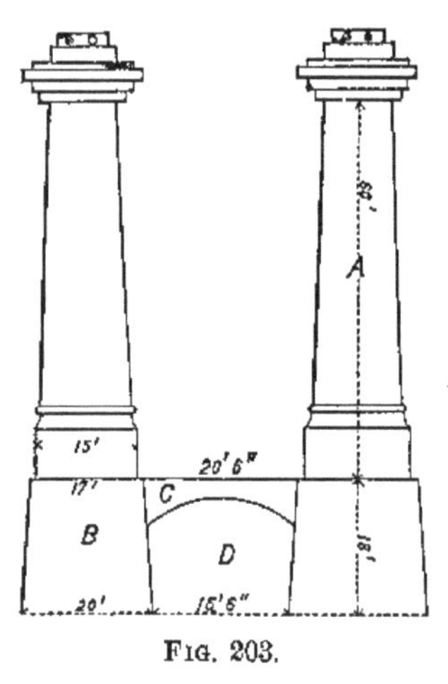

Fig. 203.

The towers (Fig. 203) are four obelisk-shaped pillars, each sixty feet high, with a square base of fifteen feet on a side, and one of eight feet at the top.

The height of the pedestals on the Canada side is eighteen feet, and on the United States twenty-eight. An arch, C, connects the two pedestals, under which is a carriage-way, D, for communicating with the lower platform.

The main cables pass over saddles on rollers placed on tops of the towers, and are fastened at their ends (Fig. 204) to chains made of iron bars attached to an anchoring plate, D, of iron, firmly secured in an anchorage of rock, B, and a mass of masonry, A.

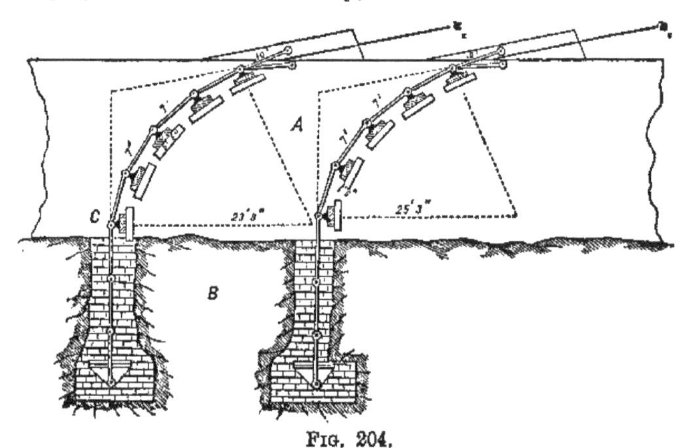

Fig. 204.

The upper cables are drawn in towards the axis of the bridge to reduce the amount of horizontal oscillations.

The following are some of its principal dimensions:
Span of the cables, 821⅓ feet.
Deflection of upper cables (mean temperature), 54 feet.
Deflection of lower cables " " 64 feet.
Length of upper cables " " 1,193 feet.
Length of lower cables " " 1,261 feet.
Ultimate strength of the four cables, 12,000 tons.
Permanent weight supported by the cables, 1,000 tons.
Tensile stress in the four cables, 1,810 tons.
Height of railroad track above mean stage of water, 245 feet.

After a constant use of over 25 years, this bridge had its anchorages reinforced (1877–8), and its superstructure renewed (1880) by substituting iron and steel for the wood. (Transactions of the Am. Society of Civil Engineers, July, 1881.)

504. **East River Suspension Bridge.**—This bridge connects the cities of New York and Brooklyn, and was thrown open to the public in 1883. Is was planned by Roebling and completed under the direction of his son, Col. W. A. Roebling. Some of its dimensions are as follows:
Length of span over the river, 1,595½ feet.
Total length of bridge, 5,989 feet.
Height of bridge at the centre over the river, 135 feet.
The ultimate strength of the cables is 49,200 tons, with a tensile stress of 11,700 tons produced by an estimated load of 8,120 tons.

CHAPTER XVIII.

V. MOVABLE AND AQUEDUCT BRIDGES.

505. **Movable bridges.**—In bridges over navigable rivers it is often necessary that one or more spans be made to move aside to allow of the passage of vessels. The term, **movable bridge**, is therefore applied to any arrangement, whatever be its nature, by means of which the roadway can at pleasure be made continuous or broken, between two points of a permanent bridge, or over a water-way. The methods used to effect this result are various.

They may be classed under five heads:

The passage may be opened or closed; 1, by turning a portion of the bridge around a vertical axis; 2, by turning it around a horizontal axis; 3, by making it roll forwards and backwards in a line with the bridge; 4, by lifting it vertically above the passage; and 5, by floating it from and into place upon the water.

506. I. **By turning around a vertical axis.**—The term, **swing-bridge,** is generally applied to a bridge which turns about a vertical axis. This form of bridge is the one most generally used when the opening is of any size. If two openings are required, the bridge rests upon a masonry pier, which is placed midway between the openings, and which supports a circular plate, whose diameter is equal, or nearly equal, to the breadth of the bridge. This plate has in the centre a pivot surrounded by a circular track with rollers. On this pivot and rollers the bridge is revolved horizontally, being turned by suitable machinery.

If only one opening is required, the abutment is generally used to support the mechanism for turning the bridge, care being taken to place the pivot far enough back from the face of the abutment so that the bridge, when open, shall not project beyond it.

In calculating the strains on the parts of such a bridge, the latter is usually considered when open, as composed of two cantilevers, each loaded with its own weight; when closed, as a bridge of two spans.

507. II. **By turning around a horizontal axis.**—Where the width of the opening is small, the moving portion of the bridge, which may be in one or two pieces, is lifted by chains attached to the extremities, the operation of lifting being assisted by counterpoises connected with the mechanism used. One of the simplest counterpoises is a lever revolving on a horizontal axis above the bridge, one end of the lever being connected with the movable end of the bridge by a chain, the other being weighted and connected with the mechanism by which the bridge is lifted.

508. III. **By moving a portion of the bridge forward and backward in a line with its axis.**—Bridges of this kind are placed upon fixed rollers, so that they can be moved forward or backward, to interrupt or open the communication across the water-way. The part of the bridge that rests upon the rollers, when the passage is closed, forms a counterpoise to the other. The mechanism usually employed for moving these bridges consists of tooth-work, and may be so arranged that it can be worked by one or more persons standing on the

bridge. Instead of fixed rollers turning on axles, iron balls resting in a grooved roller-way may be used, a similar roller-way being affixed to the frame-work beneath.

Bridges of this class are known as **rolling bridges.**

509. IV. **By lifting.**—In small bridges, like those over canals, the bridge is sometimes hung by the four corners to chains which pass over pulleys and have counterpoises at the other ends. A slight force applied to it raises the bridge to the required height, allowing the boats to pass under the bridge.

510. V. **By floating.**—A movable bridge of this kind may be made by placing a platform to form a roadway upon a boat or a water-tight box of a suitable shape. This bridge is placed in or withdrawn from the water-way, as circumstances may require.

A bridge of this character cannot be conveniently used in tidal waters, except at certain stages of the water. It may be employed with advantage on canals in positions where a fixed bridge could not be placed, in which case a recess in the side of the canal is made to receive the bridge when the passage-way is opened.

511. The general term, **draw-bridge,** is applied to all these movable bridges, although technically the term is confined to bridges of the second class, or those revolving around a horizontal axis.

Movable bridges are either simple bridges or made of truss-work belonging to one of the three systems already named.

The objections to using either a tubular, an arched, or a suspension bridge for a movable bridge are apparent. Where either of these classes is used, the passage-way can only be kept open by constructing the bridge so that a vessel can pass beneath it.

512. **Aqueduct bridges.**—In aqueducts for supplying a city with water, the volume of water conveyed is comparatively small, and the aqueduct bridge will present no peculiar difficulties except those of a water-tight channel. The latter may be made either of masonry, or of cast-iron pipes, according to the quantity of water to be delivered. If formed of masonry, the sides and bottom of the channel should be laid in the most careful manner with hydraulic cement, and the surface in contact with the water should receive a coating of the same material, particularly if the stone or brick used be of a porous nature. This part of the structure should not be commenced until the

arches have been uncentered and the heavier parts of the structure have been carried up and have had time to settle. The interior spandrel-filling, to the level of the masonry which forms the bottom of the water-way, may either be formed of solid material, of good rubble laid in hydraulic cement, or of concrete; or a system of interior walls, like those used in common bridges for the support of the roadway, may be used to sustain the masonry of the water-way.

In aqueduct bridges of masonry, supporting a navigable canal, the volume of water is much greater than in the preceding case, and every precaution should be taken to procure great solidity, and to secure the structure from accidents.

Segmental arches of medium span will generally be found most suitable for works of this character. The section of the water-way is generally of a trapezoidal form, the bottom line being horizontal. For economy, the water-way is usually made wide enough for one boat only; on one side is **a tow-path** for the horses, and on the other a narrow foot-path.

The principle of the suspension bridge is well adapted to aqueduct bridges, because, as each boat displaces its own weight of water, the only moving load is the passage of men and horses along the tow-path.

CHAPTER XIX.

BRIDGE CONSTRUCTION.

513. Before a bridge can be constructed there are three things to be considered, viz., 1st, the site; 2d, the water-way; 3d, the design or plan.

Before a bridge can be designed a thorough knowledge of the site, the amount of water-way, and the particular service required of the bridge, must all be known.

514. **Site.**—The site may already be determined, and it may not be in the power of the engineer to change it. If it is in his power to locate the site within certain limits, he will select the locality which offers the most security to the

foundations and which requires the least expense to be incurred in their construction and in that of the bridge.

In many cases it is a matter of indifference where the stream is crossed, but a careful survey of the proposed site should always be made, accompanied by borings. The object of this survey is to ascertain thoroughly the natural features of the surface, the nature of the subsoil of the bed and banks of the water-course, and the character of the water-course at its different phases of high and low water, and of freshets. This information should be embodied in a topographical map; in cross and longitudinal sections of the water-course and the substrata of its bed and banks; and in a descriptive memoir which, besides the usual state of the water-course, should exhibit an account of its changes, occasioned either by permanent or by accidental causes, as from the effects of extraordinary freshets, or from the construction of bridges, dams, and other artificial changes either in the bed or banks.

Having obtained a thorough knowledge of the site, the two most essential points next to be considered are to adapt the proposed structure to the locality, so that a sufficient water-way shall be left both for navigable purposes and for the free discharge of the water accumulated during high freshets; and to adopt such a system of foundations as will ensure the safety of the structure.

515. **Water-way.**—When the natural water-way of a river is obstructed by any artificial means, the contraction, if considerable, will cause the water, above the point where the obstruction is placed, to rise higher than the level of that below it. This difference of level is accompanied by an increase of velocity in the current of the river at this place. This damming of the water above the obstruction, and increase of velocity in the current between the level above and the one below the obstruction, may, during heavy freshets, cause overflowing of the banks; may endanger, if not entirely suspend, navigation during the seasons of freshets; and expose any structure which, like a bridge, forms the obstruction, to ruin, from the increased action of the current upon the soil around its foundations.

If on the contrary, the natural water-way is enlarged at the point where the structure is placed, with the view of preventing these consequences, the velocity of the current during the ordinary stages of the water will be decreased, and this will occasion deposits to be formed which, by gradually filling up the bed of the stream, might prove, on a sudden

25

rise of the water, a more serious obstruction than the structure itself; particularly if the main body of the water should happen to be diverted by the deposit from its ordinary course, and form new channels of greater depth near the foundations of the structure.

For these reasons, the water-way to be left after the bridge is built should be so regulated that no considerable change shall be occasioned in the velocity of the current through it during the most unfavorable stages of the water.

The beds of rivers are constantly undergoing change, the amount and nature of which depend upon the kind of soil of which they are composed, and the velocity of the current.

516. The following table shows, on the authority of Du Buat, the greatest velocities of the current close to the bed without injury to or displacement of the material of which it is composed:

Soft clay	0.25	feet per	second.
Fine sand	0.50	"	"
Coarse sand and fine gravel	0.70	"	"
Gravel, ordinary	1.00	"	"
Coarse gravel, 1 in. in diameter	2.25	"	"
Pebbles, 1½ in. in diameter	3.33	"	"
Heavy shingle	4.00	"	"
Soft rock, brick, etc.	4.50	"	"
Rock	6.00 and greater.	"	"

Knowing the material of which the bed of the river at the site is composed, and regulating the water-way so that the velocity of the current close to the bottom after the bridge has been erected, during the heaviest freshets shall not exceed the limit of safety or disturbance of the material forming the bed, the stability of the foundations is assured. If the velocity should exceed the limits here given, precautions must be taken to protect the foundations, as heretofore described.

517. **Velocity.**—The velocity of a current depends upon the slope of the bed. Since the particles of water in contact with the earth of the sides and bottom of the stream are retarded by friction, it follows that in any cross-section the velocity of the particles in the centre differs from those at the bottom and on the sides. In ordinary cases it is sufficiently exact to take the least, mean, and greatest velocities as being nearly in the proportions of 3, 4, and 5; and for very slow currents they are taken to be nearly as 2, 3, and 4.

The greatest velocity may be obtained by actual measurement, by means of floats, current metres, or other suitable apparatus, or it may be calculated from the slope of the bed of the river at and near this locality.

Having determined the greatest velocity, the mean velocity is taken as four-fifths of it. Col. Medley, in his Treatise on Civil Engineering, takes the mean velocity as nine-tenths (nearly) of the surface velocity when the latter exceeds three feet per second, and four-fifths when less than this.

Having determined the mean velocity of the natural water-way, that of the contracted water-way may be obtained from the following expression,

$$v = m\frac{S}{s}V, \quad . \quad . \quad . \quad . \quad . \quad (163)$$

in which s and v represent, respectively, the area and mean velocity of the contracted water-way; S and V, the same data of the natural water-way; and m a constant, which, as determined from various experiments, may be represented by the number 1,045.

Giving to s a particular value, that for v may be deduced, and may then be compared with the velocity allowable at this locality; or, assuming a value for v, the value of s may be deduced, and will be the area of the contracted water-way. The safest width, or area of water-way, in many cases may be inconveniently great; therefore, some risk must be run by confining the floods to more contracted limits. To reduce this risk as much as possible is the object of the engineer in seeking this information. With this information, the engineer can decide upon the number of piers, hence the number of spans of the bridge. Knowing the nature of the bottom, the character and kind of piers and abutments may be selected.

518. **Design or plan of bridge.**—Before the engineer can complete the design of the bridge, it is necessary that he should know what service it has to perform: whether it is to be a common or a railroad bridge; whether a single or double-track one. This information being given, and the knowledge acquired of the site and water-way being furnished him, he is able to decide whether the structure shall be a truss, arched, or suspension bridge; and, knowing the facilities at the place for the construction of the work, can prepare an estimate of its probable cost.

In deciding on the form of bridge which shall best combine efficiency with economy, there are many things to be considered. The cost of the superstructure, or all above the piers and abutments, increases rapidly with the length of span. Hence, economy would, as far as the superstructure is concerned, demand short spans. But short spans require an increase in the number of piers. When the height is small, the stream not navigable, and the piers easy to build, short spans may be used; but, if the foundations are in bad soils, if the river is deep, with a rapid current, or liable to great freshets, if it is navigable and requires an unobstructed water-way, the construction of piers will be very expensive, and therefore it is often desirable in these cases that there should be few or no piers in the stream; hence, long spans are necessary, even at great cost. Good judgment and accurate knowledge on the part of the engineer will be necessary, in order that these and similar questions should be decided correctly.

ERECTION OF BRIDGE.

519. The bridge having been planned, its parts all prepared and taken to the site, the abutments and piers built, the next step is to put it in position.

There are three methods, which have already been named, viz., building the bridge on a scaffolding in the position it is to occupy; building it and rolling it in position, known as *launching*; and building away from the site and then floating it to the spot, and lifting it in place.

520. **Scaffolding.**—The scaffolding is, so far as principle is concerned, the same as that already described under the head of masonry. That used for bridge construction is simply a rough but rigid trestling, resting on the ground, or on piles when the scaffolding is over water. The whole arrangement is sometimes called **staging**, and frequently **false-works.**

By means of this scaffolding the different pieces of the structure are lifted in place and fastened together. When the bridge is finished the staging is removed. This method is the one most generally used.

521. **Launching.**—This method has been used where the scaffolding would have been too great an obstruction to the stream or too costly. Deep and rapid rivers or ravines, where the bridge is erected at a very high level, or rivers with rapid currents subject to great freshets, are cases where scaffolding would be costly, and in some cases impracticable.

522. **Floating to site and lifting in place.** — This method has been used in connection with the last method.

In this method the truss or tube is placed on boats or pontons, and floated to the spot it is to occupy. Then, by cranes or other suitable lifting machinery, the truss is lifted to its place. This was the method adopted for the Britannia Tubular Bridge.

In tidal waters this method has been used with great success. The truss was put together on platforms on the decks of barges, at a sufficient height above the surface of the water, so that at high tide the truss would be above the level of its final position. The barges were then floated into position at high tide, and as the tide fell the truss was deposited in its proper place.

523. **Cost.**—The cost of erecting a bridge is divided generally into four parts: 1, Scaffolding; 2, Plant; 3, Labor; 4, Superintendence.

Scaffolding.—The cost of this forms an essential part of the estimate, and depends greatly upon the facilities for obtaining the proper materials in the vicinity of the site.

Plant.—This is a technical word used to include the tools and machinery employed in the work. The employment of steam in so many ways at the present time renders this item an important one in estimating the cost.

Labor.—The number of men, their wages, subsistence, and oftentimes their transportation, have all to be considered under this head.

Superintendence.—Good foremen and able assistants are essential to a successful completion of the work. Their wages may be included in the last item. It is usual to allow a given percentage on the estimate to include the cost of superintendence.

Summing these four items together, the cost of erecting the superstructure of the bridge may be estimated.

PART VII.

ROOFS.

CHAPTER XX.

524. The term **roof** is used to designate the covering placed over a structure to protect the lower parts of the building and its contents from the injurious effects of the weather. It consists of two distinct parts—the **covering** and the **frames** which support the covering. By some the term roof is applied only to the "covering," exclusive of the frames.

525. Roofs are of various forms—**angular, curved,** and **flat,** or nearly so.

The most common form of roof is the angular. These vary greatly in appearance and in construction. Some of the most common examples of the angular roof are the ordinary **gabled,** the **hipped,** the **curb** or **Mansard,** the **French** roof, etc.

Curved roofs and domes are frequently used. They cost more than the angular roofs, if the cost of the abutments be included. But if the abutments already exist or if for other reasons they have to be built, the curved roof, under these circumstances, in many cases, may be found cheaper and more suitable.

Flat roofs are very common, especially in hot climates. The covering of these roofs rests upon beams placed in a horizontal position, or one that is nearly so. The slope given them is generally about 4° with the horizontal.

These roofs are easy to construct, and are simple in plan, but they are heavy, do not allow the water to escape freely, and there is a waste of material in their use.

526. **Coverings.**—The coverings of roofs are made of boards, shingles, slates, mastics, the metals, or any suitable

material which will stand exposure to the weather and afford a water-tight covering. The style of the building, and the especial object to be attained, will govern their selection. The extent of surface covered by them is usually expressed in square feet. Sometimes the term **square** is only used, in which case it means an area of 100 square feet.

The weight of the materials used for the covering is about as follows:

Material.	Weight per square foot.
Copper	1 lb.
Lead	7 lbs.
Zinc	1.5 lbs.
Tin	¾ lb.
Iron (common)	3 lbs.
Iron (corrugated)	3.5 lbs.
Slates	5 to 12 lbs.
Tiles	7 to 18 lbs.
Boards, 1 inch thick	2½ lbs.
Shingles	1 lb.

These are fastened directly upon the frames, or upon pieces of scantling and boarding which rest on the frames.

527. **Frames.**—The frames which support the covering have their exterior shape to correspond to the form of the roof. These frames, known generally as **roof-trusses**, are tied together and stiffened by braces which may occupy either a horizontal or inclined position, and may be either notched upon or simply bolted to the trusses.

The trusses are placed from five to ten feet apart, depending upon the weight of the covering and the amount of load which each truss has to support. They rest usually upon pieces of timber called **wall-plates**, laid on the wall to distribute the pressure transmitted by the truss over a larger surface of the wall.

528. Although nearly the last part of a building which is constructed, the roof is one of the first to be considered in planning the building, since the thickness and the kind of wall depend greatly upon the weight of the roof. The weight of the roof and the size of the pieces to be used in its construction, when the roof is flat, are easily determined. The pieces are simple beams, subjected only to cross-strains, and the joints are of the simplest kind.

When the roof is curved or inclined, these determinations are more difficult. In these roofs the strains on the parts produced by the covering are of different kinds, and must be determined completely, both in amount and kind, before the dimensions of the different pieces can be fixed, and the best form of joints and fastenings selected.

In calculating the strains on a roof-truss, we must take into consideration, besides the weight of the covering and of the truss itself, the weight of the snow, ice, or water which may at times rest upon the covering, the effect due the action of the wind, and such extra loads as the weight of a ceiling, of machinery, of floors, etc., which may be supported by the frames.

The weight of the covering varies, as has been shown, from one pound to twenty pounds upon the square foot. The weight of the truss increases with the span, but it is only in very wide spans that the weight of the parts and of the whole truss have to be considered.

The weight of snow is assumed to be about one-tenth that of the same bulk of water. Knowing the maximum depth of the falls of snow, an approximate weight may be determined. Six pounds per square foot is the estimated weight of snow adopted by European engineers. A greater weight, even as high as twenty pounds, is recommended for the northern part of the United States.

The action of the wind is very great in some localities. Tredgold recommends an allowance of forty pounds to the square foot as an allowance for its effect.

529. **Rise and span.**—These are quantities dependent upon circumstances. The rise is dependent upon the kind of roof, the order of architecture used for the building, and the climate. The span is dependent upon the size of the building.

In gabled roofs and ordinarily angled roofs, the inclination which the sides of the roof make with the horizontal is called the **pitch**. In countries where heavy falls of snow are common the pitch is ordinarily made quite steep—although builders are now more generally inclined to a moderate pitch, even for these cases. The objections to a steep pitch are the exposing of a greater surface of the roof to the direct force of the wind, the waste of room, etc. The material of which the covering is composed affects the pitch. An ordinary roof covered with shingles should have a pitch of at least 22½ degrees; one covered with slate or tiles a pitch something greater, between 23 and 30 degrees.

The style of roof and architecture affect the pitch. Gothic styles and parts of French roofs require a pitch of 45 degrees, and even of 60 degrees.

530. Materials used in construction.—Wood and iron are the materials used for the construction of the frames. The truss may, as in other frames, be made entirely of wood, or entirely of iron, or of a combination of the two materials.

Wooden Roof-trusses.

531. The simplest wooden truss is the triangular frame. The inclined pieces are called **rafters** and the horizontal one is termed the **tie-beam.**

It is used for spans of 12 to 18 feet, and when the roof is light. For spans of 18 to 30 feet the king-post truss (Fig. 205) is used. Its component parts are:

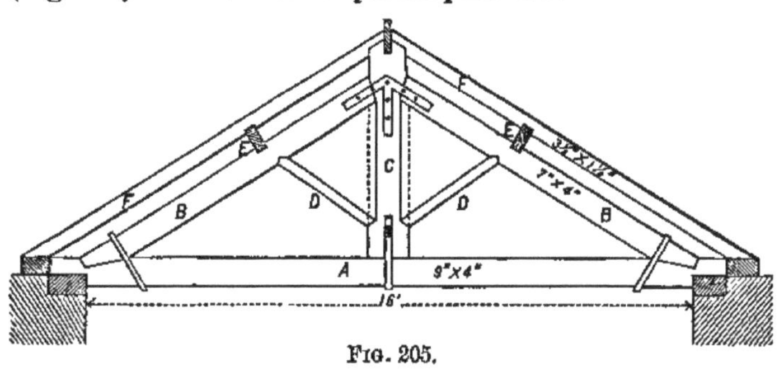

Fig. 205.

1. The **principal rafters.**—These are the inclined pieces, B B, which abut against each other or against the king-post at the top.
2. The **tie-beam.**—This is the horizontal beam, A, connected with the lower ends of the rafters to prevent their spreading out under the action of the load placed on them.
3. The **king-post.**—The upright, C, framed at the upper end upon the rafters and connected at the lower end with the tie-beam.
4. **Purlins.**—These are horizontal pieces, E, E, notched upon or bolted to the rafters to hold the frames together and to form supports for the common rafters, F, F.
5. **Common rafters.**—These are inclined pieces, F, F, of smaller dimensions than the principal rafters, placed from 1 to 2 feet apart and intended to support the covering.
6. **Struts.**—The inclined pieces, D, D, framed into the principal rafters and king-post to prevent the rafters from sagging at the middle.

If the king-post and struts be removed, the simple triangular truss is left.

532. **Queen-post truss.**—This truss is employed for spans from 30 to 45 feet long. Its parts (Fig. 206) are all shown in the figure; C, C, being the queen-posts.

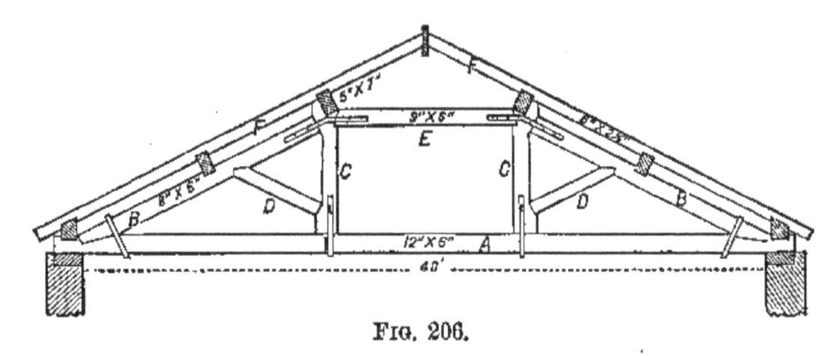

Fig. 206.

533. **Iron roof-trusses.**—Wooden roof-trusses have been used for wider spans than those named, but the use of iron in building has enabled the engineer to construct roof-trusses of wider spans which are much lighter and present a better appearance.

These trusses are sometimes made of wood and iron in combination, as we have seen in bridge-trusses, but now they are more generally made entirely of iron.

The coverings are frequently made of iron, mostly corrugated, and are fastened to the purlins by the usual methods for iron-work.

DETERMINATION OF THE KIND AND AMOUNT OF STRAINS ON THE PARTS OF A ROOF-TRUSS.

534. **Amount and kind of strains upon the different parts of the simple king-post truss.**—The method of determining the amount and kind of strains on the simple triangular frame has already been explained. (Art. 256.) It is usual, except in very short spans and where the tie-beam supports nothing but its weight, to support the middle point of this piece by a king-post. To find the strains on a triangular frame with a king-post, let A B and A C (Fig. 207) be the rafters, B C, the tie-beam, and A H, the king-post. The king-post is so framed on the rafters at A, as to hold up any load which it has to support. It is connected with the tie-beam in such a manner as to keep the middle point, H, in the same straight line with B and C.

The strains on this truss are produced most usually by a uniform load on the rafters and a load on the tie-beam.

Denote by l, the length of either rafter; by w, the load on a unit of length, including the weight of the rafters; by W',

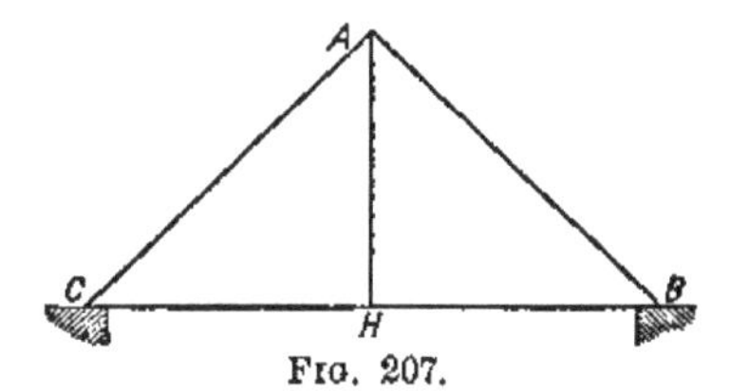

FIG. 207.

the weight of the tie-beam, including the load it has to support, as a ceiling, floor, etc., and by a, the angle A B C.

The load on one of the rafters, as A B, will be wl, and acts through the middle point, or at a distance from B equal to $\frac{1}{2}l$. The strains produced by this load are compressive on the rafter and tensile on the tie-beam, and the amount for each may be determined, as shown in Art. 254.

The king-post is used to prevent the sagging of the tie-beam at its middle point. It therefore supports, besides its own weight, $\frac{5}{8}W'$ (Art. 186), which produces a strain of tension on the king-post and which is transmitted by it to A, where it acts as a load suspended from the vertex of the frame. The strains produced by it on the rafters and tie-beam may be determined as in Art. 256.

The strains being known in amount and kind for each piece, can now be summed and the total amount on the different parts determined.

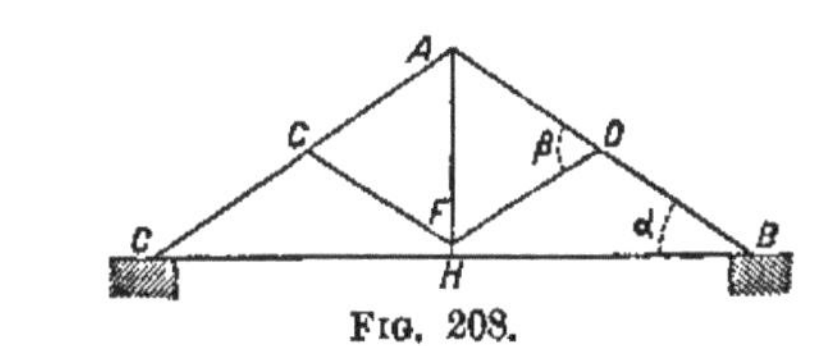

FIG. 208.

535.—Strains on a king-post truss framed with struts.—Let Fig. 208 represent an outline of this truss. Let D F and F G be the struts framed in the king-post and supporting the rafters at their middle points.

The truss is supposed to be strained by a load uniformly distributed over the rafters.

Adopt the notation used in the previous case and represent by β, the angle A D F. We may neglect without material error the weight of the struts and king-post, their weights being small compared with the load on the rafters.

The load acts vertically downwards and is equal to wl for each rafter. Acting obliquely, it tends to compress and bend them. Each rafter is a case of a beam resting on three points of support, hence the pressure on either strut is due to the action of $\frac{5}{8}wl$.

Pressure on the struts.—The pressure on the strut D F arises from the action of the component of $\frac{5}{8}wl$ perpendicular to the rafter at the point, D. Denote by P_1 the pressure on the strut in the direction of its axis. To keep the point, D, in the same straight line with A and B, the resistance offered by the strut must be equal to the force acting to deflect the rafter at that point. Hence there results,

$$P_1 \sin \beta = \tfrac{5}{8}wl \cos a. \quad . \quad . \quad . \quad (164)$$

From which we find

$$P_1 = \tfrac{5}{8}wl \frac{\cos a}{\sin \beta},$$

for the pressure on the strut, D F. In the same way the pressure on the strut F G is obtained, which in this case is exactly equal in amount.

Tension on king-post.—This pressure, P_1, is transmitted through the strut to the king-post at F. Resolving this force into its components respectively perpendicular and parallel to the axis of the king-post, we find the component in the direction of the axis to be $P_1 \sin (\beta - a)$.

The king-post supports the tie-beam at its middle point. Represent as before by W', the weight of the tie-beam and its load, and we have $\frac{5}{8}W'$ for the pull on the king-post from this source. Represent the total stress of tension by T_1, and there results,

$$T_1 = 2P_1 \sin (\beta - a) + \tfrac{5}{8}W'. \quad . \quad . \quad (165)$$

Substituting in this for P_1, its value just found, and the value of T_1 will be known.

Tension on the tie-beam.—Denote by T the tension on the tie-beam produced by the thrust along the rafters, and by Q, the vertical reaction at B caused by the load on the rafters.

The relation between the normal components to the rafter,

at B, of the three forces, Q, T, and $\frac{3}{16}wl$ acting at that point, may be expressed by this equation,

$$T \sin a = Q \cos a - \tfrac{3}{16}wl \cos a. \quad . \quad (166)$$

From which the value of T can be obtained when Q is known.

Since the truss is symmetrical with respect to a vertical through A, the sum of the reactions at B and C, due to the strains on the rafters, is 2Q, and is equal to the total load placed on the rafters, which is $2wl + \frac{5}{8}W'$. Hence

$$2Q = 2wl + \tfrac{5}{8}W',$$

and

$$Q = wl + \tfrac{5}{16}W',$$

which, substituting in equation (166), gives,

$$T \sin a = \tfrac{13}{16}wl \cos a + \tfrac{5}{16}W' \cos a,$$

and

$$T = \tfrac{13}{16}\frac{wl}{\tan a} + \tfrac{5}{16}\frac{W'}{\tan a}. \quad . \quad . \quad (167)$$

Strains on the rafters.—The forces acting in the direction of the rafters produce compressive strains, and those perpendicular, transverse strains. These are determined as previously shown.

Size of the pieces.—Having found all the strains, the limit on the unit of cross-section may be assumed and the dimensions of the pieces obtained.

Remark.—It is well to notice, that if we substitute for P_1, its value in the expression for T_1, the tension on the king-post, that we will get

$$T_1 = \tfrac{5}{8}W' + \tfrac{5}{4}wl\frac{\cos a \sin(\beta - a)}{\sin \beta},$$

which may be put under the form

$$T_1 = \tfrac{5}{8}\left[W' + 2wl \cos^2 a \left(1 - \frac{\tan a}{\tan \beta}\right)\right]. \quad (168)$$

It is seen from this value of T_1, that whenever β is equal to 90° or differs but slightly from it, the expression will reduce to the form

$$T_1 = \tfrac{5}{8}(W' + 2wl \cos^2 a).$$

536. Strains on the queen-post truss.—It is easily seen

from the foregoing how the strains on this truss may be determined. It is usual to suppose the truss (Fig. 209) separated into two parts; one the primary truss, B A C, and the other, the secondary trapezoidal truss, B D G C.

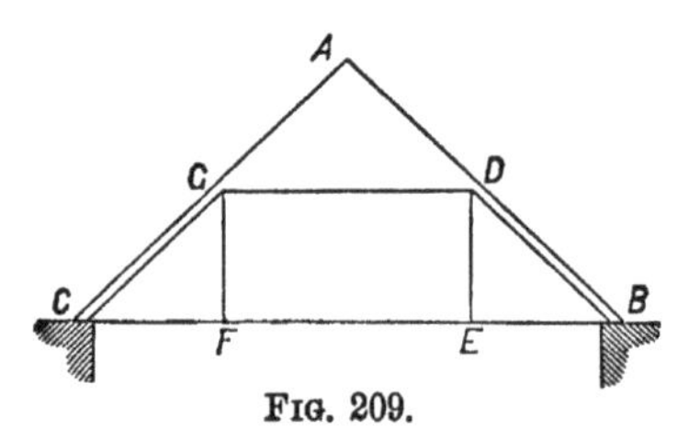

FIG. 209.

In some cases, short rafters from C to G, and B to D, are placed in contact with the principal rafters, A C and A B, which further strengthens the truss by the additional thickness given to the rafters in this part of the truss, and more fully satisfies the condition of a secondary trapezoidal truss placed within a triangular frame to increase its strength. There are various other modifications of this truss, but the method of determining the strains is not affected by them.

Iron Roof-trusses.

537. The trussing already explained under the head of Bridges enters largely into iron roof-trusses. One of the most common forms is the one in which the rafters are trussed.

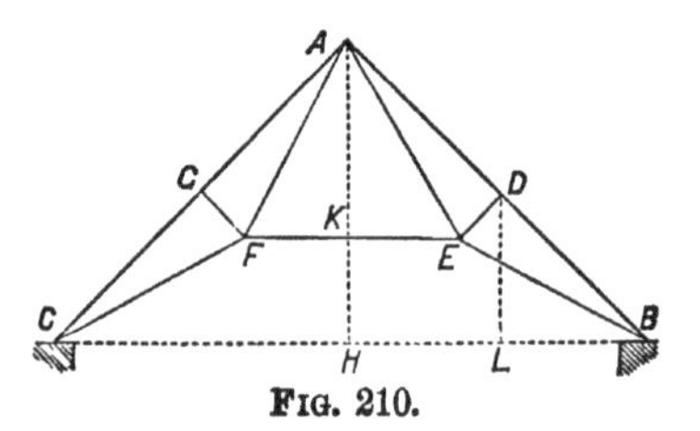

FIG. 210.

Roof-truss with trussed rafters.—A common method of supporting the middle point of a rafter is shown in Fig. 210. In this case the lower end of the strut, instead of abutting against a king-post, is held up by tie-rods joining it with the ends of the rafters.

It is seen from the figure that each rafter, with the strut and tie-rod, forms a simple king-post truss inverted. The tie-rod connecting the points, E and F, completes the truss. This tie-rod sustains the horizontal thrust produced by the strains on the rafters, preventing its action on the walls at the points of support, B and C.

In this truss the rafters are equal in length, and make equal angles with the horizon; the struts are placed at the middle points and perpendicular to the rafter; and the strains are produced by a uniform load resting on the rafters.

Use the notation of the previous cases, and denote by a the angle ABC; by β, the angle DBE; by $2b$, BC; by d, the height AH; and by d', the distance AK. The truss is symmetrical with respect to a vertical AH, through the vertex, A. Suppose the truss cut in two along this line, AH, we may preserve the equilibrium, upon removing the left half, by substituting two horizontal forces, one at A and the other at K. Suppose this done, and represent these by H and T respectively. As the weight of the tie-rods and struts is small compared with the load on the rafters, we may neglect it without material error.

The reaction at B is equal to wl.

The external forces acting on the right half of the frame are the reaction at B, the horizontal forces H and T at A and K, and the load on the rafter including its own weight. These forces act in the same vertical plane.

The analytical conditions for equilibrium are

$$\mathrm{H} - \mathrm{T} = 0, \quad \text{and} \quad wl - wl = 0,$$

and the bending moment at B is

$$wl \times \mathrm{BL} - \mathrm{H} \times \mathrm{AK} = 0.$$

We find the value of $\mathrm{H} = \frac{1}{2} \frac{wl^2 \cos a}{d'}$.

The external forces are now all known and the strains produced by them may be determined.

Pressure on the struts.—Considering the rafter as a single beam, there results

$$\mathrm{P}_1 = \tfrac{5}{8} wl \cos a,$$

for the pressure on either strut.

Tension on the tie-rods of the rafters.—Let T_1 be the tension on the tie-rod BE, and T_2 the tension on AE.

At the point, B, the normal pressure must be equal to the normal component of the resultant of the forces, wl and T_1 acting at that point, which may be expressed as follows:

$$\tfrac{3}{16}wl \cos a = wl \cos a - T_1 \sin \beta,$$

and at A, for the same reason, we have

$$\tfrac{3}{16}wl \cos a = H \sin a - T_2 \sin \beta.$$

These equations give, since H is known,

$$T_1 = \tfrac{13}{16}wl \frac{\cos a}{\sin \beta}, \text{ and } T_2 = \frac{H \sin a - \tfrac{3}{16}wl \cos a}{\sin \beta}, \quad (169)$$

for the tensions on the tie-rods B E and A E.

Tension on the main tie-rod, E F, of the truss.—From the analytical condition,

$$H - T = 0,$$

there results,

$$T = H = \tfrac{1}{2} \frac{wl^2 \cos a}{d'}. \quad . \quad . \quad . \quad (170)$$

This may be verified. The stresses, P_1, T_1, and T_2, in the pieces connected at E (Fig. 210) have been determined. These forces with T must be in equilibrium at E. Let us find the components of these forces in the direction of the strut, D E, and a perpendicular to the strut at E. (Fig. 211.)

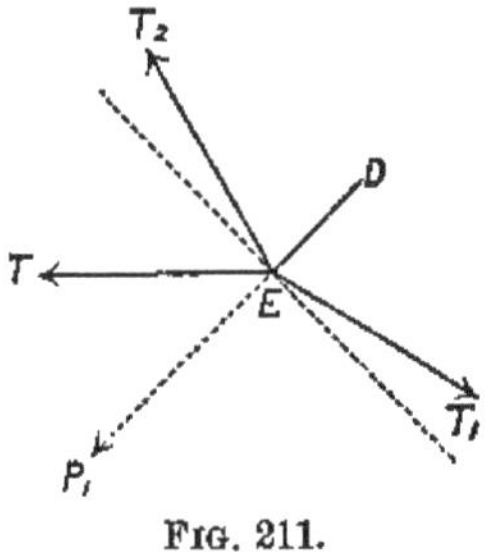

FIG. 211.

For equilibrium, we have the following:

$$(T_1 + T_2) \sin \beta - T \sin a - P_1 = 0,$$

and

$$(T_2 - T_1) \cos \beta + T \cos a = 0.$$

Substituting in the first of these equations, the values of P_1, T_1, and T_2, already obtained, there results,

$$\tfrac{10}{16}\, wl \cos a + H \sin a - \tfrac{5}{8}\, wl \cos a - T \sin a = 0, \text{ or } H = T.$$

In a similar manner, by substitution in the 2d, it can be shown that the condition is satisfied, or $H = T$.

Compression on the rafters.—The compression on the rafter at B is due to the components of the forces acting at that point parallel to the rafter. Hence

$$\text{Compression at B} = wl \sin a + T_1 \cos \beta,$$

and

$$\text{Compression at A} = H \cos a + T_2 \cos \beta. \qquad (171)$$

Frequently in the construction of this truss, the struts are extended until they meet the tie-rod joining B and C. (Fig. 212.)

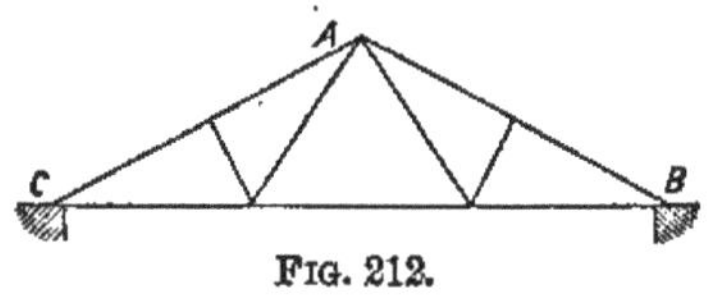

FIG. 212.

In this case the stresses are the same as those just determined in the struts and rafters, but are less in the secondary tie-rods, because of the increase in the angle β.

538. When the span is considerable, this method of trussing is oftentimes used to increase the number of supports for the rafter. By adding to the trussed rafter, the two struts, bf and cd (Fig. 213), and the two secondary tie-rods, f D and d D, two additional points of support are furnished to the rafter.

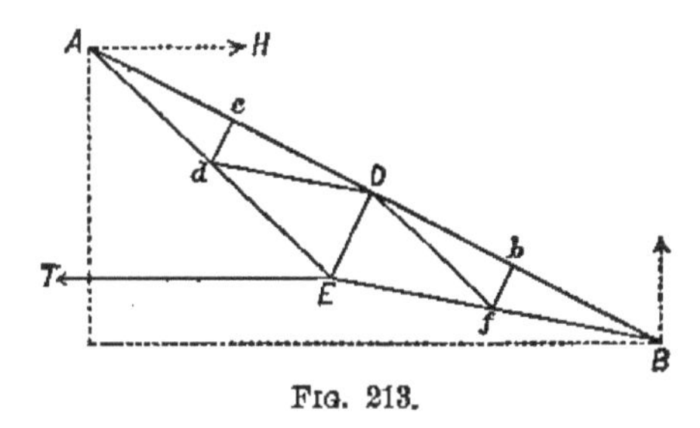

FIG. 213.

The points, b and c, are midway between B D and A D, dividing the rafter into four equal parts, and making the triangles B f D and D d A equal to each other and similar to B E A.

Using the previous notation, the reaction at B is wl, and the horizontal force at A is $\frac{1}{2}\frac{wl^2 \cos a}{d'}$, as in previous case. The external forces are all known.

Pressure on the struts.—The struts are respectively perpendicular to the rafter; the normal components of the forces acting at b, D, and c will give the amount of pressure on each strut, due to the load acting at these points. Represent this component at D by P_1, at b and c by P_2, and at A and B by P_3. Since the rafter is kept by the struts in such a position that *b*, D, and *c* are in the same straight line with A and B, it is an example of a beam resting on five supports, and we have,

$$\mathrm{P}_3 = \tfrac{11}{112} wl \cos a,\ \mathrm{P}_2 = \tfrac{2}{7} wl \cos a, \text{ and } \mathrm{P}_1 = \tfrac{13}{56} wl \cos a.$$

This value of P_2 is the amount of pressure acting on either of the struts, *bf* or *cd*, and the strain on them is determined. That on D E is still to be determined.

Tension on the secondary tie-rods.—Let T_1 be the tension on the rod, B*f*, and we have,

$$\tfrac{11}{112}\, wl \cos a = wl \cos a - \mathrm{T}_1 \sin \beta,$$

from which we get

$$\mathrm{T}_1 = \tfrac{101}{112} wl \frac{\cos a}{\sin \beta}.$$

And in the same way we find the tension T'_1 on A*d* to be

$$\mathrm{T}'_1 = \frac{\mathrm{H} \sin a}{\sin \beta} - \tfrac{11}{112} wl \frac{\cos a}{\sin \beta}.$$

Denote by T_2, T_3, T'_2, and T'_3 the tensions on *f* D, *f* E, *d* D, and *d* E respectively. Since an equilibrium exists between the forces acting at the point *f*, and the same at *d*, the components of these forces, taken respectively parallel and perpendicular to the rafter, must fulfil the following conditions:

$$\mathrm{T}_2 + \mathrm{T}_3 - \mathrm{T}_1 = 0, \text{ and } (\mathrm{T}_3 - \mathrm{T}_2 - \mathrm{T}_1) \sin \beta + \mathrm{P}_2 = 0, \text{ at } f,$$

and

$$\mathrm{T}'_2 + \mathrm{T}'_3 - \mathrm{T}'_1 = 0, \text{ and } (\mathrm{T}'_3 - \mathrm{T}'_2 - \mathrm{T}'_1) \sin \beta + \mathrm{P}_2 = 0, \text{ at } d.$$

The values of T_1, P_2, and T'_1, have already been found. The values for the others are easily deduced. They will be as follows:

$$\mathrm{T}_2 = \mathrm{T}'_2 = \frac{\mathrm{P}_2}{2 \sin \beta} = \tfrac{1}{7} wl \frac{\cos a}{\sin \beta},$$

$$\mathrm{T}_3 = \tfrac{85}{112} wl \frac{\cos a}{\sin \beta}, \text{ and } \mathrm{T}'_3 = \frac{1}{\sin \beta} (\mathrm{H} \sin a - \tfrac{27}{112} wl \cos a).$$

The strains of tension and compression on all the secondary pieces have been obtained excepting for the strut, D E, at the middle. This can now be determined.

Strain on strut, D E, at the middle.—This strain is due to the pressure, P_1, and the components of T_2 and T'_2 in the direction of the strut, or

$$\text{Compression on D E} = P = P_1 + (T_2 + T'_2) \sin \beta.$$

Substituting in this for T_2, T'_2, and P_1, their values already found, we finally obtain,

$$P = \tfrac{22}{56} wl \cos \alpha,$$

for the stress in the strut, D E.

The amount and kind of strain on each piece are now known, and the strength of the truss may therefore be determined.

539. *Roof-truss in which the rafters are divided into three segments and supported at the points of division by struts abutting against king or queen-posts.*

This form of truss shown in Fig. 214 is in common use for roofs. In this case, the rafters are trisected respectively at the points, H, D, G, and M, by the struts H K, D F, G F, and

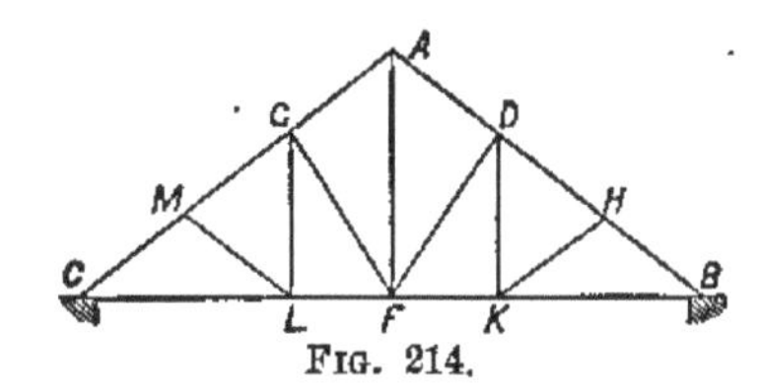

FIG. 214.

M L, which have their lower ends connected with and abutting against the vertical rods at the points K, F, and L, where these rods are fastened to the tie-rod B C.

The usual method of determining the amount of strains on the different parts of a frame of this kind is to consider it as formed of several triangular ones. In this particular case, we consider the truss A B C as made up of the secondary trusses, B H K, B D F, and B F A, on the right of A F, and a similar set on the left of it.

The strains are supposed to arise from a uniform load over the rafters, the weight of the vertical ties and the struts being neglected, as in the previous cases.

In the previous examples, the rafters have been regarded as single beams resting on two, three, five, etc., points of support, and the reactions of these points of support have been taken as the value of the load resting upon them. This process may be followed in this case and is to be preferred, whenever the rafters, A B and A C, are continuous.

In most treatises on roofs, the action of the load on the points of support is considered in a different manner. There are two general methods. Taking either half of a truss of the kind just described, one method supposes that each segment of the rafter supports one-third of the entire load on the rafter; or, each segment is considered a beam supported at its ends and uniformly loaded. According to this hypothesis, since $\frac{1}{3}wl$ is the load on the segment, $\frac{1}{3}wl$ will act at the points, H and D, and $\frac{1}{6}wl$, at B and A, of the half A B F.

The other method assumes the pressures exerted at the four points of support to be equal to each other, that is, $\frac{1}{4}wl$ to be the load acting at each of the points, B, H, D, and A. This is sometimes called "the method of equal distribution of the load."

Adopting the first method, assuming one-third of the load on the rafter as resting on each segment, let us first determine the strains in the secondary truss, B H K.

Strains on B H K.—By hypothesis, the pressure at H is $\frac{1}{3}wl$, acting vertically downwards. The problem then is the case of a simple triangular frame sustaining a load at the vertex.

Denote by a, the angle H B K; since the triangle is isosceles, the components of $\frac{1}{3}wl$ along the rafter and strut are each equal to $\frac{1}{6}\frac{wl}{\sin a}$, and develop compressive stresses in H B and H K.

The stress transmitted to B produces a vertical pressure on the point of support equal to $\frac{1}{6}wl$ and a stress of tension in B K equal to $\frac{1}{6}wl \cot \alpha$.

In like manner, the stress transmitted to K produces a vertical pull at that point equal to $\frac{1}{6}wl$ which is sustained by the tie-rod, D K, and a horizontal stress equal to and directly opposed to the stress of tension at B.

Strains on B D F.—The problem in this case is that of the simple triangular frame, sustaining a weight at the vertex.

The load acting at D is $\frac{1}{3}wl$, increased by the pull on the tie-rod, D K, or $\frac{1}{2}wl$, and is supported by the rafter B D and the strut D F. Since these pieces do not make equal angles with the vertical through D, the components of $\frac{1}{2}wl$ in the directions of these pieces are not equal. Resolving, we find the one in the direction of the rafter will be $\frac{1}{6}\frac{wl}{\sin a}$, and the other along the strut, $\frac{1}{6}\frac{wl}{\sin \beta}$; β being the angle D F K.

The first of these is transmitted to B, where it produces a vertical pressure equal to $\frac{1}{6}wl$, and a tensile stress in the tie-beam equal to $\frac{1}{6}wl$ cot α.

The other, transmitted to F, produces a pull on the king-post equal to $\frac{1}{6}wl$, and a tensile stress in the tie-beam equal to and directly opposed to that at B produced by the component acting along the rafter.

Strains on BFA.—The stress in AD is due to the assumed load, $\frac{1}{6}wl$ at A and the transmitted stress in the king-post, $\frac{1}{3}wl$, or is equal to $\frac{1}{2}wl$.

Resolving this into its components along the rafter A B and a horizontal at A, we have for the first, $\frac{1}{2}\frac{wl}{\sin\alpha}$, and for the latter, $\frac{1}{2}$ wl cot α.

The former transmitted to B produces a vertical pressure equal to $\frac{1}{2}wl$, and a tensile stress in the tie-beam equal to $\frac{1}{2}wl$ cot α.

The horizontal component at A is balanced by an equal and directly opposite component due to the half A C F.

Strains on the whole truss.—Knowing the strains on one-half, and the truss being symmetrical about the vertical through A, the stresses in all the pieces can now be determined. Summing and recapitulating, the stresses are as follows:

In B H=C M=$\frac{1}{6}\frac{wl}{\sin\alpha}+\frac{1}{6}\frac{wl}{\sin\alpha}+\frac{1}{2}\frac{wl}{\sin\alpha}=\frac{5}{6}\frac{wl}{\sin\alpha}$, compressive,

" H D=M G=$\frac{1}{6}\frac{wl}{\sin\alpha}+\frac{1}{2}\frac{wl}{\sin\alpha}=\frac{2}{3}\frac{wl}{\sin\alpha}$, "

" D A=G A=$\frac{1}{2}\frac{wl}{\sin\alpha}$, "

" H K=M L=$\frac{1}{6}\frac{wl}{\sin\alpha}$, and D F=G F=$\frac{1}{3}\frac{wl}{\sin\beta}$, "

" D K=G L = $\frac{1}{6}wl$, and A F = $\frac{1}{3}wl + \frac{1}{3}wl = \frac{2}{3}wl$, tensile,

" B K=C L = $\frac{5}{6}wl$ cot α, and K F = F L = $\frac{2}{3}wl$ cot α, "

By the use of moments.—These same values may be obtained by using the "method of sections." To apply this method to determine the stresses in the rafter, suppose a vertical section of the rafter made on the right of and consecutive to A, and take the point F as the centre of moments. Represent the compressive stress in the segment, A D, of the rafter cut by this section, by C_1. Its direction is parallel to A B, and its lever arm, which denote by p, will be equal to a perpendicular let fall from F upon the rafter. The reaction at B and the load on the rafter are known. For equilibrium we would have,

$$C_1 \times p = wl \times \text{BF} - wl \times \frac{\text{BF}}{2},$$

whence

$$C_1 = \tfrac{1}{2} wl \times \frac{b}{p}.$$

We find p to be equal to $\frac{db}{l}$, which being substituted in this expression gives

$$C_1 = \frac{wl^2}{2d}. \quad . \quad . \quad . \quad . \quad . \quad (172)$$

Substituting in (172) the value of $d = l \sin a$, we obtain

$$C_1 = \frac{wl}{2 \sin a},$$

which is the same value already determined.

If the same method be applied to the segment H B, we will find the value of C_3 to be equal to $\frac{5}{6} \frac{wl}{\sin \alpha}$.

540. In the preceding roof-truss, the inclined pieces were struts and the verticals were ties. Another form of truss is one in which the verticals are struts and the diagonals are ties. (Fig. 215.) The rafters are subdivided into a number of equal segments. At each point of division, a strut is placed, and kept in a vertical position by the main tie-beam and the inclined tie-rods, as shown in the figure.

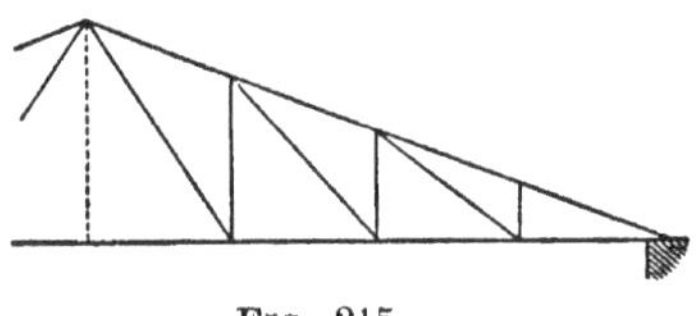

FIG. 215.

The methods previously explained will enable the student to determine the kind and amount of strains on each piece of the truss.

541. It has been recommended to check the accuracy of the calculations by some other method than the one used; the graphical method is a very convenient one for this purpose. Let us apply this method to finding the strains in the roof-truss referred to in Art. 539.

The load over the rafters is supposed to act as there taken, viz., $\frac{1}{6}$ at A and B, and $\frac{1}{3}$ at H and D, each.

Assume any point, as O. From O, on a vertical line, lay off, according to a scale, $Ob = \frac{1}{6}wl$, $bh = \frac{1}{3}wl$, $hd = \frac{1}{3}wl$, and $da = \frac{1}{6}wl$. These distances represent the loads acting at B, H, D, and A, respectively. Their sum $Oa = wl$, hence,

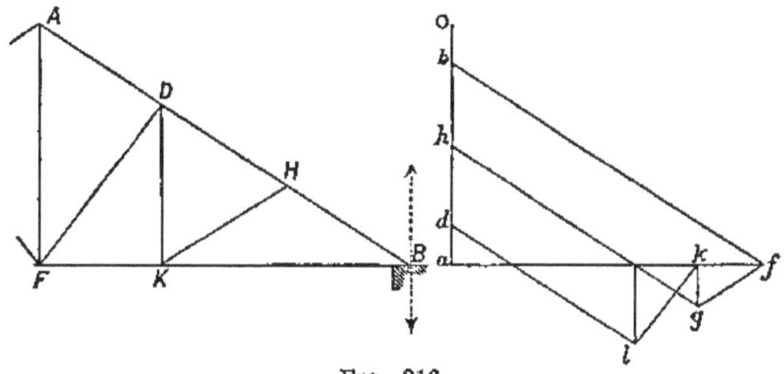

FIG. 216.

$aO = - wl$ represents the reaction at B, due to the load acting on the half A B F of the truss. The forces at B are Ob, aO, and the stresses in the pieces B H and B K. Through b, draw bf parallel to B H, and through a, draw af parallel to B K. The polygon $aObfa$ will represent the system of forces acting at B, and the lines fa and bf will represent the intensities of the stresses in K B and B H, respectively, at B, and may be taken off with the same scale used to lay off the vertical forces, Ob, bh, etc.

It is seen that the forces acting at H are the weight $\frac{1}{3}wl = bh$, the stress bf, and the unknown stresses in H K and H D. Through f, draw fg parallel to H K, and through h, draw hg parallel to H D. The polygon, $fbhgf$ will represent the intensities of the forces acting at H.

The forces acting to strain B K and H K have been determined; the forces acting at K in the directions of K D and K F are unknown. Through g, draw gk parallel to K D, and through a, draw ak parallel to K F, forming the polygon $afgka$; the lines gk and ka will represent the intensities of the stresses in D K and K F.

The stresses in the pieces K D and HD being known, the stresses in DA and DF can be determined.

In a similar way, the stresses in the other pieces can be determined.

542. **Application of graphical method to the roof with trussed rafters.**—Let us apply the same method to the trussed roof of Art. 537. Instead of the frame being uniformly loaded over the rafters, consider it as supporting a load W at the vertex A. (Fig. 217.)

The applied forces acting on the frame are the load W and the reactions at B and C. Assume a point, as O, and lay off on a vertical line the distance O*b* to represent W. The distances *bc* and *c*O will represent the reactions at B and at C. Through *b*, draw *bd* parallel to B D, and through *c*, the line *cd* parallel to B E. The triangle *bcd* will represent the

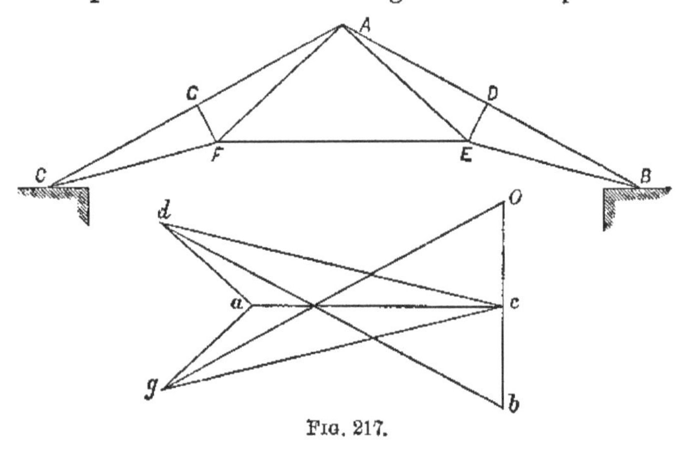

Fig. 217.

system of forces acting at B. Through O, draw the line O*g* parallel to A C, and through *c*, draw the line *cg* parallel to C F. The triangle O*cg* will represent the forces acting at C.

Going to E, since the load on the truss has been supposed to act at A, there will be no strain on D E, and the forces at E will be those acting in the direction B E already found, and the unknown forces along E A and E F. Through *d*, draw *da* parallel to E A, and through *c*, draw *ca* parallel to E F. The triangle *cda* will represent these three forces acting at E. And in the same way, the triangle *cga* would represent the strains on the pieces at F.

If there had been a force acting at E in the direction of D E, then there would have been three unknown forces acting at E, and we could not have solved the problem until one of these were known.

Purlins.

543. The purlins are simply beams, and are considered as resting on two or more supports, according to the number of frames connected by them. The strains are easily deter mined.

CONSTRUCTION OF ROOFS.

544. The most important element of the roof is the frame. The same rules given for frames, and the general methods described for their construction apply to the construction of the roof-truss.

PART VIII.

ROADS, RAILROADS, AND CANALS.

CHAPTER XXI.

ROADS.

545. A **road** is an open way or passage for travel, forming a communication between two places some distance apart.

A path or track over which a person can travel on foot is the simplest form of a road. A line, having been marked out or "*blazed*" between two places, is soon beaten into a well-defined path by constant use. A person travelling over a road like this will find nothing but a beaten path on the surface of the ground, with few or no modifications of its surface, and generally with no conveniences for crossing the streams or rivers which intersect it.

As the travel over a road of this kind increases and beasts of burden begin to be used for packing the merchandise, baggage, etc., which are to be carried over the route, modifications and improvements of the path become necessary. For convenient passage of the animals, the path must be widened, the brush and undergrowth removed, temporary bridges constructed or means of ferriage provided for crossing streams of any considerable depth, and steep ascents and descents must be modified and rendered practicable for the pack-animals. The term "**trail**" is used to designate the original path and also the path when improved so that it can be used by pack-animals.

Since transportation by wheels is cheaper and more rapid than by pack-animals, the next step will be to still further improve the road so that vehicles on wheels can be used over the route. This necessitates a still further widening of the trail, a further reduction of the slopes so as to render them practicable for carts and wagons, the providing of means to cross the streams where they cannot be forded, and the raising of the ground in those localities where it is liable to be overflowed. In this condition, the trail is called a road.

As the travel over this kind of road increases, the wants and conveniences of the community demand a further improvement of the road so that the time taken in going over it and the cost of transportation shall be reduced. This is effected by shortening the road where possible, by reducing still further the ascents and descents or by avoiding them, and by improving the surface of the road.

It has been proved that a horse can draw up a slope of $\frac{1}{20}$ only one-half the load he can draw on a level. Hence, a level road would enable one horse to do the work required of two on a road with these slopes.

It has been shown that a horse can draw over a smooth, hard road, as one of broken stone, from three to four times as much as he can draw on a soft earthen road. It therefore follows that an improvement of the surface will be accompanied by a reduction both in time and cost of the transportation.

546. The engineer may be required to lay out and make a road practicable for wagons connecting two settlements or points, in a wild, uninhabited, and therefore unmapped country, as is the case frequently on our frontier, or he may be required to plan and construct a road having for its objects the reduction of time and expense of transportation, in a country of which he has maps and other authentic information. In either case, the general principles guiding the engineer are the same. These may be considered under the following heads: 1st, **Direction**; 2d, **Gradients**; 3d, **Cross-Section**; 4th, **Road-Coverings**; 5th, **Location**; 6th, **Construction.**

DIRECTION.

547. Other things being equal, the shortest line between the two points is to be adopted, since it costs less to construct; costs less for repairs; and requires less time and labor to travel over it.

But straightness will be found of less consequence than easy ascents and descents, and as a rule must be sacrificed to obtain a level or to make a road less steep.

Good roads wind around hills instead of running over them, and this they may often be made to do without increasing their lengths. But even if the curved road, which is practically level, should be longer, it is the better; for on it a horse will draw a full load at his usual rate of speed, while on the road over the hill, the load must be diminished or the horse must reduce his rate of speed.

Roads often deviate from the straight line for reasons of economy in construction, such as to avoid swampy, marshy, or bad ground, or to avoid large excavations, or to reach points on streams better suited for the approaches of bridges, etc.

Great care must be exercised in deciding on the line which the road is to follow. If the line is badly chosen, the expense of construction and repair may be so great that it may finally be necessary to change the line and adopt a new one.

548. The considerations which should govern the selection of the line are: to connect the **termini** by the most direct and shortest line; to avoid unnecessary ascents and descents; to select the position of the road so that its longitudinal slopes shall be kept within given limits; and to so locate the line that the cost of the embankments, excavations, bridges, etc., shall be a minimum.

The wants of the community in the neighborhood of the line oftentimes affect the direction of the line, since it may be advisable and even more economical in the end to change the direction so as to pass through important points which do not lie on the general direction of the road than to leave them off the road.

GRADIENTS.

549. Theoretically, every road should be **level.** If they are not, a large amount of the horse's strength is expended in raising the load he draws up the ascent. Experiment has shown that a horse can draw up an ascent of $\frac{1}{100}$, only 90 per cent. of the maximum load he can draw on a level; up an ascent of $\frac{1}{50}$, he can draw about 80 per cent.; of $\frac{1}{30}$, he can draw only 64 per cent.; of $\frac{1}{24}$, only 50 per cent.; and of $\frac{1}{10}$, only 25 per cent.

These numbers are affected by the nature and condition of the road, being different for a rough and for a smooth road, the resistance of gravity being more severely felt on the latter.

A level road is therefore the most desirable, but can seldom be obtained. The question is to select the maximum slope or steepest ascent allowable.

An ascent affects chiefly the draught of heavy loads, as has been already shown.

A descent chiefly affects the safety of rapid travelling.

550. The slope or grade of a road depends upon the kind of vehicle used, the character of the road-covering, and the condition in which the road is kept. From the experiments above mentioned it would seem that the maximum grade for ascent should not be greater than 1 in 30, although 1 in 20 may be used for short distances.

For descent, the grade should be less than the angle of repose, or that inclination at which a vehicle at rest would not be set in motion by the force of gravity. This angle varies with the hardness and smoothness of the road-covering, and is affected by the amount of friction of the axles and wheels of the vehicles. On the best broken stone roads in good order, for ordinary vehicles, the maximum grade is taken at 1 in 35.

Steeper grades than these named produce a waste of animal power in ascending and create a certain amount of danger in descending.

551. Although theoretically the road should be level, in practice it is not desirable that it should be so, on account of the difficulty arising of keeping the surface free from water. A moderate inclination is therefore to be selected as a minimum slope for the surface of the road. This slope is taken at 1 in 125, and in a level country it is recommended to form the road by artificial means into gentle undulations approximating to this minimum.

It is generally thought that a gently undulating road is less fatiguing to a horse than one which is level. Writers who hold this opinion attempted to explain it physiologically, stating that as one set of muscles of the horse is brought into play during the ascent and another during the descent, that some of the muscles are allowed to rest, while others, those in motion, are at work. This explanation has no foundation in fact, and is therefore to be rejected. The principal advantage of an undulating road is not the rest it gives the horse, but the facilities which are afforded to the flowing of the water from the surface of the road.

CROSS-SECTION.

552. The proper width and form of roadway depend upon the amount and importance of the travel over the road.

Width.—The least width enabling two vehicles to pass with ease is assumed at 16½ feet. The width in most of the States is fixed by law.

In England, the width of turnpike roads approaching large towns, on which there is a great amount of travel, is 60 feet. Ordinary turnpike roads are made 35 feet wide. Ordinary carriage roads across the country are given a width of 25 feet; for horse-roads, the width is 8 feet; and for foot-paths, $6\frac{1}{2}$ feet.

Telford's Holyhead road is made 32 feet wide on level ground; 28 feet wide in moderate excavations; and 22 feet in deep excavations and along precipices.

In France there are four classes of main roads. The first or most important are made 66 feet wide, the middle third of which is paved or made of broken stone. The second class are 52 feet wide; the third are 33 feet wide; and the fourth are 26 feet wide. All these have the middle portion ballasted with broken stone.

The Roman military roads had their width established by law, at twelve feet when straight and sixteen when crooked.

Where a road ascends a hill by zigzags it should be made wider on the curves connecting the straight portions; this increase of width being one-fourth when the angle included between the straight portions is between 120° and 90°, and one-half when the angle is between 90° and 60°.

553. **Form of roadway.**—The surface of the road must not be flat, but must be higher at the middle than at the sides, to allow the surface water to run off freely.

If the surface is made flat, it soon becomes concave from the wear of the travel over it, and forms a receptacle for water, making a puddle if on level ground, and a gulley if the ground is inclined.

The usual shape given the cross-section of the roadway is that of a convex curve, approaching in form a segment of a circle or an ellipse. This form is considered objectionable for the reasons that water stands on the middle of the road; washes away its sides; that the road wears unequally, and is very apt to wear in holes and ruts in the middle; and that when vehicles are obliged to cross the road, they have to ascend a considerable slope.

554. The best form of the upper surface of the roadway is that of two inclined planes rounded off at their intersection by a curved surface. The section of this curved surface is a flat segment of a circle about five feet in length.

The inclination of the planes will be greatest where the surface of the road is rough and least where it is smoothest and hardest. A slope of $\frac{1}{24}$ is given a road with a broken stone covering, and may be as slight as $\frac{1}{60}$ for a road paved with square blocks. The transverse slope should always exceed the longitudinal slope of the road, so as to prevent the surface water from running too far in the direction of the length of the road.

On a steep hillside, the surface of the roadway should be a plane inclined inwards to the face of the hill. A ditch on the side of the road next to the hill receives the surface water.

555. **Foot-paths.**—On each side of the roadway, foot-paths should be made for the convenience of passengers on foot. They should be from five to six feet wide and be raised about six inches above the roadway. The upper surface should have an inclination towards the "side channels," to allow the water to flow into them and thence into the ditches. When the natural soil is firm and sandy, or gravelly, its surface will serve for the foot-paths; but if of loam or clay, it should be removed to a depth of six inches and the excavation filled with gravel.

Sods, eight inches wide and six inches thick, should be laid against the side slope of the foot-path next to the road, to prevent the wash from the water running in the side channels.

Fences, hedges, etc., where the road is to be enclosed, should be placed on the outside of the foot-paths, and outside of these should be the ditches. (Fig. 218.)

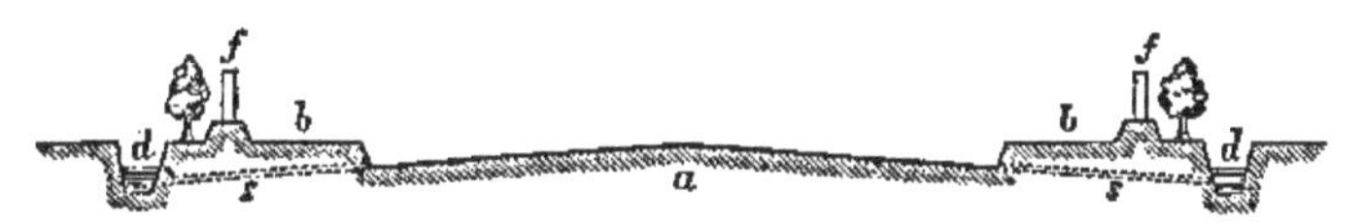

FIG. 218.—*a*, cross-section of roadway; *b*, *b*, foot-paths; *f*, *f*, fences; *d*, *d*, ditches; *s*, *s*, side drains.

556. **Ditches.**—Ditches form an important element in the construction of a good road.

The surface of the road has been given a form by means of which the water falling on it is carried off into the gutters or side channels of the road, whence it is conveyed by side drains, *s*, *s* (Fig. 218), into ditches, which immediately carry off all the water which enter them.

The ditches are sunk to a depth of about three feet below the roadway, so that they shall thoroughly drain off the water which may pass through the surface of the roadway. These ditches should lead to the natural water-courses of the country, and have a slope corresponding to the minimum longitudinal slope of the road. Their size will depend upon circumstances, being greater where they are required to carry away the water from side-hills or where they are made in wet grounds. A width of one foot at the bottom will generally be sufficient.

There should be a ditch on each side of the road, on level ground or in cuttings. One is sufficient where the road is on the side of a hill.

557. **Side-slopes.**—The side-slopes of the cuttings and embankments on each side of the road vary with the nature of the soil.

Rock cuttings may be left vertical or nearly so. Common earth should have a slope of at least $\frac{2}{3}$, and sand, $\frac{1}{3}$. Clay is treacherous and requires different slopes according to its liability to slip and the presence of water. The slope required in each case is best determined by observing the slope assumed by these earths in the locality of the work where exposed to the weather.

When the road is in a deep cutting, the side slopes should not be steeper than $\frac{1}{2}$, so as to allow the road, by its exposure to the sun and wind, to be kept dry.

Whenever the side-slopes are of **made** earth, earth removed and placed in position like that of an embankment, the slopes should be more gentle.

ROAD-COVERINGS.

558. The **road-covering** of a common country road, and most generally of all the new roads in our country, is the natural soil thrown on the road from the ditches on each side. In many cases there are even no ditches, and the road-covering or upper surface of the roadway is the natural soil as it exists on the hard subsoil beneath, when the soft material has been removed by scraping or by some other method.

Roads of this kind are deficient in the qualities of hardness and smoothness. To improve these roads, it is necessary to cover the surface with some material, as wood, stone, etc., which will substitute a hard and smooth surface for the soft and uneven earth, and which, acting as a **covering**, will protect the ground beneath from the action of the water that may fall upon it.

559. Roads may be classified from their coverings as follows:

I. Earth roads.
II. Roads of wood.
III. Gravel roads.
IV. Roads of broken stone.

V. Roads paved with stone.
VI. Roads covered or paved with other materials.
VII. Tram-roads.

I. EARTH ROADS.

560. These are the most common and almost the only kind of roads in this country. From what has been said, we know that they are deficient in hardness and generally in smoothness. In wet weather, when there is much travel of a heavy kind over them, they become almost impassable.

The principal means of improvement for these roads are to reduce the grades, thoroughly drain the roadway, and freely expose the roadway to the influence of the sun and wind. In repairing them, the earth used to fill the holes and hollows should be as gravelly as possible and free from muck or mould. Stones of considerable size should not be used, as they are liable to produce lumps and ridges, making an uneven surface disagreeable to travel upon.

II. ROADS OF WOOD.

561. **Corduroy roads.**—When a road passes over a marsh or soft swampy piece of ground which cannot be drained, or the expense of which would be too great, a corduroy road is frequently used. This kind of road is made by laying straight logs of timber, either round or split, cut to suitable lengths, side by side across the road at right angles to its length.

It is hardly worthy of the name of a road, and is extremely unpleasant to persons riding over it, but it is nevertheless extremely valuable, as otherwise, the swamp across which it is laid would at times be impassable.

562. **Plank roads.**—In districts where lumber is cheap and gravel and stone cannot be easily obtained, road-coverings of plank have been used.

The method most generally adopted in constructing a road of this class consists in laying a flooring or track, eight feet wide, of boards from nine to twelve inches in width and three inches in thickness. The boards rest upon two parallel rows of sleepers, or sills, laid lengthwise of the road, and having their centre lines about four feet apart, or two feet from the axis of the road.

The boards are laid perpendicular to the axis of the road,

experience having shown that this position is as favorable to their durability as any other and is also the most economical.

When the road is new and well made, it offers all the advantages of a good road and is a very pleasant one to use. But when the planks become worn and displaced it makes a very disagreeable and indifferent road.

Some years ago they were much used, but as a general thing they are no longer built except under very peculiar and urgent circumstances.

III. GRAVEL ROADS.

563. These are roads upon which a covering of good gravel has been laid.

The roadway is first prepared by removing the upper layer of soft and loose earth, and thoroughly draining the road. The bed is sometimes of the shape of the upper surface of the road, but more generally it is merely made level; on this a layer of gravel about four inches in thickness is laid, and when compacted by the travel over it another layer is laid, and so on until a thickness of sixteen inches at the centre has been reached.

It is advisable to compress the bed by rolling it well with a heavy iron roller before beginning to lay the gravel. In some cases a bed of broken stone has been used.

Gravel from the river shores is generally too clean for this kind of road, there not being enough clayey material mixed with it to bind the grains together. On the other hand, gravel from pits is apt to be too dirty and requires a partial cleansing to fit it for this purpose.

The gravel used should be sifted through screens, and all pebbles exceeding two inches in diameter be broken into small pieces or rejected.

The iron roller can be advantageously used to assist in compacting the layers of gravel as they are put on the road.

A gravel road carefully made, with good side ditches to thoroughly drain the road-bed, forms an excellent road.

Some gravel roads are very poor, even inferior to an earth road, caused in a great measure by using dirty gravel which is carelessly thrown on the road in spots, which cause the road to soon wear into deep ruts and hard ridges.

IV. ROADS OF BROKEN STONE.

564. The covering of roads of this class, both in this country and Europe, is composed of stone broken into small

angular fragments. These fragments are placed on the natural bed in layers, as in the gravel road, or they may be placed in layers on a rough pavement of irregular blocks of stone.

565. **Macadamized roads.**—When the stone is placed on the natural road-bed, the roads are said to be "macadamized," a name derived from Mr. McAdam, who first brought this kind of road into general use in England.

The construction of this road is very similar to that just given for a gravel road. The roadway having received its proper shape and having been thoroughly drained, is covered with a layer of broken stones from three to four inches thick. This layer is then thoroughly compacted by allowing the travel to go over it and by rolling it also with heavy iron rollers; care being taken to fill all the ruts, hollows, or other inequalities of the surface as fast as they are formed. Successive layers of broken stone are then spread over the road and treated in the same manner, until a thickness of between eight and twelve inches of stone is obtained. Care is taken that the layers, when they are spread over the surface, are not too thick, as it will be difficult, even if it be possible, to get the stone into that compact condition so necessary for a good road of this kind.

566. **Telford roads.**—This is the name given to the broken stone roads in which the stone rests on a rough pavement prepared for the bed. (Fig. 219.)

FIG. 219.

This pavement is formed of blocks of stone of an irregular pyramidal shape; the base of each block being not more than five inches, and the top not less than four inches.

The blocks are set by the hand as closely in contact at their bases as practicable; and blocks of a suitable size are selected to give the surface of the pavement a slightly convex shape from the centre outwards. The spaces between the blocks are filled with chippings of stone compactly set with a small hammer.

A layer of broken stone, four inches thick, is then laid over this pavement, for a width of nine feet on each side of the centre; no fragment of this layer should measure over two and a half inches in any direction. A layer of broken stone of smaller dimensions, or of clean coarse gravel, is spread over the wings to the same depth as the centre layer.

The road-covering, thus prepared, is thrown open to travel until the upper layer has become perfectly compact; care having been taken to fill in the ruts as fast as formed with fresh stone, in order to obtain a uniform surface. A second layer, about two inches in depth, is then laid over the centre of the roadway; and the wings receive also a layer of new material laid on to a sufficient thickness to make the outside of the roadway nine inches lower than the centre. A coating of clean coarse gravel, one inch and a half thick, is then spread over the surface, and the road-covering is considered as finished.

The stone used for the pavement may be of an inferior quality in hardness and strength to the broken stone on top, as it is but little exposed to the wear and tear occasioned by travelling. The surface-stone should be of the hardest kind that can be procured.

567. **Kind of stone** used for broken stone roads.—The stone used for these roads should be selected from those which absorb the least water, and are also hard and not brittle. All the hornblende rocks, porphyry, compact feldspar, and some of the conglomerates furnish good, durable road-coverings. Granite, gneiss, limestone, and common sandstones are inferior in this respect, and are used only when the others cannot be obtained.

568. **Repairs.**—Broken stone roads to be good must be kept in thorough repair. If the road is kept in order it will need no repairs. The difference between "kept in order" and "repairs" is that the latter is an occasional thing, while the former is a daily operation. To keep the road in order requires that the mud and dust be daily removed from the surface of the road and that all ruts, depressions, etc., be at once filled with broken stone.

It is recommended by some that when fresh material is added, the surface on which it is spread should be broken with a pick to the depth of half an inch to an inch, and the fresh material be well settled by ramming, a small quantity of clean sand being added to make the stone pack better. When not daily repaired by persons whose sole business it is to keep the road in good order, general repairs should be made in the spring and autumn by removing all accumulations of mud, cleaning out the side channels and other drains, and adding fresh material where requisite.

If practicable, the road-surface at all times should be kept free from an accumulation of mud and dust, and the surface preserved in a uniform state of evenness by the daily addition of fresh material wherever the wear is sufficient to call for it. Without this constant supervision, the best constructed road will, in a short time, be unfit for travel, and with it the weakest may at all times be kept in a tolerably fair condition.

V. ROADS PAVED WITH STONE.

569. A good pavement should offer but little resistance to the wheels, and at the same time give a firm foothold to horses; it should be durable, free from noise and dirt, and so constructed as to allow of its easy removal and replacement whenever it may be necessary to gain access to gas or water pipes which may be beneath it.

570. **Roman roads.**—The ancient paved Roman roads, traces of which may still be seen as perfect as when first made, were essentially dressed stone pavements with concrete foundations resting on sub-pavements. The entire thickness of the road-covering was about three feet, and was made as follows:

The direction of the road was marked out by two parallel furrows in the ground, and the loose earth from the space between them removed. A bed of mortar was then spread over the earth, and on this the foundation (*statumen*), composed of one or two courses of large flat stones in mortar, was laid. On this foundation was placed a course of concrete (*rudus*), composed of broken stones. If the stones were freshly broken, three parts of stone to one of lime were used; if the stone came from old buildings, two parts of lime were used. On this course a third (*nucleus*), composed of broken bricks, tiles, pottery, mixed with mortar, was placed. In this layer was imbedded the large blocks of stone (*summa crusta*) forming the pavement. These stones were irregular in form, rough on their under side, smooth on their upper, and laid so that the upper surface should be level. They were laid with great care and so fitted to each other as to render the joints almost imperceptible.

When the road passed over marshy ground, the foundation was supported by timber-work, generally of oak; the timber was covered with rushes, reeds, and sometimes straw, to protect it from contact with the mortar.

On each side of the roadway were paved foot-paths.

571. **English paved roads.**—Some of the paved roads in England are partial imitations of the Roman road. This

pavement (Fig. 220) was constructed by removing the surface of the soil to the depth of a foot or more to obtain a firm bed. If the soil was soft it was dug deeper and a bed of sand or gravel made in the excavation. On this a broken stone road-covering similar to those already described was laid. On this broken stone was spread a layer of fine clean

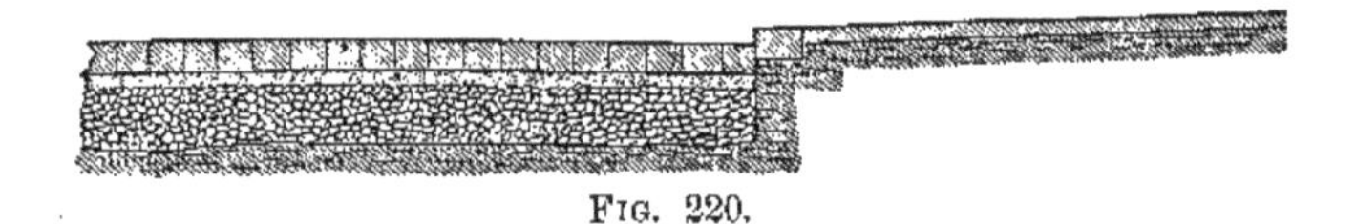

FIG. 220.

gravel, two and a half inches thick, on which rested the paving stones. The paving stones were of a square shape, and were of different sizes, according to the nature of the travel over the road. The largest size were ten inches thick, nine inches broad, and twelve inches long; the smallest were six inches thick, five inches broad, and ten inches long. Each block was carefully settled in its place by means of a heavy rammer; it was then removed in order to cover the side of the one against which it rested with hydraulic mortar; this being done, the block was replaced, and properly adjusted. The blocks of the different courses across the roadway break joints.

This pavement fulfils all the conditions required of a good road-covering, presenting as it does a hard even surface to the action of the wheels, and reposing on a firm bed formed by the broken-stone bottoming. The mortar-joints, so long as they remain tight, will effectually prevent the penetration of water beneath the pavement.

572. **Belgian pavement.**—This pavement, so named from its common use in Belgium, is made with blocks of rough stone of a cubical form measuring between eight and nine inches along the edge of the cube. These blocks are laid on a bed of sand; the thickness of this bed is only a few inches when the soil beneath is firm, but in bad soils it is increased to from six to twelve inches. The transversal joints are usually continuous, and those in the direction of the axis of the road break joints. In some cases the blocks are so laid that the joints make an angle of 45° with the axis of the roadway, one set being continuous, the other set breaking joints. By this arrangement of the joints, the wear upon the edges of the blocks, by which the upper surface soon assumes a convex shape, is diminished. It has been ascertained by experience, that when the blocks are laid in the usual manner, the

wear upon the edges of the block is greatest at the joints which run transversely to the axis.

When a bed of concrete is used, instead of or in addition to a bed of sand, and the upper surface of the blocks is rectangular instead of square, there results a pavement much used in New York City.

573. **Cobble-stone pavement.**—Rounded pebbles (**cobble stones**) are used frequently for pavements. This pavement is composed of round or egg-shape pebbles, from five to ten inches long, three to six inches wide, set on end in a bed of sand or fine gravel, and firmly settled in place by pounding with a heavy rammer. After the stones are driven, the road-surface is covered with a layer of clean sand or gravel, two or three inches thick.

The objections to this pavement are its roughness; the resistance offered to the wheels; the noise; the ease with which holes are formed in the road by the stones being pressed down in the ground by heavy loads passing over them; the difficulty of cleaning its surface; and its need of frequent repairs.

574. **Kind of stones used for pavements.**—The fine-grained granites which contain but a small proportion of mica, and the fine-grained silicious sand-stones which are free from clay, form good material for blocks for paving. Mica slate, talcose slate, hornblende slate, some varieties of gneiss, and some varieties of sand-stone of a slaty structure, yield excellent materials for pavements for sidewalks and paths.

VI. ROADS OF OTHER MATERIALS.

575. Wooden blocks have been much used recently in paving the streets of our towns and cities. Brick, concrete, asphalte, and even cast iron, are or have been used for road-coverings. Roads near blast-furnaces are frequently seen covered with the slag from the furnaces, and those near kilns where cement is burned, with cinders and clinkers from the kilns. Road-coverings of charcoal have been tried in Michigan and Wisconsin.

The wooden, brick, and asphaltic pavements are the most common of these.

Wooden pavements.—Wooden pavements are the same in principle as stone. The road-bed is formed and the blocks of wood are placed in contact with each other upon the surface of the road-bed as described for the blocks of stone pavements. The wooden blocks are parallelopipedons in form and are laid with the grain of the wood in the direction of the depth of the road. From slight differences in the details of construction of wooden pavements there has arisen quite a variety of names, as the Nicolson, the bastard Nicolson, the Stowe, the Greeley, the unpatented, etc., all using the wooden blocks, but differing slightly in other ways.

Wooden pavements offer a smooth surface; are easily kept clean; not noisy; easy for the horses and vehicles; pleasant to ride upon; and are cheaper at first cost than stone pavements. For these reasons they have been much used in the United States.

They are, however, slippery in wet weather; soon wear out; and unfit for roads or streets over which there is a heavy travel. True economy forbids their use except as temporary roads.

576. **Asphaltic coverings.**—Asphaltic roads may be composed of broken stone and this covered with asphaltic concrete, or the broken stone covered with ordinary concrete and this overlaid with a covering of asphalte mixed with sand. Asphaltic roads present a smooth surface which does not become slippery by wear; a surface free from dust and mud; not noisy; and from its imperviousness to moisture forms an excellent covering over the road-bed beneath and prevents the escape of noxious vapors from below.

Asphaltic roads properly made are growing steadily in favor and when they are better known will be more generally adopted for all streets in towns and cities, over which the travel is light.

VII. TRAM-ROADS.

577. In order that the tractive force should be a minimum, the resistance offered to the wheels of the carriage should be a minimum. In other words, the harder and smoother the road, the less will be the tractive force required. But carriages drawn by horses require that the surface of the road should be rough, to give a good foothold to the horses' feet. These two opposite requirements are united only in roads with track-ways, on which there are at least two parallel tracks made of some hard and smooth material for the wheels to run upon, while the space between the tracks is covered with a different material suitable for the horses' feet. Constructions of this class are termed "tram-roads" or "tramways." The surface of the tracks or "trams" are made flush with that of the road and are suitable for the wheels of ordinary carriages. Their construction will be alluded to in the next chapter.

CHAPTER XXII.

LOCATION AND CONSTRUCTION OF ROADS.

578. In establishing a road to afford means of communication between two given places, there are several points which must be considered by the engineer and those interested in its construction. These are the kind of road to be selected, the general line of direction to be chosen or located, and the construction of the road.

The selection of the kind of road depends upon the kind of travel which is to pass over it; the amount of travel, both present and prospective; and the wants of the community in the neighborhood of the line. The location and construction of the road depend upon the natural features of the country through which the road must pass, and as these come exclusively within the limits of the engineer's profession, they alone will be considered in this chapter.

LOCATION.

579. **Reconnoissance.**—The examination and study of the country by the eye is termed a **reconnoissance**, and is usually made in advance of any instrumental surveys, to save time and expense. The general form of the country and the approximate position of the road may frequently be determined by it.

A careful examination of the general maps of the country, if any exist, will lessen the work of the reconnoissance very much, as by this the engineer will be able to discover many of the features which will be favorable or otherwise to the location of the road in their vicinity.

Roads along the bank of a large stream will have to cross a number of tributaries. Roads joining two important streams running nearly parallel to each other must cross high ground or dividing ridges between the streams.

An examination of the map will show the position of the streams, and from these the engineer may trace the general directions of the ridges, determine the lowest and highest points, and obtain the lines of greatest and least slopes.

With this information the directions of the roads leading from one valley to another may be approximately located.

It is seen (Fig. 221) that if A and B are to be joined by a road, that the road may run direct from A to B, as shown by the dotted line joining them, or it may go, by following the

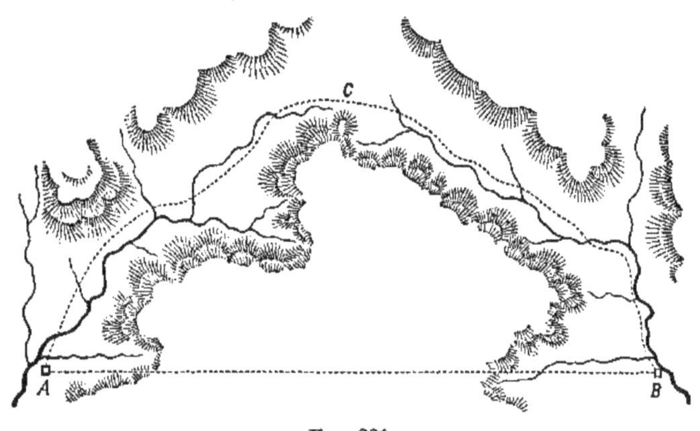

FIG. 221.

general directions of the streams, through C, as shown by the dotted line A C B. By the first route, the road would be apparently shorter, but the ascents and descents would be greater; by the second, the road would be longer, but the ascents and descents more gentle, and the total difference of level to be passed over would be less.

We can draw this conclusion from the fact that the streams have made for themselves channels which follow the lines of gentlest slope. And that if two streams flow in the same direction, the high ground or ridge separating them has the same general direction and inclination as the streams. And if two streams approach each other near their sources, as those at C in the figure, that this indicates a depression in the main ridge in this vicinity.

Hence long lines of road usually follow the valleys of streams, obtaining in this way moderate grades and crossing the ridges by the lowest passes.

The engineer having studied thoroughly the map and made himself acquainted with the natural features of the country as there indicated, proceeds to make a personal examination of the ground, to identify these natural features, and to verify the conclusions deduced from the study of the map.

In making the examination, he goes both forwards and backwards over the ground so as to see it from both directions, and in this way verify or correct the impressions he has received as to its nature.

By means of the reconnoissance he establishes "approximate" or "trial lines" for examination. These lines are marked out by "blazing" if in a wooded country, or by stout stakes driven at the important points if the country be a cleared or open one.

580. **Surveys.**—The surveys are divided into three classes: preliminary surveys, surveys of location, and surveys of construction.

The **preliminary survey** is made with ordinary instruments, generally a transit and a level, and has for its object the measurement of the length of the road, the changes of direction of the different courses, the relative heights of the different points or differences of level along the line, and of obtaining the topography of the country passed over in the immediate neighborhood of the line.

The line is run without curves, and therefore, when plotted, consists of a series of straight lines of different lengths, forming at their connection angles of varying size.

The levelling party, besides taking the measurements requisite to construct a profile of the line, make cross-section levellings at suitable points, so as to show the form of surface of the road.

The topography on each side of the line is ordinarily sketched in by eye; instrumental measurements being occasionally made to check the work.

581. **Map and memoir.**—The results of these surveys are mapped, and all the information gathered during the survey which cannot be shown on the map is embodied in a memoir.

From these trial lines thus surveyed, the engineer makes a selection, being governed by the considerations mentioned in Art. 567, viz., shortness of route, avoidance of unnecessary ascents and descents, selection of favorable grades, and economy of construction.

582. **Estimate of the cost.**—This can be made approximately after the engineer has established the grades.

The kind of road and the character of the travel over it generally fix the limits of its longitudinal slopes. To fix them exactly, the engineer constructs the profiles of the different sections of the road and draws the "grade lines" on these profiles, keeping their slopes within the general limit already assumed. Thus in a profile (Fig. 222) the grade line A B is drawn, following the mean or general slope of the ground, equalizing as far as possible the undulations of the profile above and below the grade line. The inclination of the grade line with the horizontal is then measured, and if its slope falls within the limit assumed, the grade is a satisfactory one and the amounts of excavation and embankment are nearly equal. If the inclination be found too steep, either

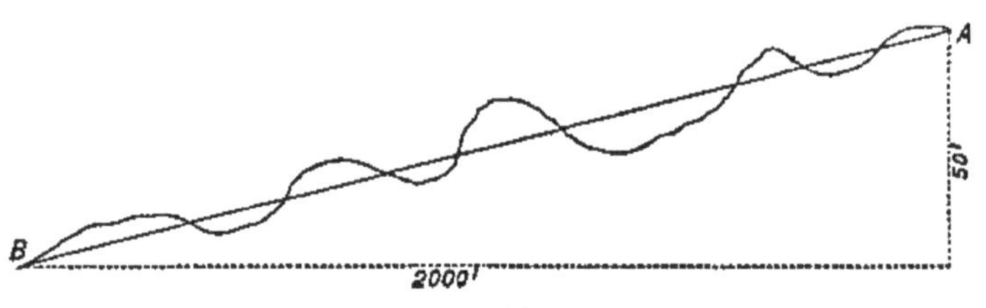

FIG. 222.

the top of the hill must be cut down or the length of the line between the two points at top and bottom be increased. The latter is the method usually adopted. Thus if the road laid out on a straight line joining C and D (Fig. 223) requires a

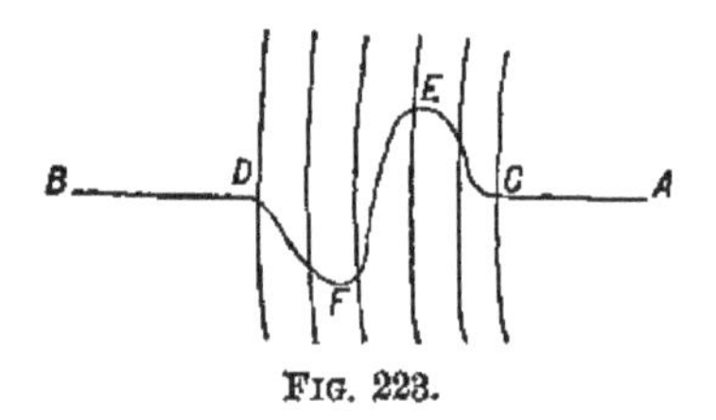

FIG. 223.

steeper grade than the maximum grade adopted, the length of the road between these points, C and D, may be increased by curving it, as shown by the line C E F D. The length to give this winding road is easily determined so that the grade of every portion of the road shall be kept within the assumed limit. The proper grade line having been determined and drawn on the profiles, the height of the embankments and the depth of the cuttings are determined.

Knowing the width of the road, the form of its surface, and the inclination of the side slopes, the cubical contents of the excavations and embankments may be calculated, and an estimate of the cost made.

The comparative costs of the routes being determined and the considerations mentioned in last article given their full weight, the engineer selects the particular line for the road.

It is well to say that it happens often that no trial lines

are necessary; the route to be followed by the road being apparent.

583. **Survey of location.**—The route being selected, it is gone over again and more accurately surveyed. It is carefully levelled at regular intervals in the direction of its length, and cross-levels at all important points are made. The angles made by the changes of direction of the line are rounded off by curves, the curves being generally arcs of circles. Advantage is taken of this survey to place the line in its best position so as to reduce to a minimum the embankments and excavation, and to give the best approaches to the points where streams are to be crossed.

The line is divided into a number of divisions, and maps of these divisions are made showing the road in plan and the longitudinal and cross-sections of the natural ground, with the horizontal and vertical measurements written upon them.

By these maps, the engineer can lay out the line on the ground and can determine the amount of excavation and embankment required for each division.

Besides these maps, detailed drawings of the road-covering, of the bridges, culverts, drains, etc., with the written specifications explaining how the work on each must be done, should be prepared.

The work is now in the condition that estimates of its cost can be accurately made and its construction begun.

584. **Survey of construction.**—The road is constructed by contract or "day labor." Whichever method is adopted, it is first necessary to "lay out the work." This laying out the work forms the third class of surveys, or survey of construction.

From the maps showing the location, the engineer proceeds to mark out the axis of the road upon the ground by means of stout pegs or stakes driven at equal intervals apart, using a transit or theodolite to keep them in the proper line. These stakes are numbered to correspond with the same points indicated on the map.

The width of the roadway and the lines on the ground corresponding to the side slopes of the excavations and embankments, are laid out in the same manner, by stakes placed along the lines of the cross profiles.

Besides the numbers marked on the stakes, to indicate their position on the map, other numbers, showing the depth of the excavations, or the height of the embankments from the surface of the ground, accompanied by the letters **Cut. Fill.** to indicate a **cutting**, or a **filling**, as the case may be, are also added to guide the workmen. The positions of the stakes on the ground, which show the principal points of the axis of the road, should be laid down on the map by bearings and distances from bench-marks in their vicinity, in order that the points may be readily found should the stakes be subsequently misplaced.

Curves.—Curves are not necessary for common roads, but it always looks better even in a common road to join two straight portions by a regular curve than by a bent line.

Curves are laid out by means of offsets from a chord or tangent, or by angles of deflection from the tangent. The latter method, using a transit or theodolite, is the one most commonly employed.

CONSTRUCTION.

585. **Earth-work.**—This term is applied to all that relates to the excavations and embankments, whatever be the material excavated or handled.

Excavations.—In forming the excavations, the inclination of the side slopes demands particular attention. This inclination will depend on the nature of the soil, and the action of the atmosphere and internal moisture upon it. In common soils, as ordinary earth formed of a mixture of clay and sand, hard clay, and compact stony soils, although the side slopes would withstand very well the effects of the weather with a greater inclination, it is best to give them a slope of $\frac{1}{2}$; as the surface of the roadway will, by this arrangement, be better exposed to the action of the sun and air, which will cause a rapid evaporation of the moisture on the surface. Pure sand and gravel require a slope of $\frac{1}{3}$. In all cases where the depth of the excavation is great, the base of the slope should be increased. It is not usual to use artificial means to protect the surface of the side slopes from the action of the weather; but it is a precaution which, in the end, will save much labor and expense in keeping the roadway in good order. The simplest means which can be used for this purpose, consist in covering the slopes with good sods, or else with a layer of mould about four inches thick, and sown with grass-seed. These means will be amply sufficient to protect the side slopes from injury when they are not exposed to any other causes of deterioration than the wash of the rain and the action of frost on the ordinary moisture retained by the soil.

The side slopes form usually an unbroken surface from the foot to the top. But in deep excavations, and particularly in soils liable to slips, they are sometimes formed with horizontal offsets, termed **benches**, which are made a few feet wide and have a ditch on the inner side to receive the surface-water from the portion of the side slope above them. These benches catch and retain the earth that may fall from the portion of the side slope above.

In excavations through solid rock, which does not disintegrate on exposure to the atmosphere, the side slopes might be made perpendicular; but as this would exclude, in a great degree, the action of the sun and air, which is essential to keeping the road-surface dry and in good order, it will be necessary to make the side slopes with an inclination, varying according to the locality; the inclination of the slope on the south side in northern latitudes being greatest, to expose better the road-surface to the sun's rays.

Embankments.—In forming the embankments, the side slopes should be made less than the natural slope; for the purpose of giving them greater durability, and to prevent the width of the top surface along which the roadway is made from diminishing by every change in the side-slopes, as it would were they made with the natural slope. To protect more effectually the side-slopes, they should be sodded or sown in grass seed; and the surface-water of the top should not be allowed to run down them, as it would soon wash them into gullies and injure the embankment. In localities where stone is plenty, a retaining wall of dry stone may be advantageously substituted for the side-slopes.

To reduce the settling which takes place in embankments, the earth should be laid in successive layers, and each layer well settled with rammers. As this method is expensive, it is seldom resorted to except in works which require great care, and are of small extent. For extensive works, the method usually adopted is to embank out from one end, carrying forward the work on a level with the top surface. In

FIG. 224.

this case, as there must be a want of compactness in the mass, it is best to form the outsides of the embankment first, and to gradually fill in towards the middle, in order that the earth may arrange itself in layers with a dip towards the centre (Fig. 224). This arrangement will in a great measure counteract the tendency of the earth sliding off in layers along the sides.

586. **Removal of the earth.**—In both excavation and embankment, the problem is "to remove the earth from the excavation to the embankment or place of deposit by the shortest distance, in the shortest time, and at the least expense." This is an important problem in practice, and its proper solution affects very materially the cost of the work.

The average distance to which the earth is carried to form the embankment is called the **lead**, and is assumed to be equal to the right line joining the centre of gravity of the volume of excavation with that of the embankment. When this lead is made the least possible, all other things being equal, the cost of removal of the earth is a minimum.

In the execution of earthwork, it is not always advisable to make the whole of an embankment from the adjoining cuttings, as the lead would be too long. In such a case, a part of the cutting is wasted, being deposited in some convenient place, forming what is known as a **spoil-bank.** The necessary earth required to complete the embankment is obtained from some spot nearer to the work, and the cutting or excavation made in supplying it is called a **borrow-pit.**

Means used to move the earth.—The earth is loosened by means of ploughs, picks, and shovels, and then thrown into wheelbarrows, carts, or wagons to be removed. A scraper drawn by a horse is frequently used to great advantage.

Resort is sometimes had to blasting to loosen the soil, when it is rock, hard clay, and even frozen earth.

The advantages of wheelbarrows over carts, and carts over wagons, etc., depend upon circumstances. When the earth is to be transported to a considerable distance, the wheelbarrow becomes too expensive. By combining the cost of filling the cart or wheelbarrow, the amount removed, and the time occupied in transporting the earth in each case, the cost of the two methods can be obtained and compared with each other.

Shrinkage.—When embankments are made in layers compacted by ramming or by being carted over, the subsequent settling is quite small. But made in the usual way, there is always a certain amount of settling which follows and which is provided for by making at first the embankment a few inches higher than it is to be. Earth occupies a less space in an embankment than in its natural state; that is, a greater number of cubic yards of excavation is required to form an embankment than there are cubic yards in its volume.

This shrinkage of the earths is about as follows:

Gravel shrinks about eight per cent.
Gravel and sand nine per cent.
Clay and clayey earth ten per cent.
Loam and light earths twelve per cent.

On the contrary, rock occupies more space when broken up than it does in its natural state, the percentage of its increase in volume varying with the way the fragments are piled together. Carelessly piled its increase of volume was found to be about seventy-five per cent., and when carefully piled, fifty per cent.

587. **Methods of obtaining the quantities to be excavated, etc.**—In comparing the costs of the routes or for rough estimates, it is sufficiently exact to take a number of equidistant profiles, and calculate the solid contents between each pair, either by multiplying the half sum of their areas by the distance between them, or else by taking the profile at the middle point between each pair, and multiplying its area by the same length as before; the first of these methods gives too large a result, and the second too small.

Where an exact estimate is to be made, the Prismoidal formula (Mensuration, p. 129) should be used. This formula gives the exact contents.

588. **In swamps and marshes.**—When the embankment is made through a swamp or marsh, many precautions are necessary.

If the bog is only three or four feet deep and has a hard bottom, it is recommended to remove the soft material and build the embankment on the hard stratum.

If it be too deep to remove the soft material, its surface, provided it be not too soft, may be covered with some substance to form an artificial bed for the embankment. Rows of turf with the grassy side downward have been used. Brushwood has also been tried.

If the swamp be deep and the material quite fluid, the first thing to do is to drain it, and then prepare an artificial bed for the embankment.

589. **Side-hill roads.**—When a road runs along the side of a hill, it is usually made half in excavation and half in embankment. But as the embankment is liable to slip if simply deposited on the natural surface of the ground, the latter should be cut into steps or offsets (Fig. 225). A low stone wall constructed at the foot of the embankment will add to its stability.

If the surface of the hill be very much inclined, the side slopes of both the excavation and the embankment should be replaced by retaining walls of dry stone (Fig. 226), or of stones laid in mortar.

The upper wall may be dispensed with when the side hill is of rock.

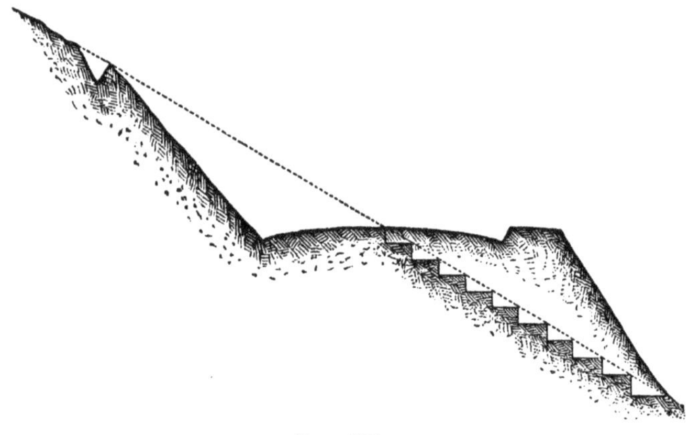

FIG. 225.

When the road passes along the face of a nearly perpendicular precipice at a considerable height, as around a projecting point of a rocky bank of a river in a mountainous

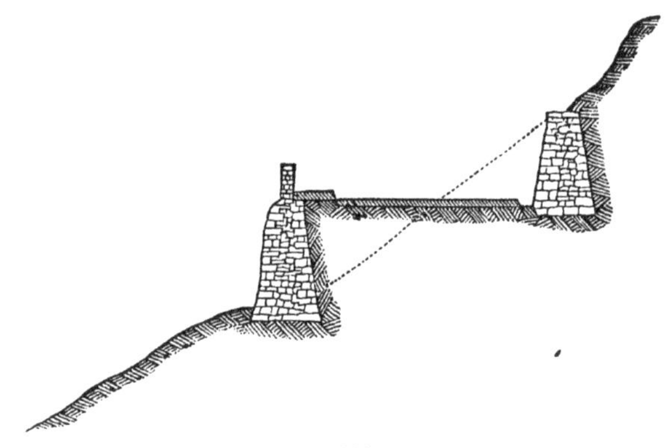

FIG. 226.

district, it may rest on a frame-work of horizontal beams let into holes drilled in the face of the precipice and supported at their outer ends by inclined struts beneath, the lower ends of which rest in notches formed in the rock.

DRAINAGE.

590. A system of thorough drainage, by which the water that filters through the ground will be cut off from the soil beneath the roadway, to a depth of at least three feet below the bottom of the road-covering, and by which the water falling upon the surface will be speedily conveyed off, before it can filter through the road-covering, is essential to the good condition of a road.

The form of the road, the side drains, and the ditches (Fig 218), are arranged and constructed with this object in view (Art. 556.)

591. **Covered drains or ditches.**—As open ditches would be soon filled by the washings of the side-slopes in certain parts of the roads, covered drains (Fig. 227) are substituted for them in these places.

Fig. 227.

They may be constructed with a bottom of concrete, flagging, or brick, with sides of the same material, or as shown in the figure, and covered with flat stones, leaving open joints of about half an inch to give free admission to the water. The top is covered with brushwood or with fragments of broken stone, or with pebbles and clean gravel, through which the water will filter freely without carrying any earth or sediment into the drain.

592. **Cross drains.**—Besides the covered drains parallel to the axis of the road in cuttings, other drains known as cross drains are made under the roadway. They should have a slope along the bottom to facilitate the escape of the water. A slope of 1 in 100 will be sufficient.

They may be constructed in the same manner as the covered drains, or trenches may be dug to the required depth with the proper slope and filled with broken stone. On the stone a layer of brushwood is placed and over this the road-covering. Drains of this kind are known as **blind ditches.** Any construction will be effective which will leave a small open waterway at the bottom of the trench which will not become choked with sediment.

If the road is level, the cross drains may run straight across, but if inclined they form a broken line, in plan the shape of the letter V, with the angular point in the centre of the road directed towards the ascent. From their form, they are termed **cross-mitre** drains.

They are placed at intervals depending upon the nature of the soil and kind of road-covering used, in some cases as much as sixty yards apart, in others not more than twenty feet.

593. **Catchwaters.**—These are broad shallow ditches constructed across the surface of the road so arranged that vehicles can pass over them easily and without shock. They are used to catch the water which runs down the length of the road and to turn it off into the side ditches. They are sometimes called water-tables.

They are necessary on long slopes, and in depressions where a descent and an ascent meet, to prevent the water from cutting the surface of the road in furrows. In a depression, they are usually placed at right angles to the road ; on a slope, they cross the road diagonally where the water is to be carried to one side ; if to both sides, their plan is that of a V with the angular point up the road.

The inclination of the bottom of the catchwater should be sufficient to carry off the water as fast as it accumulates in the trench, and where the velocity of the current flowing through them is considerable, they should be paved.

A mound of earth crossing the road obliquely is frequently used as a substitute for the catchwater. When used it should be arranged to allow carriages to pass over them without difficulty and inconvenience.

594. **Culverts.**—These structures are used to carry under the road the water of small streams which intersect it, and also the water of the ditches on the upper side of a road to the lower side, or side on which the natural water-courses lie by which the water is finally carried away.

They may be built of stone, brick, concrete, or even of wood.

Where stone is scarce, a culvert may be built of planks or slabs, forming a long box open at the ends. This is a temporary structure unless it can be kept always wet.

A small full-centre arch of brick resting on a flooring of concrete forms a good culvert.

The length of a culvert under an embankment will be equal to the width of the road increased by the horizontal distance on each side forming the base of the side-slope. At each end, wing-walls should be built, their faces having the same slope as that of the embankment. The ends of the culvert must be protected against the undermining action of the water.

The form of cross-section varies according to the circumstances of the case, depending greatly on the strength required in the structure and the volume and velocity of the water flowing through it. The dimensions of the waterway of a culvert should be proportioned to the greatest volume of water which it may ever be required to carry off, and should always be large enough to allow of a person entering it tc clean it out.

595. **Footpaths and sidewalks.**—Ordinarily, footpaths are not provided for in our country roads. They should be, however, and the remarks made in Art. 575 apply to their construction.

In cities and towns, sidewalks and crossings are arranged in all the streets. They are made of flagging-stone, brick, wood, ordinary concrete, asphaltic concrete, etc. They differ in construction only in degree from roads of the same kind.

596. **Sidewalk of flagging-stone.**—The flagstones are at least two inches in thickness, laid on a bed of gravel. The width of the sidewalk depends upon the numbers liable to use them, being wider where great crowds are frequent and less wide on streets not much used. A width of twelve feet is sufficient for most cases.

The upper surface is not level, but has a slight slope towards the street to convey the surface water to the side channels.

The pavement of the street is separated from that of the sidewalk by a row of long slabs set on their edges, termed **curb-stones,** which confine both the flagging and paving stones. The curb-stones form the sides of the side channels, and should for this purpose project six inches above the outside paving stones, and be sunk at least four inches below their top surface; they should be flush with the upper surface of the sidewalks, to allow the water to run over into the side channels, and to prevent accidents from persons tripping by striking their feet against them.

The crossings should be from four to six feet wide, and be

slightly raised above the general surface of the pavement, to keep them free from mud.

TRAM-ROADS.

Tram-roads are built of stone, of wood, or of iron.

597. **Stone tram-roads.**—The best tram-roads of stone consist of two parallel rows of granite blocks, about $4\frac{1}{2}$ feet apart from centre to centre, the upper surface of the blocks being flush with the surface of the road. The blocks should be from 4 to 6 feet long, 10 to 12 inches broad and 8 to 12 inches deep. Sometimes the upper surface is made slightly concave for the purpose of retaining the wheels on the tracks.

Stone tram-roads were used by the Egyptians, traces of them being found in the quarries which supplied stone for the pyramids.

Tram-roads of stone have been used in England, and are used at the present time in Italy.

The granite blocks used in the Italian tram-roads are from 4 to 6 feet long, about 2 feet broad, and 8 inches deep, laid on a bed of gravel 6 inches thick. The space between the "trams" is paved with cobble stones with an inclination from the outside to the middle line. The centre is therefore lower than the sides, forming a channel for the water, which flows into cross drains provided to carry it off.

In a tram-road on the Holyhead road, the granite blocks were required to be not less than 4 feet long, 14 inches broad, and 12 inches deep. The blocks were laid on a bed composed of a rough sub-pavement, similar to that used for the Telford road, on which was a layer three inches thick of small broken stone, and on top of this a layer of gravel two inches thick, compacted by a heavy roller.

The effect of this tram-road was to reduce the required amount of tractive force to less than one-half of what was required on the broken stone road.

598. **Tram-roads of wood.**—Where timber is plenty, tram-roads of wood are frequently used. They do not differ in principle of construction from the stone tramway. Since the wood is extremely perishable when buried in the damp ground, tramways of wood are used only in temporary constructions.

599. **Iron tram-roads.**—The iron tram-roads formerly used were made by covering a wooden track with flat iron bars, so as to increase the durability of the track and to lessen the resistance offered to the wheels. To keep the wheels on the track, a

flange was placed on the side of the bar (Fig. 228). The objections to these tramways were that the broad surface of the iron plate collected obstructions upon it, and that the friction of the wheels against the flange was very great.

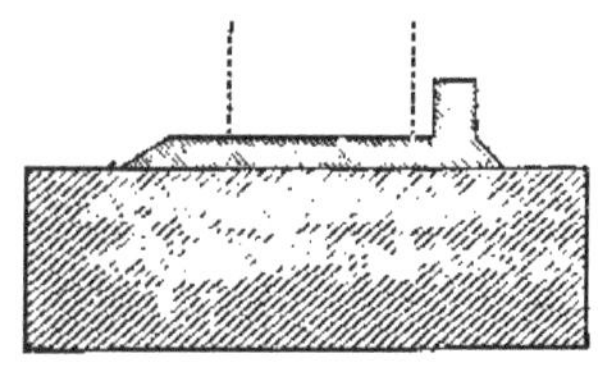

FIG. 228.

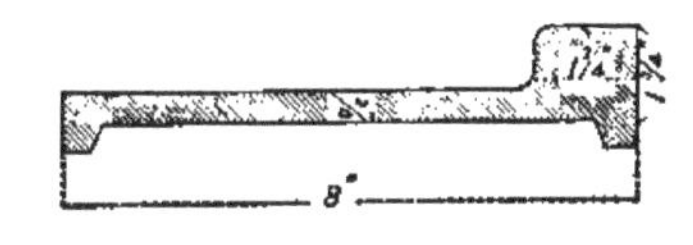

FIG. 229.

An iron plate (Fig. 229) is used quite extensively in the United States, particularly in Philadelphia, for tracks for street cars. The upper and narrower portion is used by the wheels of the car, while the wider and flat portion can be used by ordinary carriages.

CHAPTER XXIII.

RAILROADS.

600. As long as the flange attached to the bar was used to keep the wheels on the track, the road was called a tram-road. When the flange was removed from the bar and transferred to the wheel, the road became changed in character and was named a **railway** or **railroad**. The marked difference between a tram-road and a railroad is, that the former is used by all classes of carriages, while the latter can be used only by cars specially built for the purpose.

A railroad may be defined to be a track formed of iron or steel bars, called **rails**, placed in parallel lines, and upon which the wheels of vehicles run.

The general principles already alluded to as governing the location and construction of roads, apply equally to railroads, but in a higher degree. Greater importance is attached, for railroads, to straightness, to easy grades, and to using curves of larger radius where a change of direction takes place, than for any other kind of road.

601. **Direction.**—Straightness of direction is more import-

ant for railroads than for common roads, for the reasons that the shorter the line the cheaper is its cost, and that there is a greater resistance offered by curves, causing a greater expenditure of tractive force.

The same considerations which govern in determining the direction of a common road apply to the railroad, viz., cost of construction, wants of the community, etc.

602. **Grades.**—The question of grade is more one of economy than of practicability. Locomotives can be made to ascend steep grades by increasing their power and adhesion, but as the grades increase in steepness, the effective tractive force of the engine decreases. Thus with an ascent of 20 feet to the mile, an engine can draw about one-half the load which it can draw on a level; with 40 feet to the mile, about one-third, etc.

The cost of drawing a load on a railroad varies very nearly with the power employed. Hence it will cost nearly twice as much to haul a load on a grade of 20 feet to the mile as it would on a level road. This consideration will therefore justify large expenditures in the construction of the road if made with the view of reducing the grades.

The ruling or maximum grade adopted for the line depends upon the motive power used to ascend the grades and upon the avoidance of a waste of power in descending.

The steepest grade upon a given line is not necessarily the maximum inclination adopted for the road. It may be much greater than the ruling grade, and will then require special arrangements to be made to overcome it.

When the loads to be carried in one direction over the road are much heavier than those carried in the other, the ascent up which the heavy loads are to be carried should be made by easy grades, while the descent may be made by steeper ones. If the travel is equal in both directions, the ruling grades should be equal for both slopes.

The length of grades must be considered, as it is found more advantageous to have steep grades upon short portions of the line than to overcome the same difference of level by grades not so steep on longer developments.

From various experiments, it appears that the angle of repose (Art. 550) for a railroad is about $\frac{1}{280}$. But in descending grades much steeper than this, the velocity due to the accelerating force of gravity soon attains its greatest limit and remains constant, from the resistance caused by the air.

The limit of the velocity thus attained, whether the train

descends by the action of gravity alone, or by the combined action of the motive power of the engine and gravity, can be determined for any given load. It appears from calculation and experiment that heavy trains, allowed to run freely without applying the brakes, may descend grades of $\frac{1}{100}$ without attaining a greater velocity than about 40 miles an hour.

Hence, the question to be considered in comparing the advantages of different grades is one between the loss of power and speed for ascending trains on steep grades, and the extra cost of heavy excavations, tunnels, and embankments required by lighter grades.

Since locomotives are not taxed to their full extent, grades of 60 feet to the mile may be used without any practical loss of power either in the ascent or descent.

603. **Curves.**—Curves are necessary to enable the road to pass around obstacles, such as hills, deep ravines, valuable houses which cannot be removed, etc.

The objections to curves in the road are the resistances which they offer to the motion of the cars and the dangers to which the cars are exposed.

The resistances offered by the curves are chiefly due to the following causes:

1. The obliquity of the moving power while passing around the curve.
2. The friction of the flanges of the wheels against the outer rail due to the centrifugal force.
3. The friction of the flanges against the rails due to the parallelism of the axles.
4. The fastening of each pair of wheels to the same axle.

The danger of a car running off the track is much increased by curves. The car is kept on the rails while going around a curve by the flanges of the wheels and by the firmness of the outer rails. If the resistance offered by the rails and flanges should be overcome by the "quantity of motion" of the car, the latter would leave the track. Hence, where sharp curves are necessary, they should be located, if possible, near stopping places, and never at those points where the speed is to be very high or where the car will pass with great velocity, as at the foot of a steep grade.

The minimum radius of a curve depends greatly upon the speed to be employed. In France, the minimum radius allowed is 2,700 feet. In England, no curve less than 2,640 feet can be used without special permission of Parliament or the Board of Trade. The minimum radius used on the Hudson River Railroad is 2,062 feet. On the Baltimore and Ohio Railroad, the minimum radius is 600 feet, although when first constructed there were several curves of 400 feet radius, and one of 318 feet over which trains passed at a speed of 15 miles an hour.

604. **Resistances of vehicles on railroads.**—The resistance offered to the force of traction by a train of cars is due to **friction, concussion,** and the **atmosphere.** The amount of this resistance depends upon a variety of conditions, such as the condition of the road, whether well or badly constructed, in bad order, etc.; the state of the rolling machinery; the climate; the season of the year; state of the weather, etc.

In discussing the resistance, it is assumed that the cars are well made, the track in good order, and the weather moderately calm. The amount of resistance may be determined by means of a dynamometer between the engine and the train, and may be expressed either as a fraction or as a certain number of pounds per ton, the latter being generally used.

That part of the resistance offered by the train due to friction is constant at all speeds; that due to concussion and the atmosphere varies with the velocity, increasing with the speed. The law of increase is not fully known.

605. **On a level and straight road.** — The resistance offered by a train running on a level and straight road, nearly as possible under the conditions in ordinary practice, has been determined by experiment to be nearly that given by the following formula:

$$r = \frac{v^2}{171} + 8, \quad . \quad . \quad . \quad . \quad . \quad (173)$$

in which r is the resistance in pounds per ton of the engine, tender, and train; and v the velocity in miles per hour.

Hence it is seen, that for a train moving at the rate of 20 miles an hour, the resistance would be 10.33 pounds per ton of the entire train.

If the road is in bad repair, the values obtained by this formula should be increased 40 per cent.; for strong side winds, 20 per cent.

606. **Resistance due to grades.**—The resistance due to a grade is found by multiplying the whole weight of the train by the difference of level and dividing this product by the length of the slope. By this rule it is found that the resistance per ton due to a grade of 24 feet in a mile is

$$2{,}240 \times \frac{24}{5{,}280} = 10.2 \text{ pounds,}$$

or about the same as that on a level with the speed of 20 miles an hour. Therefore, if the train runs over this grade at

20 miles an hour, the resistance would be just double, or it would require the same power to run one mile on the grade that would draw the same load at the same speed two miles on a level road.

607. **Resistance due to curves.**—The resistance due to curvature is much affected by the gauge of the road, the elevation of the outer rail, the form of surface of the tires and the size of the wheels, the speed and length of the train, etc. Hence, experiments made to obtain this resistance will be found to vary greatly for the same curve on different roads. The point to be gained, however, is to find the amount of curvature which will consume an amount of power sufficient to draw a train one mile on a straight and level road.

It is assumed that the resistance from curvature is inversely as the radius; that is, the resistance offered by a curve of 2° is double that of a curve of 1°.

From experiments made under his direction, Mr. Latrobe deduced the resistance upon a curve of 400 feet radius to be double that upon a straight line.

Upon averaging a large number of experiments made for this purpose, it is found that a radius of 574 feet, or curve of 10°, offers a resistance to a train travelling at the rate of 20 miles an hour, double that on a straight and level line, at the same speed. Hence a curve of ten degrees causes a resistance of ten pounds to the ton. Knowing this resistance, that for any other curve is easily obtained.

If we desire to make the resistance uniform upon any system of grades and curves, it will be necessary, whenever a curve occurs upon a grade, to reduce the latter to an amount sufficient to compensate for the resistance caused by the curve.

608. **Mr. Scott Russell's formula.** — Formula (173) gives the value of the total resistance without separating it into its parts.

The formula of Mr. Russell and Mr. Harding gives separate expressions for each resistance. This formula is as follows:

$$r = 6\text{W} + \tfrac{1}{3}v\text{W} + \frac{v^2\text{A}}{400}, \quad . \quad . \quad . \quad (174)$$

in which r and v are the same as in (173), W, the weight of the train in tons, and A, the area of frontage of the train in square feet.

This formula may be expressed in words, as follows:

1. Multiply the weight in tons by 6. The product will be the amount in pounds due to friction.

2. Multiply the weight in tons by the velocity in miles per

hour and divide the product by 3. The result will be the amount in pounds due to concussion.

3. Multiply the square of the velocity in miles per hour by the frontage of the train in square feet and divide the product by 400. The result will give the resistance in pounds due to the atmosphere.

4. Add these three results, and the sum is the total resistance. Divide the total resistance by the weight, and the quotient is the resistance per ton.

The foregoing results corresponded closely with the experiments for speed from 30 to 60 miles per hour. At lower rates of speed, the rule gave too great results.

Another formula has been used in which the resistance of the atmosphere is assumed to be proportional to the volume of the train. It is as follows:

$$r = \left(6 + \frac{v}{15}\right) W + \frac{v^2 B}{50,000}, \quad . \quad . \quad (175)$$

in which B is the volume of the train, the other quantities being the same as in (174).

609. **Tractive force.**—The forces employed to draw the cars on railroads are gravity, horses, stationary engines, and locomotive engines.

610. **Gravity.**—Gravity either assists or opposes the other kinds of motive power on all inclined parts of a railroad. It may be used as the sole motive power on grades which are sufficiently steep. In this case the loaded cars descending the grade draw up a train of empty ones. The connection is made between the trains by means of a wire rope which runs over pulleys placed along the middle of the track.

611. **Horses.**—Horses are frequently used to draw cars on a railroad.

The power of a horse to move a heavy load is ordinarily assumed at 150 pounds, moving at the rate of 2½ miles an hour for 8 hours a day. At greater speeds his power of draught diminishes; for example to half that load at 4 miles an hour, etc.

The power of the horse is rapidly diminished upon ascents. On a slope of 1 in 7 (8¼°) he can carry up only his own weight (Gillespie).

612. **Stationary engines.**—These are employed sometimes where the speed is to be moderate, the grade steep, and the distance short.

The power is usually applied by means of an endless wire rope running on pulleys, like that employed where gravity is the only motive power. And as in that case, the descent of one train is generally made to assist in the drawing up of another to the top of the inclined plane.

613. **Locomotive engines.**—The principal motive power on railroads is the locomotive engine.

The locomotive is a non-condensing, high-pressure engine, working at a greater or less degree of expansion according to circumstances, and placed on wheels which are connected with the piston in such a manner that any motion of the latter is communicated to them.

The power exerted in the cylinder and transferred to the circumference of the driving wheel is termed "**traction**;" its amount depends upon the diameter of the cylinder, the pressure of the steam, the diameter of the driving wheel, and the distance, called the **stroke**, traversed by the piston from one end of the cylinder to the other.

The means by which the traction is rendered available for moving the engine and its load is the friction of the driving wheels on the rail; this is called the "**adhesion**," and its amount varies directly with the load resting on the wheels, and with the condition of the surface of the rails, varying from almost nothing when ice is on the rails, up to as much as one-fifth of the weight on the driving wheels when the surface of the rail is clean and dry.

The speed of the engine depends also upon the rapidity with which its boiler can generate steam. One cylinder full of steam is required for each stroke of the piston. Each double stroke corresponds to one revolution of the driving wheels and to the propulsion of the engine through a space equal to their circumference.

Steam-production, adhesion, and **traction**, are the three elements which determine the ability of a locomotive engine to do its work. The work required of the engine depends upon the nature and amount of the traffic over the road and the condition of the road. Hence, engines of different proportions are employed on the same road, one set to haul heavy loads at low velocities and another set to move light loads at high rates of speed.

Stronger and more powerful engines are needed on a road with steep grades and sharp curves than on roads with easy grades and large curves.

Locomotive engines may be so proportioned as to run at any speed from 0 to 60 miles an hour; to ascend grades even as steep as 200 feet in the mile; and to draw from 1 to 1,000 tons.

The weight and speed of the trains, and the ruling grades of the road determine the amount of power required of the engine. This power depends, as has just been stated, upon the steam-producing capacity of the boiler, upon the leverage with which the steam is applied, and upon the adhesion.

614. **Gauge.**—The width of a railroad between the inner sides of the rails is called the **gauge.**

The question as to what this width should be has been a subject for discussion and of controversy among engineers.

The original railroads were made of the same width as the tram-roads on which the ordinary road wagon was used. It happened that the width of the tram-road was 4 feet 8½ inches; this was adopted for the railroad, and soon became universal. In a few cases, other widths were adopted, but the advantages of uniformity so far exceed all other considerations, that the width of 4 feet 8½ inches is now generally adopted for main lines or roads of the first class.

For branch lines, a still narrower gauge is recommended; a width of 3 feet, and even of 2 feet 6 inches, has been employed. A road of this narrow gauge costs less to construct and admits of steeper grades and sharper curves being used.

Railroads may have either a single or a double track. When first constructed and where the traffic is light, a single track is used, but even then it is recommended to secure ground sufficient for a second track when the latter becomes necessary.

The New York Central Railroad has four tracks, two of which are used for passenger traffic and two for movement of freight.

LOCATION AND CONSTRUCTION OF RAILROADS.

615. **Location.**—Location of railroads is guided by the same principles as that of ordinary roads and is made in the same manner. The greater importance to railroads of easy grades and straightness justifies a greater expenditure for surveys, which are more elaborate than those required for common roads.

616. **Construction.**—This may be divided into two parts: forming the "**road-bed,**" and the "**superstructure.**"

The remarks already made concerning the "construction of roads" apply to "forming the road-bed of a railroad."

The excavations and embankments are generally much greater on railroads than for any other of the roads usually constructed. Where, for instance, an ordinary road would wind around a hill, a railroad would cut through it, in this way obtaining straightness and avoiding curves.

The sides of an excavation are often supported by retaining walls in order to reduce the width of the cutting at the top.

617. **Tunnels.**—When the depth of excavation is very great it will frequently be found cheaper to make a passage under ground called a **tunnel.**

The choice between deep cutting and tunnelling will depend upon the relative cost of the two and the nature of the ground. When the cost of the two methods would be about equal, and the slopes of the deep cut are not liable to slips, it is usually more advantageous to resort to deep cutting than to tunnelling. So much, however, will depend upon local circumstances, that the comparative advantages of the two methods can only be decided by a careful consideration of these circumstances for each particular case. Where a choice may be made, the nature of the ground, the length of the tunnel, that of the deep cuts by which it must be approached, and also the depths of the working shafts, must all be well studied before any decision can be made. In some cases it may be found that a long tunnel with short deep cuts will be most advantageous in one position, and a short tunnel with long deep cuts in another. In others, the greater depth of working shafts may be more than compensated for by the obtaining of a safer soil, or a shorter tunnel.

As a general rule tunnelling is to be avoided if possible.

The dimensions and form of the cross-section will depend upon the nature of the soil and the object of the tunnel as a communication. In solid rock, the sides of the tunnel are usually vertical, the top curved, and the bottom horizontal. In soils which require to be sustained by an arch, the excavation should conform as nearly as practicable to the form of cross-section of the arch.

In tunnels through unstratified rocks, the sides and roof may be left unsupported; but in stratified rocks there is danger of blocks becoming detached and falling: wherever this is to be apprehended, the top of the tunnel should be supported by an arch.

In choosing the site of a tunnel, attention should be had, not only to the nature of the soil, and to the shortness and straightness of the tunnel, but also to the facilities offered for getting access to its course at intermediate points by means of shafts and drifts.

618. **Shafts.**—Vertical pits which are sunk to a level with the crown or top of the tunnel are known as **shafts.**

There are three kinds: **trial, working,** and **permanent shafts.**

Trial shafts are, in general, sunk at or near the centre line of the proposed tunnel to ascertain the nature of the strata through which the tunnel is to be excavated. Their dimensions and shape are regulated by the uses to which they are to be put.

Working shafts are used to give access to the tunnel, for the purpose of carrying on the work and removing the material excavated, for admitting fresh and discharging foul air, and for pumping out water.

Their dimensions will be fixed by the service required of them. Their distance apart varies between 50 and 300 yards, although in some cases they are only from 20 to 30 yards apart, and in others none are used.

They may be located along the centre line of the tunnel or they may be on a line parallel to it.

Permanent shafts are generally working shafts that have been made permanent parts of the tunnel for the purposes of ventilation and of admitting light.

619. **Drifts.**—Small horizontal or slightly inclined underground passages made for the purpose of examining the strata, for the purpose of drainage, of affording access to the tunnel for the workmen and for transport of materials, etc., are termed **drifts** or **headings.**

Their least dimensions are those in which miners can conveniently work, or from 4½ to 5 feet high and 3 feet wide.

Headings are almost always used to connect the working shafts, running along the centre line or parallel to the line of the tunnel. In soft ground, the heading is at or near the bottom of the tunnel; in rock or hard and dry material at or near the top.

620. **Laying out tunnels.**—The establishment of a correct centre line for a tunnel and the fixing of the line at the bottom of the shafts are most important operations and require the utmost care.

The work is commenced by setting out, in the first place, with great accuracy upon the surface of the ground, the profile line contained in the vertical plane of the axis of the tunnel, and at suitable intervals along this line, sinking working shafts. At the bottom of these shafts the centre line is marked out by two points placed as far apart as possible. By these the line is prolonged from the bottom of the shaft in both directions.

In constructing the Hoosac Tunnel, so accurate were the alignments, that the heading running eastward from the central shaft for a distance of 1,563 feet met the heading from the eastern end with an error of but five-sixteenths of an inch; and the heading running westward for 2,056 feet met the heading from the western end with an error of but nine-sixteenths of an inch.

An elaborate trignometrical survey was used to lay out the Mont Cenis Tunnel, which was 7.5 miles long, with no working shafts.

621. **Operation of tunnelling.**—The shafts and the excavations which form the entrances to the tunnel are connected by a drift, usually five or six feet in width and seven or eight feet in height, made along the crown of the tunnel when the soil is good. After the drift is completed, the excavation for the tunnel is gradually enlarged; the excavated earth is raised through the working shafts, and at the same time carried out at the ends. The speed with which the drift is driven determines the rate of progress of the whole.

If the soil is loose, the operation is one of the most hazardous in engineering construction, and requires the greatest precautions against accident. The sides of the excavations must be sustained by strong rough frame-work, covered by a sheathing of boards to secure the workmen from danger. When in such cases the drift cannot be extended throughout the line of the tunnel, the excavation is advanced only a few feet in each direction from the bottom of the working shafts, and is gradually widened and deepened to the proper form and dimensions to receive the masonry of the tunnel, which is immediately commenced below each working shaft, and is carried forward in both directions towards the two ends of the tunnel.

In some cases, two headings were run forward and the side walls of the tunnel were built before the remainder of the section was excavated.

The ordinary difficulties of tunnelling are greatly increased by the presence of water in the soil through which the work is driven. Pumps, or other suitable machinery for raising water, placed in the working shafts, will, in some cases, be requisite to keep them and the drifts free from water until an outlet can be obtained for it at the ends, by a drain along the bottom of the drift.

622. **Drainage and ventilation of tunnels.**—The drainage of a tunnel is effected either by a covered drain under the road-bed at the centre or by open drains at the sides.

Artificial ventilation is found not to be necessary in ordinary tunnels, and the permanent shafts constructed for the purpose have been considered detrimental rather than beneficial in getting rid of the smoke. The passage of the train appears to be the best ventilator; the air being thoroughly disturbed and displaced by the quick motion of the train through the tunnel.

623. **Ballast.**—The tops of the embankments and the bottom of the excavations are brought to a height called the "formation level," about two feet below the intended level of the rails. The remaining two feet, more or less, is filled up with gravel, or gravel and sand, or broken stone, or similar material, through which the water will pass freely. This layer is called the "**ballast,**" and the material of which it is composed should be clean and hard, so as not to pack into a solid mass preventing the water from passing through it.

The object of the ballast, besides allowing the water to run off freely, is to hold the sleepers firmly in their places and to give elasticity to the road-bed.

624. **Cross ties.**—The cross ties or "sleepers" are of wood, hewn flat on the top and bottom; they are from 7 to 9 feet long for the ordinary gauge, 6 inches deep, and from 6 to 10 inches wide. The distance between the ties depends upon the weight of the engines used on the road and the strength of the rail; 2½ feet from centre to centre is about the usual distance. The nearer the sleepers are to uniformity in size and to being equidistant from each other, the more uniform will the pressure from the passage of the train be distributed over the ground.

The sleepers may be of oak, pine, locust, hemlock, chestnut, etc. They last from 5 to 10 years, depending upon their positions and the amount of travel over them. Their duration may be increased by using some of the preservative means referred to in Art. 25.

625. **Rails.**—The rails are made of wrought iron, or of wrought iron with a thin bar of steel forming the top surface, or entirely of steel.

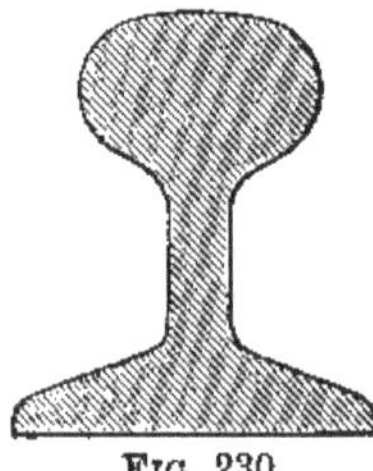

Fig. 230.

Since the rail acts as a support for the train between the ties, and as a lateral guide for the wheels, it must possess strength and stiffness to a marked degree. The top surface should be of sufficient size and hardness to withstand the action of the rolling loads, and the bottom surface should be wide enough to afford a good bearing upon the tie. The rail should have that form which gives the required strength with the least amount of material. The form of cross-section in most general use at the present time in the United States is shown in Fig. 230. This particular rail is 4½ inches high and 4 inches wide at the bottom. The width of the head varies from $2\frac{1}{4}$ to $2\frac{1}{2}$ inches the top surface having a convex form, circular in cross-section, described with a radius double the height of the rail. The thickness of the rib or stem is generally from $\frac{1}{2}$ to $\frac{3}{4}$ of an inch, although recent experiments would indicate that a less thickness might be used with safety.

The rails are rolled in lengths varying from 15 to 21 feet, and when laid are connected by fish-joints and fastened to the cross-ties by spikes. The method of fastening formerly used was to confine the ends of the rails in a cast-iron chair which rested on the cross-ties. This method may be seen on some of the older railroads, but is fast going out of use on all first-class roads.

626. **Coning of the wheels.**—The wheel running on the outer rail of a curve has to pass over a greater distance than the one running on the inner rail. Since the wheels and axles are firmly connected, some arrangement must be made to keep the wheels from dragging or slipping on the rails and to reduce the twisting strain brought on the axles. This is usually effected by making the tread of the wheel conical instead of cylindrical, so that the tendency of the car to press against the outer rail brings a larger diameter upon the outer and a smaller diameter on the inner rail. The difference between these diameters must be proportioned to the distance to be traversed by the wheels, and must depend, therefore, upon the radius of the curve and the gauge. The sharper the curve, the greater should be the difference between the diameters. Upon many roads it is customary to widen the gauge from 4 feet $8\frac{1}{2}$ inches to 4 feet 9 inches on sharp curves, thus allowing more play for the wheels and giving a greater difference in the diameters of those parts of the wheel in contact with the rails. As the tread of the wheel is conical, the tops of the rails are inclined, or given a "**cant**" to fit this cone. The amount of inclination depends upon the amount of conical form given to the tread of the wheel. For the common gauge, this inclination is taken at about $\frac{1}{20}$.

627. **Elevation of the outer rail.**—When the track is straight, a line drawn in the cross-section made by a plane perpendicular to the axis of the road, tangent to the upper surfaces of the rails, is horizontal. On the curved portions of the track the centrifugal force tends to throw the car against the outer rail. This tendency is resisted by raising the outer rail to a certain height above the inner one. The rule for obtaining this height is expressed as follows:

$$h = \frac{v^2 g}{32R'} \quad . \quad . \quad . \quad . \quad . \quad (176)$$

in which h is the elevation above inner rail in inches; v, the velocity in feet per second; g, the gauge of the road in inches; and R, the radius of the curve in feet.

628. **Crossings, switches, etc.**—To enable trains to pass from one track to the other, **crossings** are arranged as shown in Fig. 231. The connection between the crossing and the track is made by a **switch.**

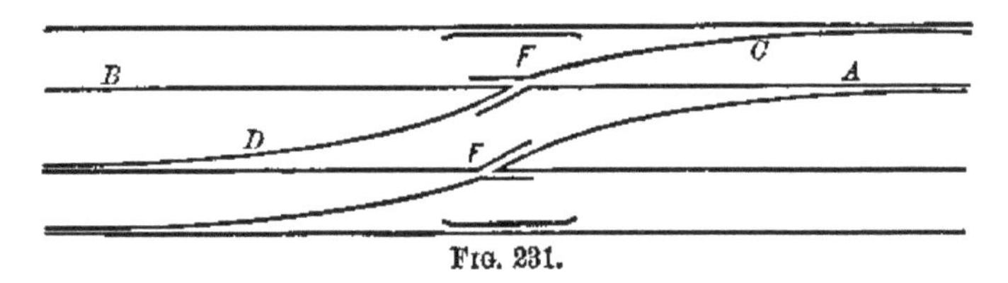

FIG. 231.

The switch consists of one length of rails, movable around one of the ends, so that the other can be displaced from the line of the main track and joined with that of the crossing, or the reverse, depending upon which line of rails the train is to use. A vertical lever is attached to the movable end by means of which the ends of the rails are pushed forward or shoved back, making the connection with the tracks. The handles of the lever should be so fashioned and painted that their position may be seen from a considerable distance.

Where one line of rails crosses another, an arrangement called a **crossing-plate, or frog** (Fig. 232), is used to allow free passage of the wheels.

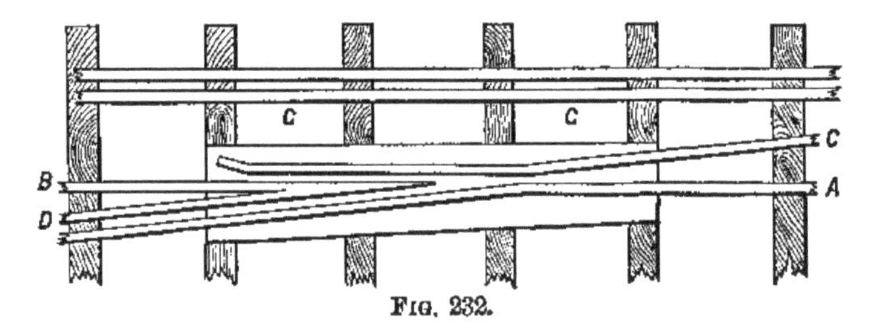

FIG. 232.

In order that the wheels should run smoothly on the rail A B, the rail C D must be cut at its intersection with the former; for a similar reason, the rail A B must be cut at its intersection with C D.

A guard-rail, G G, is used to confine the opposite wheel for short distance and prevent the wheel running on A B from leaving the rail at the cut. This guard-rail is parallel to the outer rail and placed about two inches from it. It extends a short distance beyond the opening in both directions and has its ends curved slightly, as shown in Fig. 231.

The angle between the lines of the main track and the crossing should be very small, not greater than 3°.

629. **Turn-tables.**—When the angle is too great to use the crossing, the arrangement called a turn-table is employed. This consists of a strong circular platform of wood or iron, movable around its centre by means of conical rollers beneath it running upon iron roller-ways. Two rails are laid upon the platform to receive the car, which is transferred from one track to the other by turning the platform sufficiently to place the rails upon it in the same line with those of the track upon which the car is to run. The greater the proportion of the weight borne by the pivot at the centre and the less that borne by the rollers, the less will be the friction.

630. **Telegraph, mile-posts, etc.**—On all well managed railroads, telegraph lines are essential to the safe working of the road. These should be connected with every station. By their use, the positions of the different trains at all hours are made known.

Mile-posts, numbered in both directions, should be placed along the sides of the road. Posts showing the grades, the distance to crossings of roads, to bridges, etc., should be used wherever necessary.

CHAPTER XXIV.

CANALS.

631. A **canal** is an artificial water-course. Canals are used principally for purposes of inland navigation; for irrigation; for drainage; for supplying cities and towns with water, etc.

NAVIGABLE CANALS.

632. **Navigable canals** may be divided into three classes; **level canals**, or those which are on the same level throughout; **lateral canals**, or those which connect two points of different levels, but have no summit level; and **canals with a summit level**, or those connecting two points which lie on opposite sides of a dividing ridge.

I. **Level canals.**—In canals of this class, the level of the water is the same throughout. As in roads, straightness of direction gives way to economy of construction, and the economical course will be that which follows a contour line, unless a great saving may be made by using excavation or embankment. Where changes of direction are made, the straight portions are connected by curved ones, generally arcs of circles, of sufficient curvature to allow the boats using the canal to pass each other without sensible diminution in their rate of speed.

II. **Lateral canals.**—In these canals, the fall of water is in one direction only. Where the difference of level between the extreme points is considerable, the canal is divided into a series of levels or ponds, connected by sudden changes of level. These sudden changes in level are overcome by means of **locks** or other contrivances by which the boat is transferred from one level to the other.

III. **Canals with summit levels.**—These are canals in which the points connected are lower than the intermediate ground over which the canal has to pass, and in consequence the fall is in both directions. As the water for the supply of the summit level must be collected from the ground which lies above it, it follows that the summit level should be at the lowest point of the ridge dividing the two extremes of the canal.

633. **Form and dimensions of water-way.**—The general **width** of a canal should be sufficient to allow two boats to pass each other easily. Where great expense would be incurred in giving this width, like that of a bridge supporting a canal, short portions may be made just wide enough for one boat.

The **depth** should be such as not to materially increase the resistance to the motion of the boat beyond what is felt in open water.

The bottom of the canal is generally made horizontal. The sides are inclined, and when of earth should not be steeper than one upon one and a half; if of masonry, the sides may be vertical or nearly so. In the latter case a greater width must be given to the bottom of the canal.

The water-way is usually of a trapezoidal form, in cross-section (Fig. 233) with an embankment on each side, raised above the general surface of the country and formed of the material from the excavation for the canal.

The **relative dimensions** of the parts of the cross-section may be generally stated as follows:

The width of the water-way, at bottom, should be at least twice the width of the boats used in navigating the canal.

The depth of the water-way should be at least eighteen inches greater than the greatest draft of the boat.

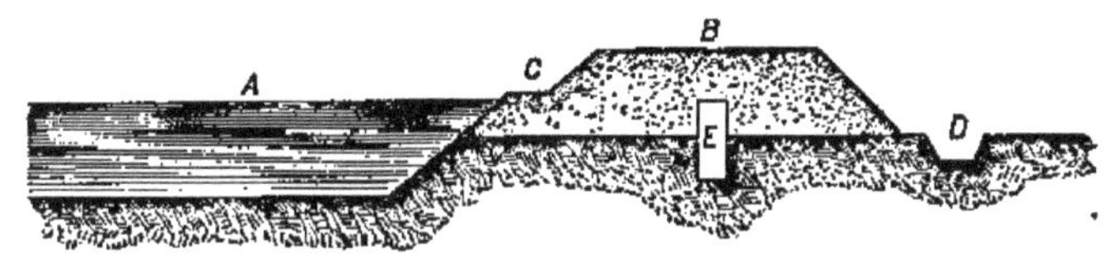

FIG. 233.—A, water-way. B, towpath. C, berm. D, side-drain. E, puddling of clay.

The least area of water-way should be at least six times the greatest midship section of the boat.

634. **A towpath** for horses is made on one of the embankments and a footpath on the other. This footpath should be wide enough to serve as an occasional towpath.

The towpath should be from ten to twelve feet wide, to allow the horses to pass each other with ease; and the footpath at least six feet wide. The height of the surfaces of these paths, above the water surface, should not be less than two feet, to avoid the wash of the ripple; nor greater than four feet and a half, for the facility of the draft of the horses in towing. The surface of the towpath should incline slightly outward, both to convey off the surface water in wet weather and to give a firmer footing to the horses, which naturally draw from the canal.

The width given to these paths will give a sufficient thickness to the embankments to resist the pressure of the water against them, and to prevent filtration through them, provided the earth is at all binding in its composition.

635. **Construction.** — All canal embankments should be carefully constructed. The earth of which they are formed should be of a good binding character, and perfectly free from mould and all vegetable matter, as the roots of plants, etc. In forming the embankments, the mould should first be removed from the surface on which they are to rest, and the earth then spread in uniform layers, from nine to twelve inches thick, and well rammed. If the character of the earth, of which the embankments are formed, is such as not to present entire security against filtration, a puddling of clay, two or three feet thick, should be laid in the interior of the mass, extending from about a foot below the natural surface up to the same level with the surface of the water. Sand is useful in stopping leakage through the holes

made in the embankments near the water surface by insects, moles, rats, etc.

The side slopes of the embankment vary with the character of the soil: towards the water-way they should seldom be less than two base to one perpendicular; from it, they may be less. The interior slope is usually not carried up unbroken from the bottom to the top; but a horizontal space, termed a **bench** or **berm**, about one or two feet wide, is left, about one foot above the water surface, between the side slope of the water-way and the foot of the embankment above the berm. This space serves to protect the upper part of the interior side slope, and is, in some cases, planted with such shrubbery as grows most luxuriantly in moist localities, to protect more efficaciously the banks by the support which its roots give to the soil. The side slopes are better protected by a revetment of dry stone, from six to nine inches thick. Aquatic plants of the bulrush kind have been used, with success, for the same purpose; being planted on the bottom, at the foot of the side slope, they serve to break the ripple, and preserve the slopes from its effects.

Side drains must be made, on each side, a foot or two from the embankments, to prevent the surface water of the natural surface from injuring the embankments.

636. Slight leakage may sometimes be stopped by sprinkling fine sand in small quantities at a time over the surface of the

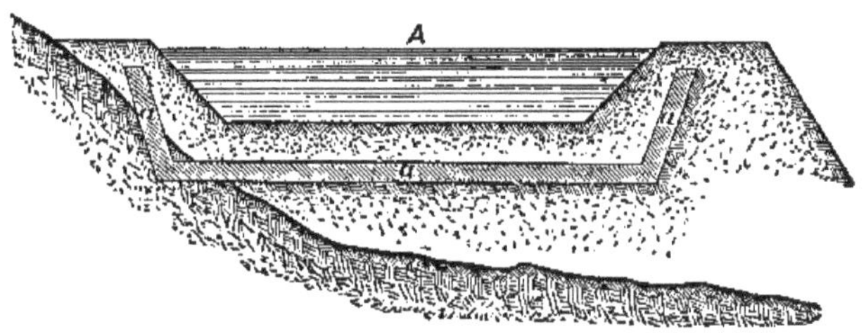

Fig. 234.

water in the vicinity of the leaks. The sand settling to the bottom gradually fills the crevices in the sides and bottom of the canal through which the water escapes.

The leakage may be so great that it may be necessary, in certain cases, to line the canal with masonry, concrete, or to face the sides with sheet-piling to retain the water.

When the bottom of the canal is composed of fragments of rock forming large crevices, or composed of marl, it has been frequently found necessary to line the water-way in such localities with masonry (Fig. 234) or with concrete.

In a lining of this kind, the stone used was about four inches thick, laid in cement or hydraulic mortar, and covered with a coating of mortar two inches thick, making the entire thickness of the lining six inches. This lining was then covered, both at bottom and on the sides, by a layer of earth, at least three feet thick, to protect it from the shock of the boats striking against it.

637. **Size of canals.**—The size of a canal depends upon the size of the boats to be used upon it. The dimensions of common canal boats have been fixed with a view of horses being used to draw them. The most economical use of horse-power is to draw a heavy load at a low rate of speed. Assuming a speed of from two to two and a half miles an hour, a horse can draw a boat with its load, in all about 170 tons. This requires a boat of the ordinary cross-section to be about twelve feet wide, and to have a draught of four and a half feet when fully loaded.

Boats of greater cross-section are frequently used, and are drawn by various applications of steam as well as by horse-power. The methods used are various, as the screw propeller, stationary engines with endless wire ropes, etc. Canals are sometimes made only twelve feet wide at bottom, with a draught of four feet; common canals are from twenty-five to thirty feet wide at bottom, with a depth of from five to eight feet; ship or large canals are fifty feet wide at bottom, and have a depth of twenty feet. These are the minimum dimensions.

638. **Locks.**—An arrangement termed a **lock** is ordinarily used to pass a boat from one level to another.

A lock is a small basin just large enough to receive a boat, and in which the water is usually confined on the sides by

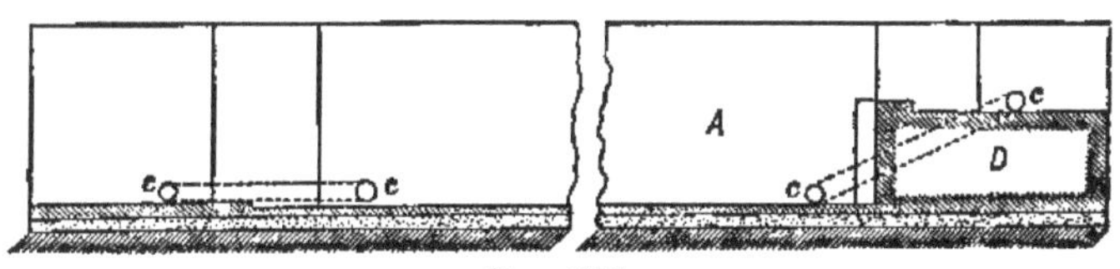

FIG. 235.

two upright walls of masonry, and at the ends by two gates; the gates open and shut, both in order to allow the passage of the boat and to cut off the water of the upper level from the lower, or from the water in the lock.

A lock (Figs. 235 and 236) may be divided into three distinct parts: 1st. The part included between the two gates, which is termed the **chamber.** 2d. The part above the upper gates, termed the **fore** or **head-bay**. 3d. The part below the lower gates, termed the **aft** or **tail-bay.**

Fig. 235 shows a vertical longitudinal section through the axis of a single lock built on a foundation of concrete, and Fig. 236 represents the plan.

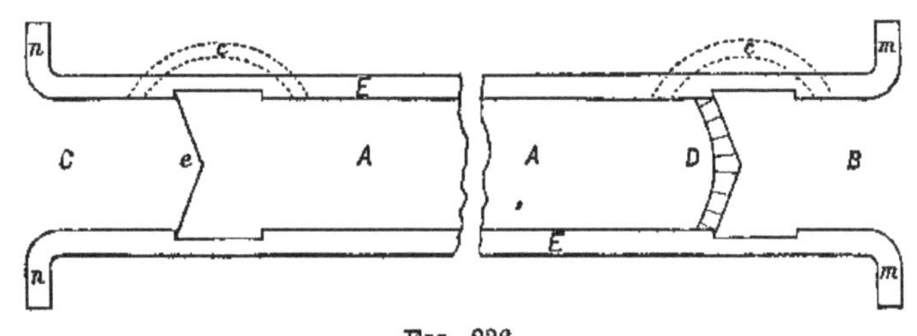

FIG. 236.

In these figures, A is the lock-chamber; E, E, the side walls; B, the head-bay; C, the tail-bay; and D, the **lift-wall.**

The lock-chamber must be wide enough to allow an easy ingress and egress to the boats commonly used on the canal; a breadth of one foot greater than the greatest breadth of the boat is deemed sufficient for this purpose. The length of the chamber is regulated by that of the boats; it should be such that when the boat enters the lock from the lower level, the tail-gates may be shut without requiring the boat to unship its rudder.

The plan of the chamber is usually rectangular, the sides receiving a slight batter; as when so arranged they are found to give greater facility to the passage of the boat than when vertical. The bottom of the chamber is either flat or curved; more water will be required to fill the flat-bottomed chamber than the curved, but less masonry will be required in its construction.

The chamber is terminated just within the head-gates by a vertical wall, the plan of which is usually curved. As this wall separates the upper from the lower level, it is termed the **lift-wall**; it is usually of the same height as the lift of the levels. The top of the lift-wall is formed of cut stone, the vertical joints of which are normal to the curved face of the wall; this top course projects from six to nine inches above the bottom of the upper level, presenting an angular point for the bottom of the head-gates, when shut, to rest against. This projection is termed the **mitre-sill.** Various degrees of opening have been given to the angle between the two branches of the mitre-sill; it is, however, generally so determined, that the perpendicular of the isosceles triangle, formed by the two branches, shall vary between one-fifth and one-sixth of the base.

The side-walls sustain the pressure of the embankment against them, and when the lock is full the pressure from the water in the chamber. The former pressure is the greater and the more permanent of the two and the dimensions of the wall are determined to resist this pressure. The usual manner of doing this is to make the wall four feet thick at the water line of the upper level, to secure it against filtration; and then to determine the base of the batter, so that the mass of masonry shall present sufficient stability to resist the thrust of the embankment. The spread and other dimensions of the foundations will be regulated according to the nature of the soil, as in other masonry structures.

The bottom of the chamber, as has been stated, may be either flat or curved. The flat bottom is suitable to firm soils, which will neither yield to the vertical pressure of the chamber walls nor admit the water to filter from the upper level under the bottom of the lock. In either of these cases, where yielding or undermining may be expected, the bottom should be an inverted arch. The thickness of the masonry of the bottom will depend on the width of the chamber and the nature of the soil. Were the soil a solid rock, no bottoming would be requisite; if it is of soft material, a very solid bottoming, from three to six feet in thickness, may be necessary. Great care must be taken to prevent the water from the upper level filtering through and getting under the bottom of the lock.

The lift-wall may have only the same thickness as the side walls, but unless the soil is very firm, it would be more prudent to form a general mass of masonry under the entire head-bay, to a level with the base of the chamber foundations, of which mass the lift-wall should form a part.

The head-bay is enclosed between two parallel walls, which form a part of the side walls of the lock. They are terminated by two wing walls, *m*, *m*, at right angles with the side walls. A recess, termed the **gate-chamber**, is made in the wall of the head-bay; the depth of this recess should be sufficient to allow the gate, when open, to fall two or three inches within the facing of the wall, so that it may be out of the way when a boat is passing; the length of the recess should be greater than the width of the gate. That part of the recess where the gate turns on its pivot is termed the **hollow quoin**; it receives what is termed the **heel** or **quoin-post** of the gate, which is made to fit the hollow quoin. The distance between the hollow quoins and the face of the lift-wall will depend on the pressure against the mitre-sill, and the strength of the stone; eighteen inches will generally be found sufficient.

The side walls need not to extend more than twelve inches beyond the other end of the gate-chamber. The wing walls may be extended back to the total width of the canal, but it will be more economical to narrow the canal near the lock, and to extend the wing walls only about two feet into the banks or sides. The dimensions of the side and wing walls of the head-bay are regulated in the same way as the chamber walls. The top of the side walls of the lock may be from one to two feet above the general level of the water in the upper level.

The bottom of the head-bay is flat, and on the same level with the bottom of the canal; the exterior course of stones at the entrance to the lock should be so jointed as not to work loose.

The side walls of the tail-bay are also a part of the general side walls, and their thickness is regulated as in the preceding cases. Their length will depend chiefly on the pressure which the lower gates throw against them when the lock is full, and partly on the space required by the lockmen in opening and shutting the gates. These walls are also terminated by wing walls, *n*, *n*, similarly arranged to those of the head-bay. The points of junction between the wing and side walls should, in both cases, either be curved or the stones at the angles be rounded off. One or two perpendicular grooves are sometimes made in the side walls of the tail-bay, to receive stop-planks, when a temporary dam is needed, to shut off the water of the lower level from the chamber, in case of repairs, etc.

The gate-chambers for the lower gates are made in the chamber walls; the bottom of the chamber, where the gates swing back, should be flat, or be otherwise arranged so as not to impede the play of the gates.

The bottom of the tail-bay is arranged, in all respects, like that of the head-bay.

639. Those parts of the lock where there is great wear and tear, as at the angles generally, should be of cut-stone; or where an accurate finish is indispensable, as at the hollow quoins. The other parts may be of brick, rubble, concrete, etc., but every part should be laid in cement or the best hydraulic mortar.

The mitre-sills are generally faced with timber, to enable them to withstand better the blows which they receive from the gates, and to make a tighter joint.

640. The locks are filled and emptied through sluices in the head and tail-gates, opened and closed by slide valves, or by culverts made of masonry or iron pipe placed as shown in the figures at *c*, *c*, *c*, etc. The latter is the method generally recommended. From the difficulty of repairing the sluices when out of order, many prefer the use of valves in the gates.

The bottom of the canal below the lock should be protected by what is termed an **apron**, which is a covering of plank laid on a grillage, or of dry stone. The length will depend upon the strength of the current; generally a distance of from fifteen to thirty feet will be sufficient.

641. **Lock gates.**—The gates may be made of wood or of iron. Each gate is ordinarily composed of two leaves, each leaf consisting of a framework, covered with planking or iron plates. The frame, when of timber, consists usually of two uprights, connected by horizontal pieces let into the uprights with the usual diagonal bracing.

In gates of this kind, each leaf turns about an upright, which is called the **quoin** or **heel-post.** This post is cylindrical on the side next to the hollow quoins, which it exactly fits when the gate is shut. It is made slightly eccentric, so that when the gate is opened it may turn easily without rubbing against the quoin. At its lower end it rests on a pivot, and its upper end turns in a circular collar which is strongly anchored in the masonry of the side walls. One of the anchor-irons is usually placed in a line with the leaf when shut, the other in a line with it when open; these being the best positions to resist most effectually the strain produced by the gate. The opposite upright, termed the **mitre-post,** has one edge bevelled off, to fit against the mitre-post of the other leaf of the gate, forming a tight joint when the gate is shut.

A long, heavy beam, termed a **balance beam** from its partially balancing the weight of the leaf, is framed upon the quoin-post, and is mortised into the mitre-post. The balance beam should be about four feet above the top of the lock; its principal use being to bring the centre of gravity of the leaf near the heel-post and to act as a lever to open and shut the leaf.

Sometimes this bar is dispensed with, and the leaves are supported on rollers placed under the lower side to assist the pivot in supporting their weight. These rollers run on iron rails placed on the floor of the gate-chamber. In these cases the gates are ordinarily opened and shut by means of windlasses and chains. This is the method generally used for very large gates. Gates formed of a single leaf moving on a horizontal axis are frequently used.

642. **Inclined planes.**—Instead of locks, inclined planes are sometimes used, by means of which the boats are passed from one level to another. In these cases, water-tight caissons or cradles, on wheels are used.

At the places where the levels are to be connected, the canal is deepened to admit of the caisson or the cradle to run in under the boat to be transferred. Two parallel lines of rails start from the bottom of the lower level, ascend an inclined plane up to a summit a little above the upper level, and then descend by a short inclined plane into the upper level. Two caissons or cradles, one on each set of rails, are connected by a wire rope, so that one ascends while the other descends. Power being applied, the boats are transferred to the appropriate levels.

The caissons are preferred because they balance each other at all times on the inclined plane, whether the boats are light or heavy, as they displace exactly their own weight of water in the caisson. In some cases, the caissons have been lifted vertically instead of being drawn up inclined planes.

643. **Guard lock.**—A large basin is usually formed at the outlet, for the convenience of commerce; and the entrance from this basin to the canal, or from the river to the basin, is effected by means of a lock with double gates, so arranged that a boat can be passed either way, according as the level in the one is higher or lower than that in the other. A lock so arranged is termed a **tide** or **guard lock,** from its uses. The position of the tail of this lock is not indifferent in all cases where it forms the outlet to the river; for were the tail placed up stream, it would generally be more difficult to pass in or out than if it were down stream.

644. **Lift of locks.**—The vertical distance through which a boat is raised or lowered by means of the lock is called the "**lift.**" This vertical distance between two levels may be overcome by the use of a single lock or by a "flight of locks." The lift of a single lock ranges from two to twelve feet, but generally in ordinary canals is taken at about eight feet. Where a greater distance than twelve feet has to be overcome, two or more, or a flight of locks, are necessary.

In fixing the lengths of the levels and the positions of the locks, the engineer, if considering the expenditure of water, will prefer single locks with levels between them, to a flight of locks.

In most cases, a flight is cheaper than the same number of single locks, as there are certain parts of the masonry which can be omitted. There is also an economy in the omission of the small gates, which are not needed in flights. It is, however, more difficult with combined than with single locks to secure the foundations from the effects of the water, which forces its way from the upper to the lower level under the locks. Where an active trade is carried on, a double flight is sometimes arranged, one for the ascending, the other for the descending boats. In this case the water which fills one flight may, after the passage of the boat, be partly used for the other, by an arrangement of valves made in the side wall separating the locks.

The engineer is not always left free to select between the two; for the form of the natural surface may require him to adopt a flight at certain points. In a flight the lifts are made the same throughout, but in single locks the lifts vary according to circumstances. Locks with great lifts consume more water, require more care in their construction, and require greater care against accidents than the smaller ones, but cost less for the same difference of level.

645. **Levels.**—The position and the dimensions of the levels must be mainly determined by the form of the natural surface. By a suitable modification of its cross-section, a level can be made as short as may be deemed desirable; there being but one point to be attended to in this, which is, that a boat passing between the two locks, at the ends of the level, will have time to enter either lock before it can ground, on the supposition that the water drawn off to fill the lower lock, while the boat is traversing the level, will just reduce the depth to the draught of the boat.

646. **Water supply.**—Two questions are to be considered: the quantity of water required, and the sources of supply.

The **quantity of water required** may be divided into two portions: 1st. The quantity required for the summit level, and those levels which draw from it their supply. 2d. The quantity which is wanted for the levels below those, and which is furnished from other sources.

The supply of the first portion, which must be collected at the summit level, may be divided into several elements: 1st. The quantity required to fill the summit level, and the levels which draw their supply from it. 2d. The quantity required to supply losses, arising from accidents; as breaches in the banks and the emptying of the levels for repairs. 3d. The supplies for losses from surface evaporation, from leakage through the soil, and through the lock gates. 4. The quantity required for the service of the navigation, arising from the passage of the boats from one level to another.

The quantity required to fill the summit level and its dependent levels will depend on their size, an element which can be readily calculated; and upon the quantity which would soak into the soil, which is an element of a very indeterminate character, depending on the nature of the soil in the different levels.

The supplies for accidental losses are of a still less determinate character.

The supply for losses from surface evaporation may be determined by observations on the rain-fall of the district, and the yearly amount of evaporation. Losses caused by leakage through the soil will depend on the greater or less capacity which the soil has for holding water. This element varies not only with the nature of the soil, but also with the shorter or longer time that the canal may have been in use; it having been found to decrease with time, and to be, comparatively, but trifling in old canals. In ordinary soils it may be estimated at about two inches in depth every twenty-four hours, for some time after the canal is first opened. The leakage through the gates will depend on the workmanship of these parts.

In estimating the quantity of water expended for the service of the navigation, in passing the boats from one level to another, two distinct cases require examination: 1st. Where there is but one lock; and 2d. Where there are several contiguous locks, or, as it is termed, a **flight** of locks between two levels.

To pass a boat from one level to the other—from the lower to the upper end, for example—the lower gates are opened, and the boat having entered the lock they are shut, and water is drawn from the upper level to fill the lock and raise the boat; when this operation is finished, the upper gates are opened and the boat is passed out. To descend from the upper level, the lock is first filled; the upper gates are then opened and the boat passed in; these gates are next shut, and the water is drawn from the lock until the boat is lowered to the lower level, when the lower gates are opened and the boat is passed out.

Hence, to pass a boat, up or down, a quantity of water must be drawn from the upper level to fill the lock to a height which is equal to the difference of level between the surface of the water in the two; this volume of water required to pass a boat up or down is termed the **prism of lift.** The calculation, therefore, for the quantity of water requisite for the service of the navigation, will be simply that of the number of prisms of lift which each boat will draw from the summit level in passing up and down.

An examination of the quantity of water used in passing from one level to another, will show that the quantity required for a flight of locks is greater than that required for isolated locks.

The **source of supply** of water is the rain-fall. The rain-water which escapes evaporation on the surface and absorption by vegetable growth, either runs directly from the surface of the ground into *streams*, or sinks into the ground, flows through crevices of porous strata and escapes by *springs*, or collects in the strata, from which it is drawn by means of *wells*.

647. In whatever way the water may be collected, the measurement of the rain-fall of the district from which it comes is of the first importance. To make this measurement, the area of the district called the **drainage area** or **catchment basin**, and the depth of the rain-fall for a given time must be determined.

Drainage area.—This area is generally a district of country enclosed by a ridge or **water-shed** line which is continuous except at the place where the waters of the basin find an outlet. It may be divided by branch ridges or spurs into a number of smaller basins, each drained by a stream which runs into the main stream.

Depth of rain-fall.—The depth is determined by establishing rain-gauges in the district and having careful observations made for as long a period as possible.

The important points to be determined are: 1. The least annual rain-fall; 2. The mean annual rain-fall; 3. The greatest annual rain-fall; 4. The distribution of the rain-fall throughout the year; 5. The greatest continuous rain-fall in a short period.

For canal purposes, the least annual rain-fall and the longest drought are the most important points to be known.

Knowing the depth of the rain-fall and the area of the catchment basin, an estimate of the amount of water which may be available for the canal may be made. Theoretically considered, all the water that drains from the ground adjacent to the summit level, and above it, might be collected for its supply; but it is found in practice that channels for the conveyance of water must have certain slopes, and that these slopes, moreover, will regulate the supply furnished in a certain time, all other things being equal. The actual discharge of the streams should be measured so as to find the actual proportion of available to total rain-fall, and the streams should be measured at the same time the rain-gauge observations are made.

The measurement of the quantity of water discharged by a

stream is called "**gauging**," and to be of value should be made with accuracy and extend through some considerable time.

648. **Feeders and reservoirs.**—The usual method of collecting the water, and conveying it to the summit level, is by feeders and reservoirs. The **feeder** is a canal of a small cross-section, which is traced on the surface of the ground with a suitable slope, to convey the water either into the reservoir, or direct to the summit level. The dimensions of the cross-section, and the longitudinal slope of the feeder, should bear certain relations to each other, in order that it shall deliver a certain supply in a given time. The smaller the slope given to the feeder, the lower will be the points at which it will intersect the sources of supply, and therefore the greater will be the quantity of water which it will receive. The minimum slope, however, has a practical limit, which is laid down at four inches in 1,000 yards, or nine thousand base to one altitude; and the maximum slope should not be so great as to give the current a velocity which would injure the bed of the feeder. Feeders are furnished, like ordinary canals, with contrivances to let off a part, or the whole, of the water in them, in cases of heavy rains, or for making repairs.

A **reservoir** is a place for storing water to be held in reserve for the necessary supply of the summit level. A reservoir is usually formed by choosing a suitable site in a deep and narrow valley, which lies above the summit level, and erecting a dam of earth, or of masonry, across the outlet of the valley, or at some more suitable point, to confine the water to be collected. The object to be obtained is to collect the greatest volume of water, and at the same time present the smallest evaporating surface, at the smallest cost for the construction of the dam.

649. **Dams.**—The dams of reservoirs have been variously constructed: in some cases they have been made entirely of earth; in others, entirely of masonry; and in others, of earth packed in between parallel stone walls. It is now thought best to use either earth or masonry alone, according to the circumstances of the case; the comparative expense of the two methods being carefully considered.

Earthen dams should be made with extreme care, of the best binding earth, well freed from everything that might cause filtrations.

The foundation is prepared by stripping off the soil and excavating and removing all porous materials, such as sand, gravel, and fissured rock, until a compact and water-tight bed is reached.

A culvert for the outlet-pipes is next built. This should rest on a foundation of concrete and should have the masonry laid in cement or the best of hydraulic mortar. It should be well coated with a clay puddling. Frequently the inner end of the culvert terminates in a vertical tower, which contains outlet-pipes for drawing water from different levels, and the necessary mechanism by means of which the pipes can be closed or opened. Sometimes a cast-iron pipe alone is laid without any culvert.

The earth is then carefully spread in layers not over a foot thick and rammed. A "puddle-wall" with a thickness at the base of about one-third its height and diminishing to about half this thickness at the top, should form the central part of the dam. Care should be taken that it forms a water-tight joint with the foundation and also with the puddle coating of the culvert.

The dam may be from fifteen to twenty feet thick at top. The slope of the dam towards the pond should be from three to six base to one perpendicular; the reverse slope need only be somewhat less than the natural slope of the earth.

The outer slope is usually protected from the weather by being covered with sods of grass. The inner slope is usually faced with dry stone, to protect the dam from the action of the surface ripple.

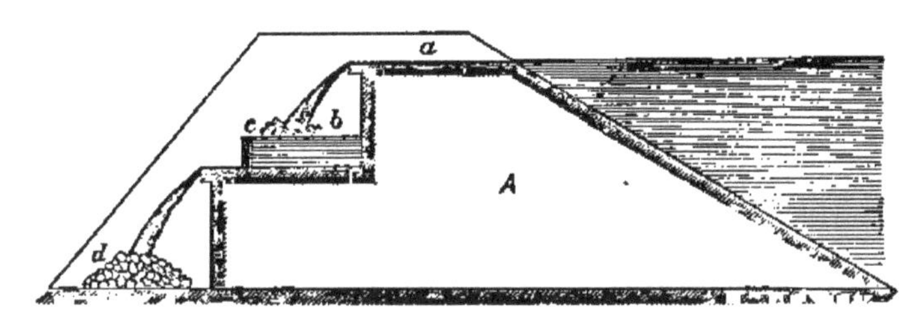

FIG. 237.—A, body of the dam.
a, top of the waste-weir.
b, pool, formed by a stop-plank dam at *c*, to break the fall of the water.
d, covering of loose stone to break the fall of the water from the pool above.

Masonry dams are water-tight walls, of suitable forms and dimensions to prevent filtration, and to resist the pressure of water in the reservoir. The cross-section is usually that of a trapezoid, the face towards the water being vertical, and the exterior face inclined with a suitable batter to give the wall sufficient stability. The wall should be at least four feet thick at the water line, to prevent filtration, and this thickness may be increased as circumstances may require.

650. **Waste-weirs.**—Suitable dispositions should be made to relieve the dam from all surplus water during wet seasons. For this purpose arrangements should be made for cutting off the sources of supply from the reservoir; and a cut, termed a **waste-weir** (Fig. 237), of suitable width and depth, should be made at some point along the top of the dam, and be faced with stone, or wood, to give an outlet to the water over the dam. In high dams the total fall of the water should be divided into several partial falls, by dividing the exterior surface over which the water runs into offsets. To break the shock of the water upon the horizontal surface of the offset, it should be covered with a sheet of water retained by a dam placed across its outlet.

In extensive reservoirs, in which a large surface is exposed to the action of the winds, waves might be forced over the top of the dam, and subject it to danger; in such cases the precaution should be taken of placing a parapet wall towards the outer edge of the top of the dam, and facing the top throughout with flat stones laid in mortar.

651. **Water-courses intersecting the line of the canal.**—The disposition of the natural water-courses which intersect the line of the canal will depend on their size, the character of their current, and the relative positions of the canal and stream.

Small streams which lie lower than the canal may be conveyed under it through an ordinary culvert. If the level of the canal and stream is nearly the same, it may be conveyed under the canal by an inverted syphon of masonry or iron, usually termed a **broken-back** culvert, or if the water of the stream is limpid, and its current gentle, it may be received into the canal. Its communication with the canal should be so arranged that the water may be shut off or let in at pleasure, in any quantity desired.

In cases where the line of the canal is crossed by a torrent, which brings down a large quantity of sand, pebbles, etc., it may be necessary to make a permanent structure over the canal, forming a channel for the torrent; but if the discharge of the torrent is only periodical, a movable channel may be arranged, for the same purpose, by constructing a boat with a deck and sides to form the water-way of the torrent. The boat is kept in a recess in the canal near the point where it is used, and is floated to its position, and sunk when wanted.

When the line of the canal is intersected by a wide water-course, the communication between the two shores must be effected either by a canal aqueduct bridge, or by the boats descending from the canal into the stream.

652. **Dimensions** of canals and their locks in the United States.—The original dimensions of the New York Erie Canal and its locks have been generally adopted for similar works subsequently constructed in most of the other States. The dimensions of this canal and its locks were as follows:

Width of canal at top	40 feet.
Width at bottom	28 "
Depth of water	4 "
Width of tow-path	9 to 12 "
Length of locks between mitre-sills	90 "
Width of locks	15 "

For the enlargement of the Erie Canal, the following are the dimensions:

Width of canal at top	70 feet.
Width at bottom	42 "
Depth of water	7 "
Width of tow-path	14 "
Length of locks between mitre-sills	110 "
Width of lock at top	18.8 "
Width of lock at bottom	14.6 "
Lift of locks	8 "

Between the double locks a culvert is placed, which allows the water to flow from the level above the lock to the one below, when there is a surplus of water in the former.

IRRIGATING CANALS.

653. Canals belonging to this class are used to bring from its source a supply of water, which, when reaching certain localities, is made to flow over the land for agricultural purposes. This kind of canal is practically unknown in the United States, as the farmer depends almost entirely on the rain-fall alone for the requisite amount of moisture for his crops.

Irrigation canals of large size have been used in India for hundreds of years; they are also found in Italy. Rude imitations, of small size, are to be seen in Mexico, the territory of New Mexico, lower part of California, and other parts of the United States.

In certain parts of our country they could be used to great

advantage, and since in the future they may be used, it is thought advisable to allude briefly to them in this treatise.

The special difference between a navigable and an irrigation canal is that the former requires that there should be little or no current in the canal, so that navigation may be easy in both directions, while the latter requires that the canal should be a running stream, fed by continuous supplies of water at its source, to make up the losses caused by the amounts of water drawn off from the canal for the purposes of irrigation.

Hence, for two canals of the same size, the navigable canal will require a less volume of water than the irrigation canal, and is more economically constructed on a low level.

The irrigation canal should be carried at as high a level as possible, so as to have sufficient fall for the water which is to be used to irrigate the land on both sides of it and at considerable distances from it. This irrigation is effected by means of branch canals leading from the main one, whence the water is carried by small channels on the fields.

654. The problem of an irrigation canal is to so connect it with the stream furnishing the supply of water, and to so arrange the slope of the bed of the canal, that the canal shall not become choked with silt.

A canal opening direct into the stream which supplies it with water, if proper arrangements are not made, will be liable to have the volume of water greatly increased in time of freshets, and at other times have the supply entirely cut off. In the first case, large quantities of silt would be washed into the canal, choking it up as the water receded to its proper level. In the second case, the supply would probably fail at the critical period of the growing crops when water was greatly needed.

A good selection of the point where the canal joins the stream, and the use of sluices to govern the supply of water, will greatly prevent the occurrence of either of these conditions.

To prevent the silting up of the canal, the slope of the bed is so fixed that the water shall have a uniform velocity throughout. It is therefore seen that, as the water is drawn off at different points for the irrigation of the land, on the right and left of the canal, the volume of water is reduced. The portions of the canal below these points must then be so fixed as to preserve the same rate of motion in the water. This is done by decreasing the width and depth of the canal, and increasing the slope of the bed. Thus starting with a water-way 100 feet wide, 6 feet deep, having a slope of 6 inches to the mile, the width of water-way, as the water is drawn off, may be contracted to 80, 60, 40, and 20 feet with the corresponding depths, 5½, 5, 4½, and 4 feet; to keep the velocity uniform the bed should have slopes of 6.4, 7, 7.9, and 10.3 inches per mile.

655. An irrigation canal may be used for the purposes of navigation. In this case the principles already laid down for navigable canals equally apply, with the condition, however, that the velocity of the current in the canal should not be so slight as to injure its uses as an irrigation canal, nor so swift as to offer too great a resistance to the boats using it as a navigable canal.

DRAINAGE CANALS.

656. Canals of this class are the reverse of irrigation canals. They are used to carry off the superfluous water which falls on or flows over the land.

The water-levels of canals for drainage, to be effective, should at all times be at least three feet below the level of the ground.

Each channel for the water should have an area and declivity, when subjected to the most unfavorable conditions, sufficient to discharge all the water that it receives as fast as this water flows in, without its water-level rising so high as to obstruct the flow from its branches or to flood the country.

Hence, to plan such a system the greatest annual rain-fall of the district, and the greatest fall in a short period or flood must be known.

Where the land to be drained is below the level of high water, the area to be drained must be protected by embankments. The canals are then laid off on the plan just given, and the water from the main canals is removed by pumping.

Drainage canals may be divided into two classes: **open** and **covered**. Where pure water is to be removed, the former are used; when filthy water, or foul materials, are to be removed, the latter are used, and are known then as **sewers**. **Sewerage** is the special name used to designate the drainage of a city or town, in which the foul waters and refuse are collected and discharged by sewers.

As far as the principles of construction are concerned sewers do not differ from the works already described. Especial attention must be paid to prevent the escape of the foul gas and disagreeable odors from the drains.

CANALS FOR SUPPLYING CITIES AND TOWNS WITH WATER.

657. As sewers are only particular cases of drainage canals, so canals for supplying cities with water are only particular cases of irrigation canals, and are therefore governed by the same general principles in their construction.

The canals of this class are usually covered, and receive the general name of **aqueducts.**

658. The health and comfort of the residents of cities and towns are so dependent upon a proper supply of water and a good system of sewerage that the greatest care must be taken by the engineer that no mistakes are made by him in planning and constructing either of these systems. The principles which regulate in deciding upon the quantity of water required, the means and purity of the supply, the location of the reservoirs, the method of distribution, etc., form a subject which can be considered in a special treatise only. The same remark applies also to sewerage.

CPSIA information can be obtained
at www.ICGtesting.com
Printed in the USA
LVHW040229240820
663982LV00001B/15